CISM COURSES AND LECTURES

The series presents lecture notes, monographs, edited works and proceedings in the field of Mechanics, Engineering, Computer Science and Applied Mathematics.
Purpose of the series is to make known in the international scientific and technical community results obtained in some of the activities organized by CISM, the International Centre for Mechanical Sciences.

INTERNATIONAL CENTRE FOR MECHANICAL SCIENCES

COURSES AND LECTURES - No. 353

EDDY STRUCTURE IDENTIFICATION

EDITED BY

J.P. BONNET
UNIVERSITY OF POITIERS

SPRINGER - VERLAG WIEN - NEW YORK

Le spese di stampa di questo volume sono in parte coperte da
contributi del Consiglio Nazionale delle Ricerche.

This volume contains 139 illustrations

Printed in Italy

In order to make this volume available as economically and as rapidly as possible the authors' typescripts have been reproduced in their original forms. This method unfortunately has its typographical limitations but it is hoped that they in no way distract the reader.

ISBN 3-211-82802-8 Springer-Verlag Wien New York

PREFACE

This book presents the talks given during the joint CISM/ERCOFTAC advanced course on "Eddy Structure Identification Techniques in Free Turbulent Shear Flows" held in Udine (Italy), May 23-26, 1994.

This course was a unique opportunity to bring together information that is needed for both experimentalists, theoreticians and computationalists for the analysis of eddy structures in turbulent shear flows.
It is now widely recognized that Large Scale Structures or Eddies in turbulent flows are essential for a physical understanding of these flows, for developing new prediction codes, for analyzing experimental and computational data and for developing new research areas such as flow control. With the exception of some particular situations, like transitional or artificially excited flows, these structures are embedded inside the overall chaotic behavior of the turbulent field. The researcher or the engineer is often faced with the problem of capturing, defining and analyzing the statistics and dynamics of these structures. Several approaches are now available for such identification, due to improvement of theoretical approaches as well as of experimental and numerical techniques. Identification techniques should be chosen after analysis of the geometrical flow properties, and closely depend upon the nature of the available information from experiments or calculations. The present course mainly focused on identification techniques specifically developed for free turbulent flows, but some applications to wall bounded flows were also addressed.

The different detection methods are now well qualified and mature enough to be used individually or concurrently. The researcher can choose the required method according to the objectives of his study and the available information, or the computational facilities and/or the time allowed for these studies.

In addition to the statistical analysis usually performed, new areas are now open which stem from the various concepts linked with the Structure Identification methods. As an example, harmonic analysis can be advantageously replaced by other decompositions. These new aspects allow to re-visit the usual concepts of flow control, an essential topic for future fluid mechanic applications.

The aim of this book is to provide the basis needed for structure characterization in free turbulent flows. Emphasis is placed on practical aspects of properly designed measurements and related algorithms. The topics covered included: Identification and basic concepts of Coherent Structures, multiple decomposition, experimental requirements (multiprobe concepts) and limitations, signal preprocessing. The aspects linked with numerical simulations are also considered. The major identification techniques are described in detail: Eduction Schemes, Proper Orthogonal Decomposition, Stochastic Estimation, Pattern Recognition Analysis, Wavelet Transform based techniques.

The experimental constraints and limitations are stated and the state-of-the-art of experimental methods which permit the application of CS detection are presented. The possibilities of CS detection from numerical results are also discussed.

In addition, basic concepts on Coherent Structures or eddies are presented. Topological concepts and their relation to dynamics, fractals and multi-fractals are introduced. In order to illustrate the applications of the detection processes, several parts are devoted to vortex dynamics with typical applications to experimental and numerical data for several flows.

Practical applications are presented, such as turbulence management, active control for drag reduction, mixing, combustion, ... Geophysical and industrial aspects are also analyzed. Lastly, the predictive possibilities of the techniques and some directions for future work are discussed.

Examples of applications of several detection methods to test flows based on experiments and numerical simulations allow for comparison of the possibilities, limitations and possible developments. These real examples also help the readers to select the most appropriate approach for their own problem, taking into account the aims of their study and the specificity of the available data.

Some of the topics addressed here have already been partially published elsewhere but new researchers are now faced with the problem of performing a vast bibliographical survey in order to cover the basic topics of this field. Therefore, this book is quite useful for beginners as well as for trained scientists because it covers not only the classical theories and results, but also more updated concepts and applications of Structure Eduction. It provides examples of applications of experimental and computational methods. It also opens new roads based on these analyses.

J.P. Bonnet

CONTENTS

GENERAL CONCEPTS ON STRUCTURE IDENTIFICATION

J.P. Bonnet and J. Delville
University of Poitiers, Poitiers, France

ABSTRACT

The notions of Coherent Structures and their detection in turbulent flows are introduced, and some applications in far wake flows are discussed. In a second section, considerations on the requirements for experiments in view of structure eduction are presented, in connection with techniques such as Fourier analysis and Proper Orthogonal Decomposition (POD). The third part of this talk presents two techniques used for structure detection: The Variable Integration Time Average (VITA) method and the Pattern Recognition Approach (PRA). The last part will be devoted to the presentation of some typical results obtained from applications of various CS eduction techniques on the same data measured in a plane mixing layer. The following techniques are then discussed: Pseudo-Flow-Visualisation (PFV), Proper Orthogonal Decomposition, Linear Stochastic Estimation (LSE) and some complementary techniques.

1 INTRODUCTION.

The existence of so-called Coherent Structures (CS) in turbulent flows is now well accepted. These structures play a crucial role in mixing, development, stability, radiated noise etc... CS in transitional or artificially excited flows are generally easy to identify. However, in ordinary turbulent flows, these structures are embedded inside the overall chaotic behavior of the turbulence field and are then much more difficult to detect. The experimentalists or numerical analysts are faced with a first question: how can these structures be extracted from the background turbulence and how can their average characteristics (in the sense of probability or dominance) be determined?

This is a non trivial task because the coherent structures are generally hidden in a random field and because the technique used to determine when and where certain structures are passing or present is *directly related to the definition* of the coherent structure itself. Several techniques for structure identification are available and the choice of a particular one depends not only on the definition of the structure, but also on the experimental and numerical capabilities available. For example, from an experimental point of view, some approaches require multiple sensors in order to obtain simultaneous instantaneous information at several spatial locations. Other approaches, however, only require knowledge at one or only a few spatial locations. The nature of the information is also crucial; velocity components, pressure, contaminants, etc... If numerical simulations are concerned, it is necessary to choose and save only the appropriate relevant parameters.

It is always necessary to choose the optimal experimental arrangement of probes or location of data in order to characterize the structures. The problem is equivalent to trying to find the simplest (minimum) probe arrangement that gains maximum significant information about the structures. The recent advances in data collection, instrumentation, digital data processing, and increased computational and storage capabilities of computers, including graphic software, have resulted in an exponential increase in the application of these methods.

Indeed, it is first necessary to try to define the Coherent Structure. The term itself is controversial. Sometimes called Organized Motion, Large Scale Structure or Coherent Eddy, these events should share a minimum set of characteristics. For the sake of simplicity we will hereafter employ the generic term CS to denote these events whatever the corresponding definition that is assumed.

Within the last few years, several meetings devoted to topics directly related to CS have been held. Detailed proceedings are available (for these meetings) [65, 68, 66, 67, 69, 70].

Among others, let us recall some available definitions:

From Lesieur [45, 72]:

Suppose that, at a given time t_0 a vorticity concentration exists within a domain $D_{t_0} \subset R^3$. Let D_t be the image of D_{t_0} at $t \geq t_0$.

D_t is a *Coherent Structure (coherent vortice)* if :

i) D_t keeps a recognizable pattern for time delays greater (typically 5 times) than the turnover time $(\omega_0{}^{-1})$.

ii) D_t is unpredictable (i.e. extremely sensitive to small perturbations of initial conditions). More details on this definition will be given by Metais during this course.

From Hussain [39, 40, 70]:

"A *Coherent Structure* is defined as a flow module with instantaneous phase-correlated vorticity". This definition and related concepts and illustrations will be provided also by Hussain during this course.

Berkooz *et al.* in [8] define CS as "organized spatial features which repeatedly appear (often in flows dominated by local shear) and undergo a characteristic temporal life cycle". A rational method for extracting such "features" was proposed by Lumley [47] and will be described later in this course. This author [49] introduces the concept of "Building Blocks" (basis of non-specified functions) based on the notion of more or less "energetic modes" on which the velocity field is projected. Using the Shot-Noise Decomposition, typical representations of the CS can be provided.
El Tahry reports several other proposals in [66] :
Perry defines a C.S. as a pattern which occurs periodically but is not necessarily ordered (scale and convection velocities being random), and whose orientation is fixed.
According to Kline, C.S. are recursive events contributing to the dynamics of the flow, such as the production of Reynolds stress.
Wygnanski states that the CS can be associated with the dominant modes of the instability. This proposal is somewhat contradicted by Jimenez.
Stull defines CS as entities which contribute to transport of momentum and passive scalars over given distances without considerable diffusion.
In addition, let us recall that Hunt classifies CS in a topological way. Several aspects of this approach will be presented by Vassilicos during the present course.
The concept of a "Conditional Eddy" introduced by Adrian in some sense avoids the problems of defining CS, since the conditional eddies are defined by the event itself. This concept does not imply quasi-periodicity of the flow, nor any mechanism related to the generation of Reynolds stress (see for example [2, 70]).

Each of the Structure Eduction methods correspond to one (or several) of these definitions.

In the case of numerical approaches such as DNS or LES, the available information is fully detailed by definition, so that any definition can be retained and any structure detection method is applicable. An illustration of the multiple possibilities offered by numerical data can be found in [50], [3], [70], or in the TCFD special issue [71].

Contrarily, in the case of experimental approaches, it is first necessary to define the available or accessible information.

Historically, one of the first and, in a way, most simple methods to detect CS was to visualize these events. Several visualization methods are available, the most popular being to use dyes or smoke in incompressible flows, or schlieren in compressible flows. The following plate 1 recalls some "morceaux choisis" of spectacular visualizations in free shear flows [64], [73, 74, 11, 14, 19] obtained from experiments. Two other examples obtained from numerical methods [60, 18, 72] are shown in Fig. 2.

Although apparently basic, the visualization process is not obvious. When tracers are used, the location of seeding should be chosen with particular care, so as to ensure that the tracer particles follow the vorticity. Also, it is necessary to assume that the

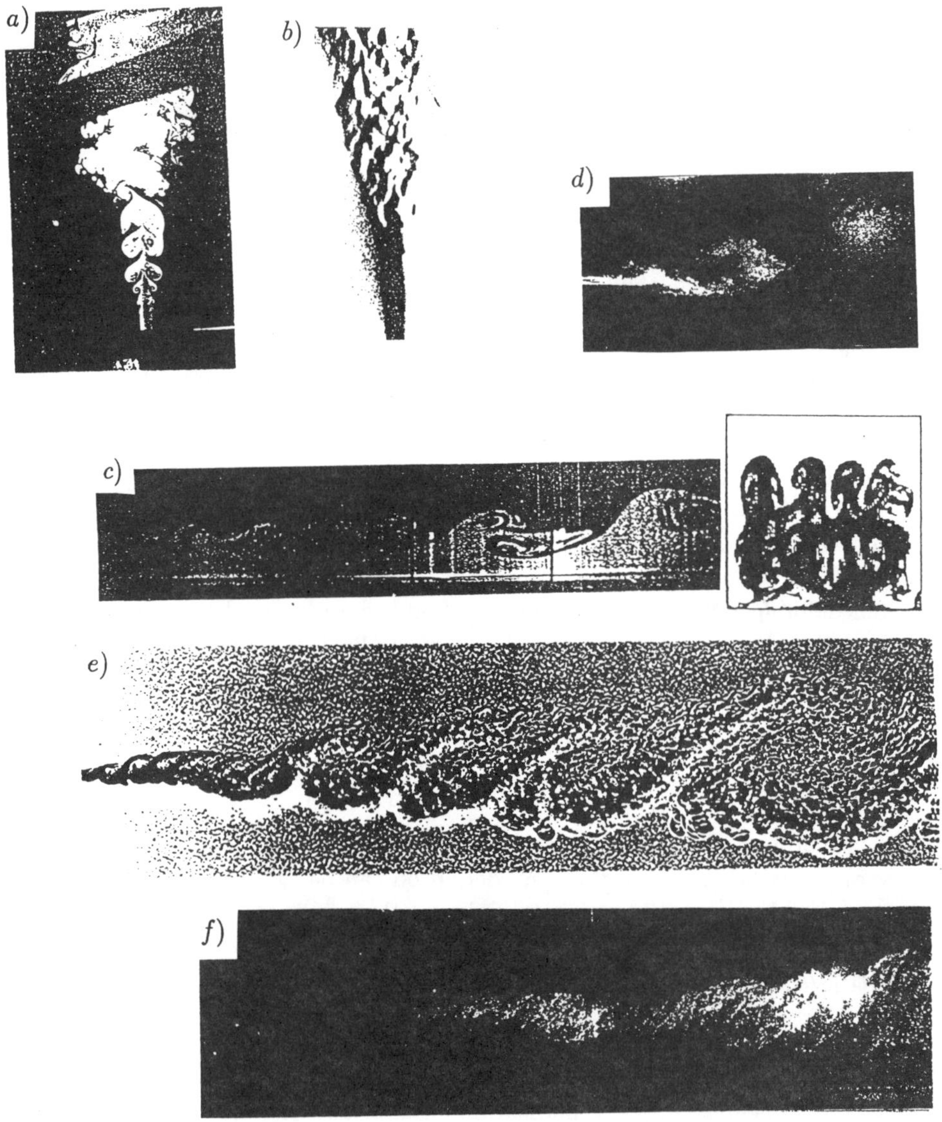

Figure 1: Typical visualizations in free shear flows from experiments. a)Free jet, Laser Induced Fluorescence [73] ; b) Free jet, Schlieren [74]; c) Mixing layer, Fluoresceine (longitudinal and front views) [11]; d) Subsonic mixing layer, Smoke [38]; e) Two-gases mixing layer, Schlieren [14]; f) supersonic/subsonic mixing layer, Laser planogram [19]

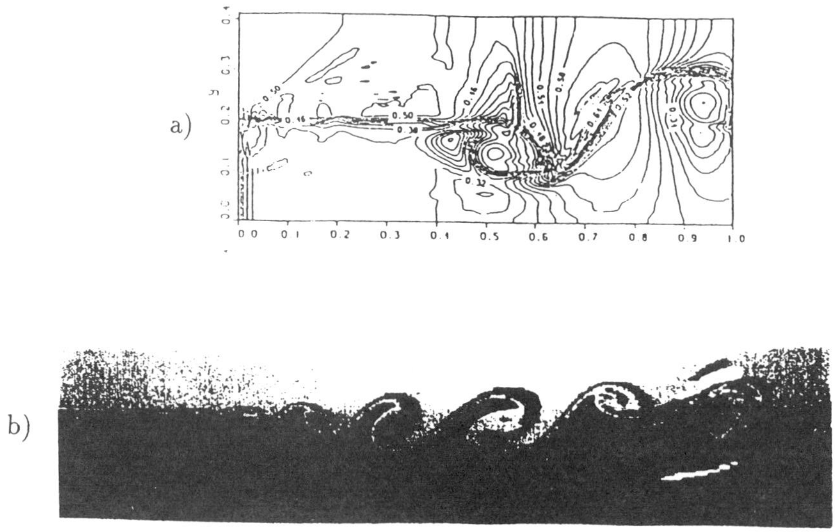

Figure 2: Typical visualizations in free shear flows from numerical simulations in supersonic mixing layers. a) Semi-deterministic modeling [60] ; b) LES from [18]

molecular diffusion is negligible.

The visualizations of Cimbala *et al.* [16], reproduced on Fig. 3, show examples of misleading results obtained when the seeding (here smoke) is introduced in different positions in a wake with a smoke wire. In this case, when the tracer is introduced far from the initial roll-up of vortices or observed far from the seeding location, erroneous interpretations can be made. This effect can be seen clearly when comparing the patterns obtained at the same streamwise distance with various seeding locations (consider for example the location $x/d = 250$ in the three photographs of Fig. 3).

More quantitative analysis can be performed with conventional measurement methods. Up to now, hot wire anemometry is the privileged method due to two main properties: it delivers velocity signals which are continuous in time and, due to its relatively low cost, it allows multiple point measurements.

PIV (2D or, better Holographic) associates visualization and multi-point velocity measurement, which allows the user to obtain visual information and potentially to use any detection method. These approaches are at present limited to research Labs, but are under rapid development so that systematic use is probable in the near future (v. Adrian or Hussain in this course).

Laser-based quantitative measurements are generally not used in CS eduction processes, due to limitations in the number of measurement locations and due to the random occurrence of the signals. However, a few attempts have been made to use LDV for CS characterization purpose [44, 43]. In such cases, the non-intrusive character of

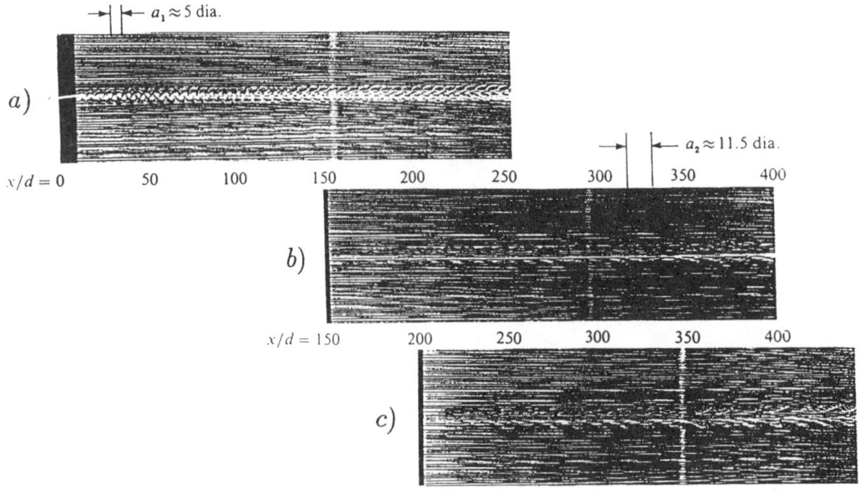

Figure 3: Comparison of visualizations obtained for different seeding locations in the wake of a cylinder ($R_e = 130$) from Cimbala *et al.* [16]. Smoke wire at a) $x/d = 8$; b) $x/d = 150$; c) $x/d = 200$;

LDV is a crucial advantage.

Except for the excited case, usual CS detection requires at least 2 point information, particularly when velocity is concerned. The early use of two point measurements was to compute the space-time correlation between two signals (see for example [24]). Later, instantaneous analysis was made possible by use of more complex data storage and processing.

1.1 A simple example of correlation analysis

We recall here, from the pioneering work of Townsend [58], a simple example that exhibits the information that can be gained from the analysis of correlations.

Despite their global, stochastic character, space-time correlations indeed contain relevant information on the large scale characteristics of turbulent flows. This example can be derived from a simple analytical approach. Let us consider an isolated eddy corresponding to the following velocity distribution:

$$\begin{aligned} u_1 &= -\frac{\partial}{\partial x_2}\left[e^{-\frac{1}{2}\alpha^2 x^2}\right] \\ u_2 &= \frac{\partial}{\partial x_1}\left[e^{-\frac{1}{2}\alpha^2 x^2}\right] \end{aligned} \tag{1}$$

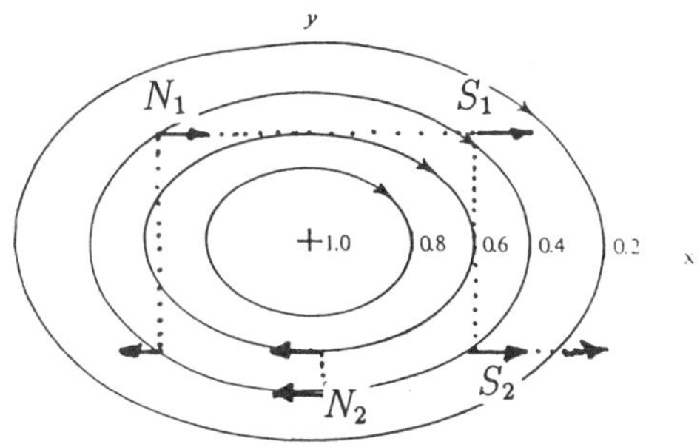

Figure 4: Basic eddy and typical measurements for "streamwise" separations (S) or "normal" separation (N)

$$u_3 = 0$$

Where $\alpha^2 x^2 = \alpha_1^2 x_1^2 + \alpha_2^2 x_2^2$.

This is a simple 2D eddy in which circulation lies in the $x_1 - x_2$ plane. The parameter α defines the shape of the eddy; an example is given in Fig. 4. From this repartition, Townsend proposes to compute the correlation by considering that the flow is built from a homogeneous random spatial distribution of eddies in the $x_1 - x_2$ plane. The ensemble average is then replaced by a space average:

$$\begin{aligned} r_{ij}(r_1, r_2) &= \overline{u_i(\vec{x} - \vec{r}/2).u_j(\vec{x} + \vec{r}/2)} \\ &= \iint_{-\infty}^{+\infty} u_i(x_1 - \tfrac{r_1}{2}, x_2 - \tfrac{r_2}{2}).u_j(x_1 + \tfrac{r_1}{2}, x_2 + \tfrac{r_2}{2}) dx_1 dx_2 \end{aligned} \tag{2}$$

Let us restrict the analysis to the x_1 component in order to compute only r_{11}. The "instantaneous" product can be written from (1,2) in simplified form:

$$u(\vec{x} - \vec{r}/2).u(\vec{x} + \vec{r}/2) = A(x_1).e^{\frac{1}{4}\alpha^2 r^2}.(x_2^2 - \frac{r_2^2}{4}).e^{-\alpha_2^2 x_2^2} \tag{3}$$

Where A depends on the characteristics of the eddy ($\alpha^2 r^2 = \alpha_1^2 r_1^2 + \alpha_2^2 r_2^2$).
The integration over the $x_1 - x_2$ plane then yields:

$$r_{11}(r_1, r_2) = B.e^{-\frac{1}{4}\alpha^2 r^2}. \int_{-\infty}^{+\infty} (x_2^2 - r_2^2/4).e^{-\alpha_2^2 x_2^2}.dx_2 \tag{4}$$

The final form of the correlation is:

$$r_{11}(r_1, r_2) = C.e^{-\frac{1}{4}\alpha^2 r^2}.(1 - \frac{1}{2}\alpha_2^2 r_2^2) \tag{5}$$

(B and C are constants).
Two simple illustrations can then be given in order to illustrate direct analysis from the

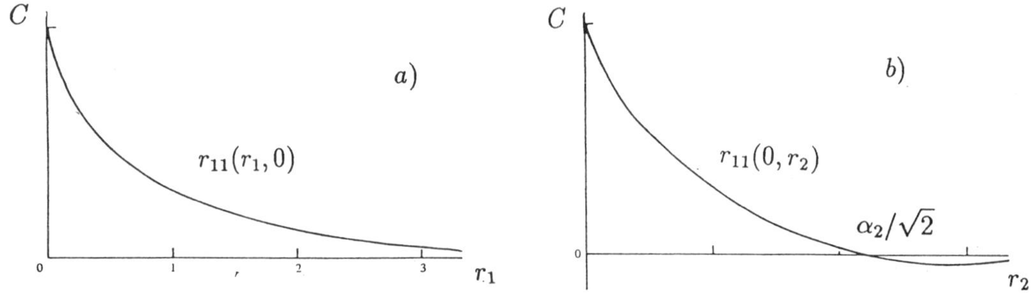

Figure 5: Longitudinal (a) and lateral (b) correlations

correlations. First, let us consider the correlation in the x_1 direction ($r_1 \neq 0; r_2 = 0$). It follows that:

$$r_{11}(r_1, 0) = C.e^{-\frac{1}{4}\alpha_1^2 r_1^2}$$

This function is strictly positive with an exponential decay, as schematically plotted in Fig. 5-a).

If we analyze the correlation for a separation perpendicular to the observed velocity, i.e. x_2 direction ($r_1 = 0; r_2 \neq 0$), the correlation has a zero crossing behavior:

$$r_{11}(0, r_2) = C.e^{-\frac{1}{4}\alpha_2^2 r_2^2}.(1 - \frac{1}{2}\alpha_2^2 r_2^2)$$

In this last case, negative correlation occurs, as shown on Fig. 5- b).

In the examples provided, it is easy to justify a-posteriori the global distinct behavior of the two correlation coefficients for the two space directions. If we suppose that two sensors are separated in the concerned velocity direction ($\equiv$ "streamwise" direction, here x_1), the velocity product will be positive whatever the distance and the location are. This is schematically illustrated in plots S of Fig. 4. Contrarily, if the transverse separation is considered, for some space separation associated with some locations in space (for example when two probes are located on each side of a passing structure), the velocity product can be negative, and thus the correlation can have negative lobes as illustrated in the plots noted N in Fig. 4.
It is therefore clear that the correlation between two points in space (or time) contains relevant information on the structure of the flow. Townsend [57] has applied this approach to the far wake velocity correlations. The use of complex "model eddies", instead of the crude one given above, has proven to be of limited efficiency. More sophisticated analysis of the correlations is required to extract unambiguously the relevant information.

1.2 Overview of the detection and analysis of CS

It is well admitted that multiple (2 or more) point measurements are the most fruitful methods for educing and characterizing the CS in any turbulent natural (non-excited) flow.

Although any classification remains arbitrary, data processing in the case of simultaneous measurements at 2 or more points can be classified into two main groups, according to the approach used. The first one, that we can call the "conditional approach" consists of detecting, from some condition, the passage or presence of some coherent events. This detection allows one to collect events and then to make averages in order to retrieve the main characteristics of the CS. This first approach requires variable degrees of subjectivity or pre-knowledge of the flow characteristics. The second approach consists of using statistical information (such as space-time correlations for example) to derive the characteristics of the CS. This approach is entirely "objective" or, at least, more "objective" than the first one. However, it is limited to stochastic information, although under certain conditions it can provide instantaneous characteristics and can be associated to the conditional approach, making the two approaches complementary. From both approaches, in addition to the information on the morphology of the CS, it is possible to determine the dynamics of the CS, their contribution to the turbulent processes, etc... These characteristics are essential for the purposes of analysis and control of the flows, for the development of new computational methods, and for the validation of numerical simulations.

<table>
<tr><th>CONDITIONAL</th><th>Non-CONDITIONAL</th></tr>
<tr><td>• Detection: (⇔ Structure definition)
- Fluctuation Levels
- 4 Quadrants
- VITA
- Gradients
- TPAV
- Tracers (temperature ...)
- Visualizations</td><td>• Space-Time Correlation
& Spectral Analysis

• Orthogonal Decomposition</td></tr>
<tr><td colspan="2">• Wavelets
• Stochastic Estimation
• Pattern Recognition</td></tr>
<tr><td colspan="2">STATISTICS:</td></tr>
<tr><td>- Ensemble Average
- Multiple Decomposition
- CS dynamics</td><td>- More Probable Characteristics
- Modal Decomposition
- Dynamical Systems</td></tr>
<tr><td colspan="2">- Statistical Properties</td></tr>
</table>

However, in a recent paper, Berkooz *et al.* [8, 70] have shown how much these different methods are interdependent and Kevlahan *et al.* [70] have compared the application of several of these methods to the same synthetic signal.

Considerations on the nature of the information obtained by each method, the complexity of use, the type of information required and mathematical and/or computational associated tools should be taken into account before any choice. This choice will guide or will be guided by the experimental method available. These different aspects will be discussed further in this course.

In the case of numerical approaches such as DNS or LES, the researcher is also faced with a problem of the same nature as are the experimentalists, except that the researcher knows all the variables of the flow. The analysis of the results in terms of CS help to decide what quantity should be stored precisely (v. Savill in [70]).

In the last few years, rapid improvements of the capabilities of multi-point measurement techniques have been observed. The cost of anemometers have decreased and probe technology has been improved. The data storage process has been improved considerably and computer memory space has increased. It is thus possible to apply most of the CS detection methods to both experiments and computations. However, the approach to CS detection can be quite different. Computations provide generally better spatial resolution than experiments. Contrarily, the time steps obtained from experiments can be as small as needed.

The aim of this course is to provide the basis needed for structure characterization in free turbulent flows. Emphasis will be placed on practical aspects of properly designed measurements and related algorithms. The topics covered will include: evidence and basic concepts of CS, multiple decomposition, experimental requirements (multiprobe concepts) and limitations, signal preprocessing. Linkage to numerical simulations will also be considered.

The experimental constraints and limitations will be discussed and the state-of-the-art of experimental methods that permit the application of CS detection techniques will be presented. CS detection utilizing numerical results will also be considered.

In addition, basic concepts related to CS or eddies will be given. Topological concepts and their relations to dynamics, fractals and multi-fractals will be introduced. In order to illustrate the applications of the detection processes, several courses will be devoted to vortex dynamics and will include typical applications to experimental and numerical data bases for several flows. Practical applications will be presented, such as turbulence management, active control, drag reduction, mixing, combustion, aeroacoustics, ... geophysical and industrial aspects will also be analyzed. The predictive abilities of the techniques and some directions for future work will be discussed.

Applications of several detection methods to test flows obtained from both experiments and numerical simulations will be given with the purpose of highlighting the strengths and the weaknesses of these methods. These real examples will also help the participants to select the appropriate approach to their own problem, taking into account the aims of their study and the data available to them. Alternately, this course

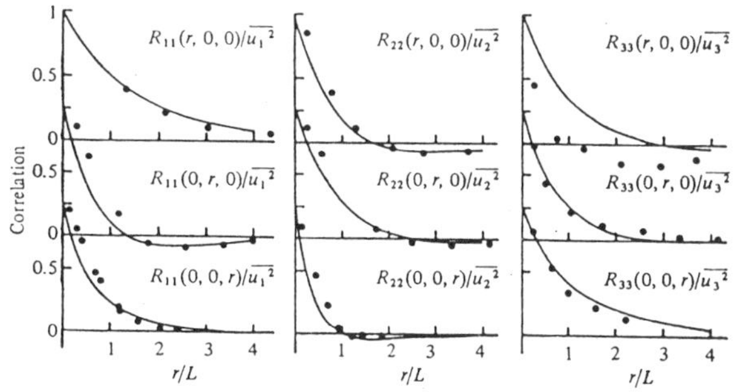

Figure 6: Normal components of the correlation functions in a wake measured by Grant [36]

will serve in designing either the experimental arrangement and associated measurement apparatus, or the optimal organization of numerical results for more efficient application of a particular CS detection technique.

1.3 The example of the far wake

Although geometrically simple, the wake of a cylinder is a flow which has been analyzed by various authors during several decades. In an early study, the two-point space-time correlation was measured by Grant [36]. He measured nine components of the correlation tensor which are presented in Fig. 6.

Later on, multi-point measurements were performed with arrays of, typically, 8 wires lying either in the transverse (Y) or in the spanwise (Z) direction. Combinations of transverse and spanwise arrays, and the introduction of temperature as a passive contaminant, have also been involved in the various investigations.

It is not the purpose of this section to treat in detail the different results. Instead, we will provide examples of the global information that has been published on this flow configuration to serve as illustrations of the applications of some CS characterization methods. We will focus on the far wake results, omitting all the near wake results.

• Payne and Lumley [53] applied Proper Orthogonal Decomposition for the first time to the measurements of Grant. They used the Gram-Charlier expansion for retrieving all the components of the cross-correlation tensor. The result is a pattern with a double-roller form; the axes of the rollers tend to be parallel to the direction of the mean shear (see Fig. 7).

• From other measurements with a single HWR, Delville [23] performed a Proper Orthogonal Decomposition of the Space-Time-Correlations in order to analyze the individual mode contributions. A typical reconstruction from the first mode is given in Fig. 8. More details can be found in section 4.5.

• From two-point correlations, Townsend obtained the "double-roller eddy" model

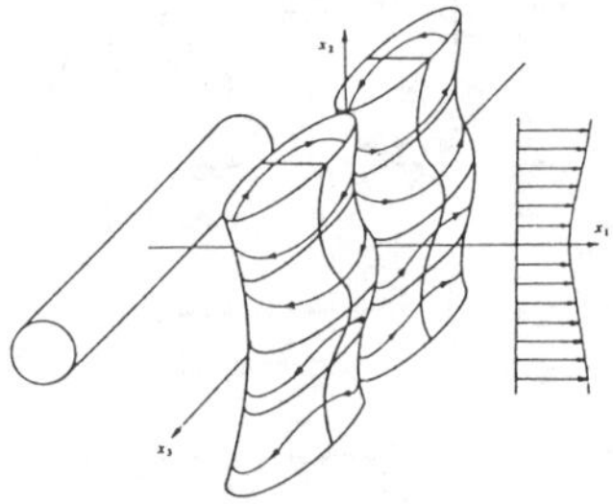

Figure 7: Sketch of the velocity pattern obtained from the POD method from the measurements of Grant [53]

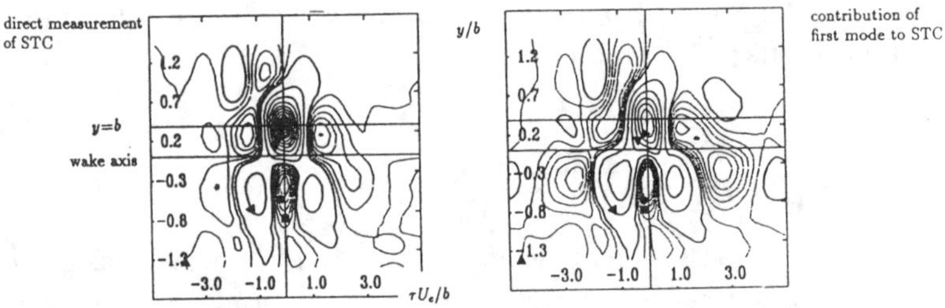

Figure 8: Comparison between raw data and first POD mode in a far wake from Delville [23]

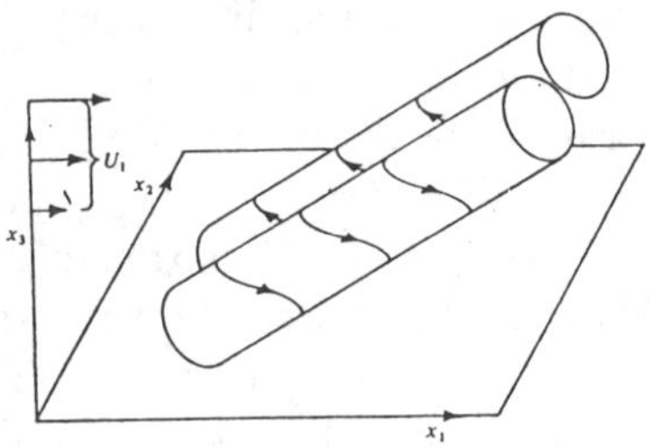

Figure 9: Sketch of Townsend's double-roller eddy (the arrows indicate the eddy streamlines) [57].

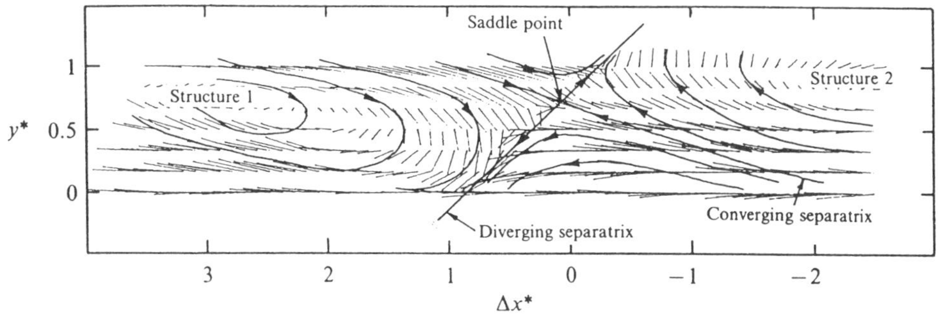

Figure 10: Velocity vector representation of the structures (in a reference frame moving at the structure average convection velocity). After [5]

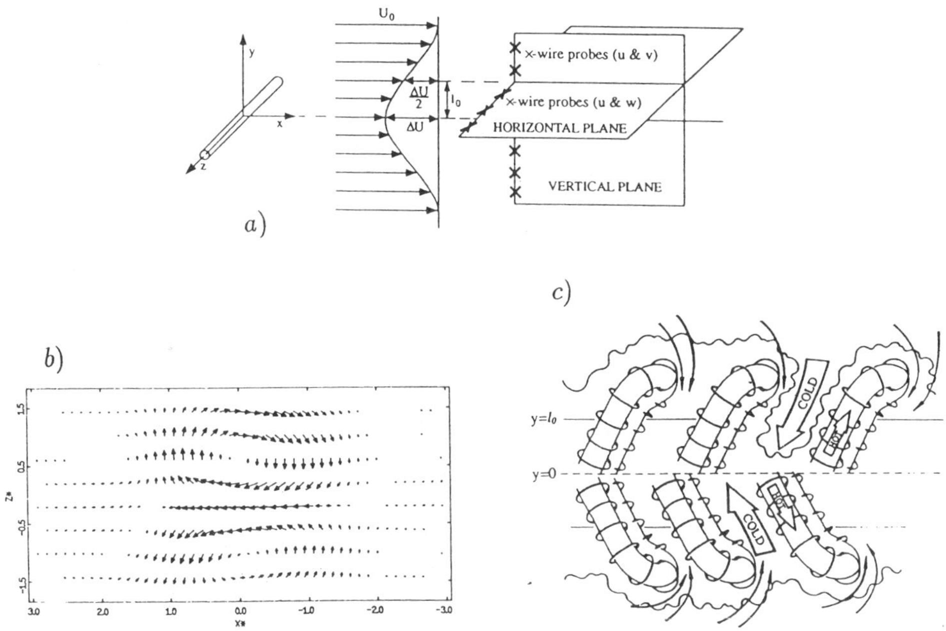

Figure 11: Sketch of the entraining motions from Ferré *et al.* and Giralt and Ferré [34]. a) experimental arrangement; b) a double-roller eddy; c) instantaneous 3D sketch; d) instantaneous u and w in the horizontal plane at $y = y_0$

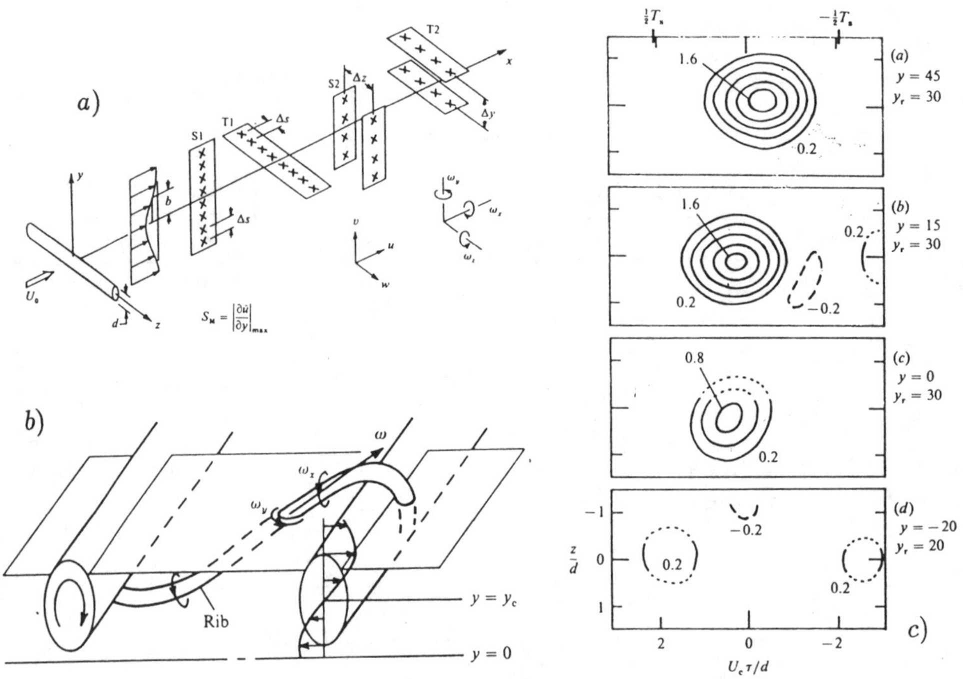

Figure 12: Characterization of CS in the intermediate wake from Hayakawa and Hussain [41]. a) experimental arrangement; b) illustration of the detection of the ribs; c) typical ensemble averages of ω_y for various T_2 rake locations

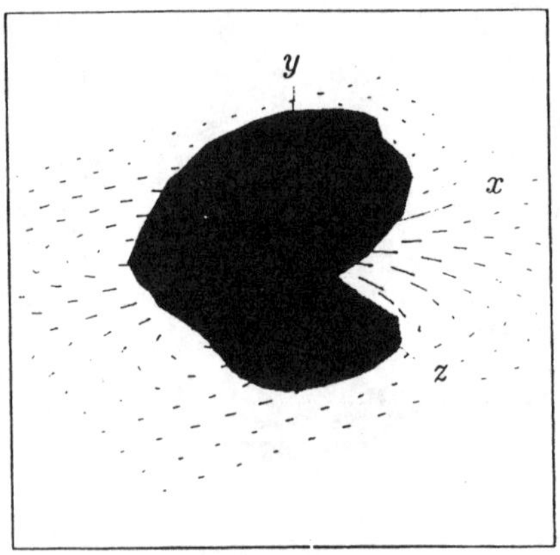

Figure 13: An oblique view of constant vorticity magnitude for the "Q4-Q1" event from Gieseke and Guezennec [70]

from a direct analysis of the correlations based on the generalization of the analysis described in section 1.1. Fig. 9 presents the schematic description of the pattern.

• Townsend [58] and Mumford [52] made use of PRA to educe the u footprints of the double-roller eddy. The results were more precise and quantitative than those obtained initially by Townsend.

Several other authors have analyzed the far wake by other means. Some typical examples are:

• With temperature used as tracer, Antonia *et al.* [5] performed a topological analysis of the signals. A typical vector representation of the structures is given in Fig. 10.

• With two hot wire rakes, Ferré *et al* [25] used temperature as a tracer to retrieve from PRA a 3D sketch of the CS. The result was confirmed by later measurements [28, 34]. Fig. 11 shows the typical eddies that can be deduced from these methods.

• Hayakawa and Hussain, in the case of an intermediate wake, applied conditional methods to educe the CS [41]. These authors, with a sophisticated probe arrangement (see Fig. 12-a)), associated with a specific detection scheme, see Fig. 12-b), were able to compute the 3D vortical characteristics of the wake flow, as illustrated by Fig. 12-c).

• An illustration of the use of 4-quadrant analysis is given by Gieseke and Guezennec [70]. They made use of conditions on the sign of the velocity component to estimate, from a Linear Stochastic Estimation, the field of the associated events. Fig. 13 presents a 3D view of the selected event.

2 SOME EXPERIMENTAL CONSIDERATIONS.

As discussed previously, the CS detection methods generally involve at least two point measurements, and continuously time evolving signals. Taking into account the recent progress in hot-wire anemometry, digital sampling, data storage, pre- and post-processing, we will focus our attention on multi-point measurements. Indeed, it is not surprising to encounter experiments dealing with tens of sensors.

In this section, we consider the requirements and constraints which govern the proper design of experiments in view of structure detection and analysis. These discussions will be devoted mainly to approaches involving hot-wire anemometry; however, most of the features described herein also apply to other techniques. General considerations prevailing for any conventional (one-point) measurement technique are recalled first, followed by specific concerns for multi-point measurements.

2.1 Sensor requirements.

2.1.1 General considerations.

As for any turbulence measurement, experiments dealing with CS detection methods should satisfy the usual requirements such as frequency response, spatial resolution, ... It is not the purpose of the present paper to analyze in detail these requirements. The reader can refer to the archival literature on this field [29]. However, especially when discussing the use of multi-point sensors, some additional constraints will appear, requiring the knowledge of some basic concepts that we will recall in the next section. In particular, the considerations related to the data collection, such as the phase between signals, quantization, aliasing and sampling rate, careful attention, and will be addressed.

2.1.2 Data collection.

Phase requirements: In the case of multi-point and/or multi-component measurements, the use of simultaneous sampling is recommended to avoid phase errors. With hot-wire measurements, the signals being continuous in time, this can be performed by using sample-hold amplifiers on each channel or by using individual synchronous converters for each channel. Any existing phase differences can be measured and *a posteriori* corrected by a phase rotation in Fourier space. However in the latter case, the zero time delay information cannot be retrieved.

Quantization: The resolution of A/D converters is limited to a range related to the number N_b of bits used for the conversion. The resolution should be high enough so that the quantization noise falls below the required level. Since this noise is generally a white noise, it affects the frequencies where the signal is of small amplitude compared to other frequencies. In turbulent flows, the spectra generally drop off rapidly at high frequencies and thus quantization effects occur in this domain. This effect can be of importance if the dissipative scales are considered. In order to avoid this effect, two solutions can be proposed: N_b can be increased (eg. the use of 16 bits A/D conversion instead of 12 bits provides an increase in the dynamic range of nearly 2 decades); another solution is to split the spectral analysis into two or more parts by performing analog filtering (low, high or band pass filtering) before the data acquisition. Indeed, this last approach is useful only when spectral descriptions are concerned.

Sampling rate and record length requirements: For most CS analysis, the time histories of the signals have to be considered. In this case, the usual Nyquist criterion (requiring low pass filtering at half the sampling frequency) is not strict enough. For such purposes, the sampling frequency should be about 5-10 times higher than the Nyquist frequency. On the other hand, in the case where only statistical moments are to be considered, aliasing is not a problem and in fact it is better for statistical reasons to use sampling rates lower than the Nyquist rate [30].

The choice of the appropriate record length is also related to the purpose of the study. If spectral estimations are required (or related quantities), the record length has to be long compared to the largest integral time scale. In any case (even if the spectral characteristics are not required) the length of the records has to be long enough to obtain accurate statistics, and it is recommended to use independent samples separated by many integral scales rather than a single sample. Obviously, in the latter case, statistical independence between the samples is not ensured. Finally let us recall that the total length of the record is a function of the quantity that has to be characterized. In the case of conventional statistics, the precision of the estimators can be computed (the length of the sample being larger when higher order moments are requested). On the contrary, if conditional sampling is used, the length of the records cannot be predicted without a minimum knowledge of the flow characteristics. In this case, an *a posteriori* convergence criteria should be used. If the events appear rarely, the necessary amount of data can be huge.

2.2 Particular requirements for multi-point measurements.

As stated previously, multi-point measurement techniques are developing rapidly. The main tool used in such techniques is the hot-wire rake, hereafter denoted HWR. Specific requirements and problems have to be taken into account when using HWR, most of which are discussed in [33] or [32].

2.2.1 Flow perturbation

The first point that has to be addressed is the perturbation of the flow due to the presence of a HWR. Indeed, this problem is real for any hot wire measurement but is exaggerated when rakes are used.
It is thus mandatory to miniaturize the probe holder to limit rake/flow interferences, and to design a probe body and prongs that are long enough to reduce rake holder/sensor perturbations. An inevitable compromise must be made between mechanical strength and perturbations. Vibrations must be avoided, the length to wire diameter ratio should be kept high enough, and perturbations (e.g. heat exchange) as well as parasitic electrical phenomena between neighboring wires should be avoided.
Relatively dense spatial meshes with minimum perturbation effects can be obtained with specially designed HWR: figure 14 shows an example of a 48 single wire HWR where the probes spacing is 1.5mm.

In any case, checks have to be done for each flow in order to quantify flow perturbations. Two checks need to be performed:
• First, single probe measurements should be compared with the equivalent results obtained with the corresponding sensor in the rake. Velocities, spectral behavior and high order moment measurements should be validated.
• Second, two-point measurements should be compared between rake and equivalent 2

individual probe methods. These comparisons can be made for space-time correlations and/or on cross-spectra.
It should be emphasized that these checks have to be made for each rake configuration, because the perturbation effects differ from one probe arrangement to an other. For example, in the case of a plane mixing layer with quasi two-dimensional CS, the rake/flow interference is different if the rake lies in the transverse (perpendicular to the shear plane) or parallel direction. In the latter case, the perturbations of the large scale structures can be greater than for the first configuration.
Indeed, these effects are more pronounced when two (or more) rakes are used simultaneously.

2.2.2 Dynamic range

When using a rake of probes in a sheared flow, each probe is placed in a different flow regime: the mean flow velocity, mean flow direction, turbulence levels, Reynolds numbers, flow frequency ranges, and wall proximity, can all vary from one sensor to the other. These parameters can induce very different dynamic responses of the sensors. Each anemometer should be tuned individually in order to obtain the correct frequency response for the local flow regime. Calibration of the entire set of sensors can be difficult *in situ* due to turbulence. Specific calibration wind tunnels may be necessary. These problems may become very complex in some configurations. As an example, a mixing layer between supersonic and subsonic flows was studied in [19]. In such a flow case, each probe is in a different Mach number regime and has a very different dynamic response.
The dynamic range has to be adjusted to obtain the information of interest. If only large scales (or low frequencies) are to be analyzed, a crude calibration can be sufficient and, after checks, to quantify flow perturbations calibration of the whole rake can be performed in a single properly chosen regime.
However, it has to be recalled that if the higher moments of the signal are considered as relevant parameters, the dynamic response of the probe has to be improved. This must be kept in mind in the study of transient flows where, even if the typical scales of the flow are large, the transient evolution of the boundary conditions can be orders of magnitudes faster than these scales.
In addition to the abovementioned effects, spatial and temporal resolution must be analyzed, as in any turbulence measurement. The size of the measurement volume induces a spatial integration effect which limits the bandwidth of the sensor. This low-pass filtering impedes detection of the smallest dynamic scales of the flow. It is impossible to obtain any relevant information about scales typically smaller than twice the size of the measurement volume. The temporal resolution is directly related to the dynamic behavior of the measurement device. It is also limited by the spatial resolution of the sensor. For example, for a hot wire of length l and a flow with a

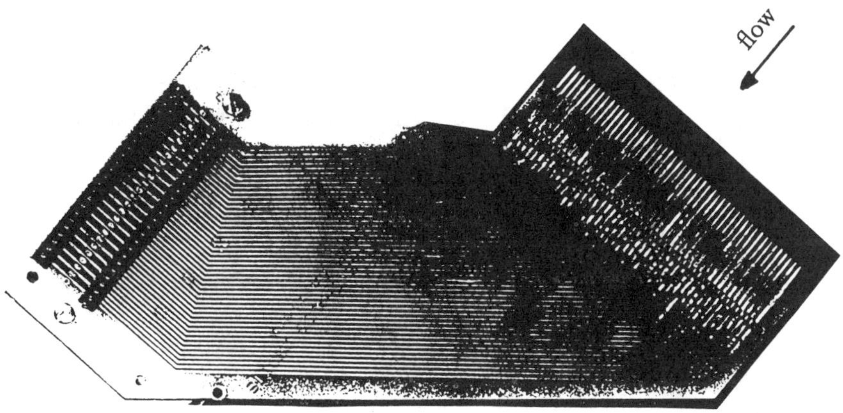

Figure 14: An example of a multi-probe rake: Rake of 48 single wire probes. Probe spacing 1.5mm, wire length 0.5mm, wire diameter 2.5μm, thickness of the rake 1.6mm. Used by Delville in [22].

convective velocity U_c, the resolvable frequencies are in the range:

$$f \leq 0.5U_c/l. \tag{6}$$

2.2.3 Data preprocessing.

Several authors have detailed other problems encountered when using multiple probes.

The characterization of structures requires in general the acquisition of a large amount of data (for statistical accuracy) involving experiments of long duration. The evolution of calibration characteristics is not identical for all the probes, and therefore small but non-negligible discrepancies may exist from probe to probe. When tens of probes are used at the same time, it is hardly possible to perform individual recalibrations.

One method to overcome this problem is to use inverse-calibration procedures. After a first accurate calibration of each probe and qualification of the flow, the data obtained from the experiment during one run are used to obtain new calibration laws [37, 22]. This pre-processing rescales properly the measurements.

In addition, long time records imply that during the data-acquisition process, some of the samples may be spoiled by parasitic effects such as electronic spurious noise, sudden flow perturbations, ... It then may become necessary to select the data records before applying the final data processing. This can lead to rejection of some samples whose statistics deviate too much from the estimator. In order to automatize the process, the selection can be based on statistical concepts. For example, the Probability Density Function of the mean-squared value or of higher moments can be computed and analyzed in terms of their gaussian character, recalling that, even if a process is not gaussian, the probability distribution of estimations of its averaged quantities have to be gaussian.

2.3 Multi-point configuration requirements.

Some considerations on the spatial arrangement of multipoint measurements will now be presented. The different features that can be found in the analysis of time varying signals summarized above have their counterpart if spatially varying information is to be analyzed. Generally, in stationary turbulent flows, the time dependency can be related to the Fourier analysis. However, in the case of spatially varying signals, two typical flow cases must be considered:
When the spatial direction of the measurement corresponds to either a homogeneous, or a periodic direction, Fourier analysis naturally provides a basis for analyzing the measurement requirements. Alternatively, when this direction corresponds to an inhomogeneous direction (this kind of configuration is naturally very important from an engineering point of view), a Fourier analysis cannot be performed. For the latter case the POD method is used to specify the measurement requirements.

It will be hereafter assumed that all the concerns addressed above are satisfied. We will focus now on the aspects related to the design of multi-point measurements. For both flow cases mentioned above, the problem is to decide upon the number of probes, as well as their location, spacing and the extent of their span.

2.3.1 Homogeneous/periodic fields.

In the case of either a homogeneous or a periodic field, the natural tool that can be shown to provide the optimal flow representation is the Fourier decomposition[1] (v. Lumley [48], or George [31]). Hence, the multipoint measurements must fulfill the conventional Fourier analysis criteria (in term of extent and probe separation). In the following, a homogeneous direction will be called z and the corresponding wavenumber k_z, while a direction of periodicity will be called θ and the corresponding Fourier mode number m. Let us consider that the number of points in the direction under consideration is N and its extent is L.

For a periodic direction (eg. azimuthal configuration of an axisymmetric flow), the appropriate basis is built from the modes m, so that the flow can be reconstructed

[1] Note that even if the intent is not to decompose the results of the experiment in terms of Fourier modes, **any** attempt to process the data by involving any multipoint ($N \geq 2$) dependency in this direction will implicitly depend on the Fourier coefficients.

from a *Fourier series*:

$$u(\theta) = \sum_{m=-\infty}^{+\infty} a_m \exp(jm\theta) \quad \text{where the Fourier coefficients are:} \tag{7}$$

$$a_m = \frac{1}{2\pi}\int_0^{2\pi} u(\theta)\exp(-jm\theta)\,d\theta \quad \text{note that } a_m \text{ are complex} \tag{8}$$

However, since measurements are performed for a limited number of points, the summation involves only $N = N_\theta$ points. Therefore, we obtain $N_\theta/2+1$ modes to describe this direction. This is in fact equivalent to what would be obtained by applying a Fourier discrete decomposition to a periodic temporally evolving signal.

In the case of a statistically homogeneous field, the same representation can be obtained in terms of the *Fourier transform* $\widehat{u}(k_z)$:

$$u(z) = \int_{-\infty}^{+\infty} \widehat{u}(k_z)\exp(jk_z z)\,dk_z \quad \text{where the Fourier transform is given by:} \tag{9}$$

$$\widehat{u}(k_z) = \int_{-\infty}^{+\infty} u(z)\exp(-jk_z z)\,dz. \tag{10}$$

Similar to the periodic configuration, an experiment involving $N = N_z$ points in the homogeneous direction provides $N_z/2+1$ spatial modes[2].

Two main effects can result: aliasing and the window effect

- Aliasing: Within an experimental context, there is no fundamental difference between a Fourier series and a Fourier transform since the experiment is always performed on a finite extent with a limited number of probes. The range of independent coefficients extends from wavenumbers $2\pi/L$ for the lowest one to $\pi/(NL)$ for the highest one. A consequence of this is that one given mode can be aliased into lower modes[3]. Consider for example samples, equally spaced, where M modes are needed to represent the signal and where only N measurement locations are used ($M > N/2$). Then $N/2$ modes can be computed. The information in the mode m ($m > N/2$) appears in the calculated mode $N-m$. There is then a fold at the cut-off wavenumber $N/2$.

  The aliasing effect may be negligible if the spectrum varies monotically and drops rapidly as the mode number increases. This feature is generally found in the one-dimensional spectra of time varying signals. On the other hand, examples can be found of spatially sampled data where the contribution to the lowest wavenumbers of the spectra is small compared to that of the higher modes. In the latter case, a significant aliasing towards the lowest modes may be present [32] (v. Fig. 15).

  For time varying signals, a low pass filtering procedure can be used **before sampling** in order to avoid aliasing effects; however, this cannot be applied easily

[2] This is the counterpart of what can be obtained in case of Fourier analysis for statistically stationary temporally varying signals.

[3] Note that once aliased, the original data never can be recovered.

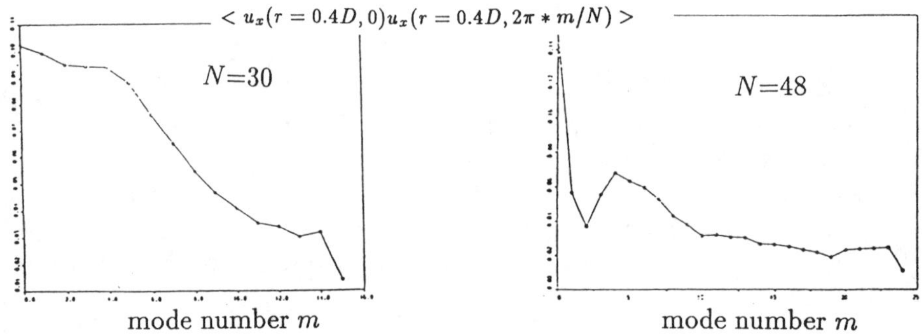

Figure 15: Modal decomposition of the mixing layer of an axisymmetric jet using both 30 and 48 positions around the circle. From George & Glauser [32]

to spatial sampling. An alternative could be to use very closely spaced probes, and then to perform a spatial average between neighbouring probes in order to remove the highest spatial wavenumbers. However, practical limitations do not allow such procedures. Thus, in most experiments the aliasing problem is neglected, in the hope that this effect will not occur. A quite trivial way to avoid spatial aliasing is to use sensors of sufficient length so that the highest spatial modes are suppressed by the effects of integration along the measurement volume [32]. It would be therefore best to use sensors whose length is comparable to the separation between the sensors. Hot wires are very well suited for this kind of approach and can provide a way to determine the influence of the aliasing effect on the results. A less satisfactory alternative could be to use a temporal low pass filter at a frequency related to the convective velocity and to the small scales that need to be removed (equation 6).

- Window effect: A window effect is imposed on spatial measurements due to the finite extent of the measurement field. This phenomenon is the counterpart of what can be found in time varying signals. In fact, results provided by measurements actually correspond to the convolution of the flow field with the spectral window $W(k) = L\text{sinc}(\pi kL)$. A spectral leakage appears whose effects increase if peaks are present in the spectrum. In general, this problem is significant primarily in homogeneous directions, where only one portion of the flow field can be considered. In the case of periodic directions (eg. axisymmetric configuration), this phenomenon does not occur as long as the entire flow field is considered.

Finally let us recall some problems related to the use of two point measurements. These problems may be encountered in applications where the spatial correlation is measured by two point procedures and then Fourier transformed in order to obtain the spectral representation.
For the homogeneous direction z the spectral correlation is defined by:

$$\Psi(k_z) =< \hat{u}(k_z)\hat{u}^{(*)}(k_z) > . \tag{11}$$

Ψ can be estimated via this equation only if the Fourier transforms can be used before averaging. For two-point measurements, the spatial correlation[4] $R(\delta_z) =< u(0)u(\delta_z) >$ must be estimated and the spectral correlation is then computed using:

$$\Psi(k_z) = \int_{-\infty}^{+\infty} R(\delta_z) \exp(-jk_z\delta_z)\, d\delta_z \tag{12}$$

However with such an approach, due to statistical errors in the determination of R, it is possible to obtain negative values of the auto-spectra $\Psi(k_z)$ at some wave-numbers, which is physically impossible. In order to avoid this problem it may be necessary to smooth the data, or even to use special algorithms for the Fourier Transform.

2.3.2 Inhomogeneous directions.

As in the previous section, the same questions can be asked concerning the number of probes, separations, ... to be used in order to infer, the flow organization for inhomogeneous directions from the measurements. Unlike homogeneous or periodic configurations, there is no general counterpart to the Fourier modes from which guidelines can be found. However the same kind of phenomena (aliasing,...) have their counterpart in data sampled in inhomogeneous directions.

The natural basis to use in order to quantify these problems is formed by the eigenfunctions obtained from the Proper Orthogonal Decomposition (v. Lumley this course) (POD)[5] of the flow. We consider here a one-dimensional decomposition of the inhomogeneous direction y:

$$\int_L R(y, y')\phi^{(n)}(y')\, dy' = \lambda^{(n)}\phi^{(n)}(y) \tag{13}$$

where $\phi^{(n)}$ and $\lambda^{(n)}$ are the nth eigenvector and eigenvalue and $R(y, y')$ is the two-point correlation. Among the properties of this decomposition, which are described in the

[4]here R is a function of δ_z only because the direction is homogeneous.

[5]Whether or not interested by the POD itself, because the POD is optimal for the energy representation and reduces to the Fourier decomposition in case of homogeneity. The field could also be expanded in Fourier modes (even less optimally) and the windowing and aliasing effects might be extrapolated from the previous analysis. However the use of POD can lead to less severe requirements.

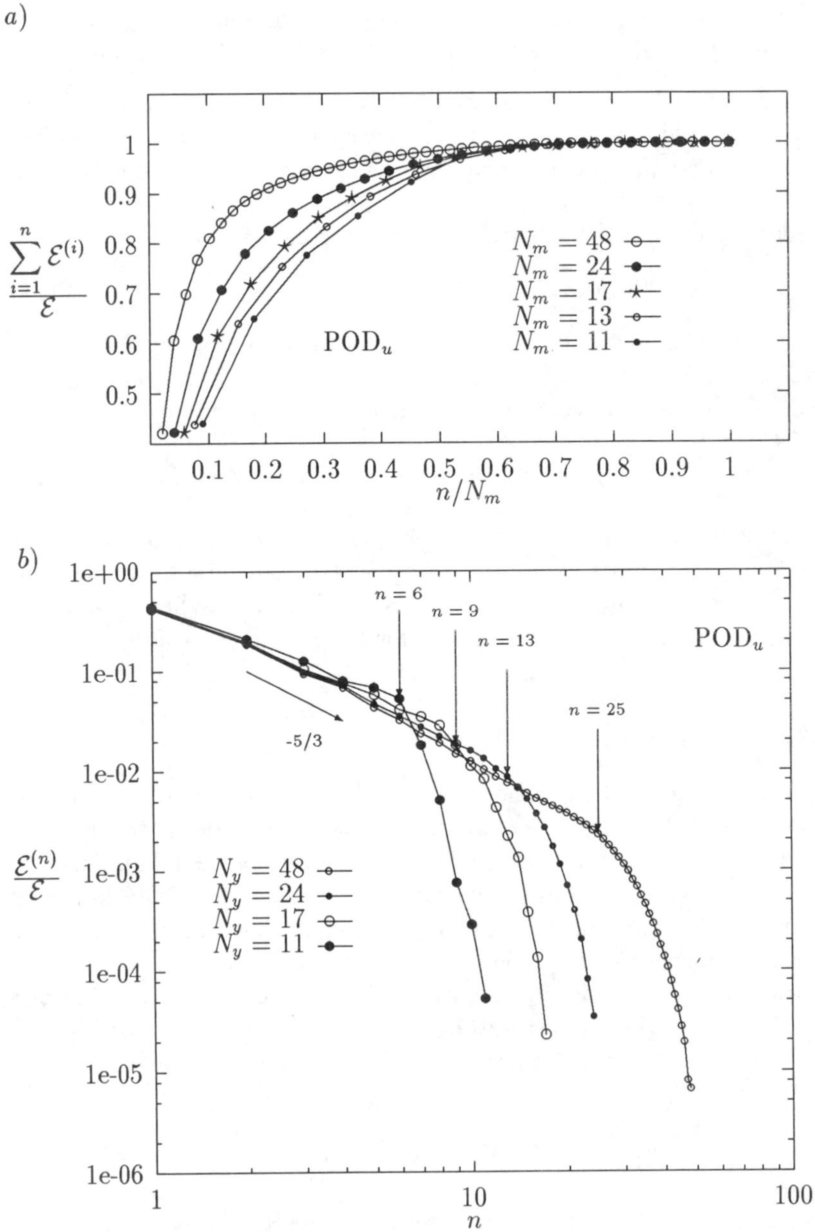

Figure 16: *a*) Influence of the number of grid points on the POD convergence ; *b*) Modal distribution of the modes of the POD for various mesh sizes.

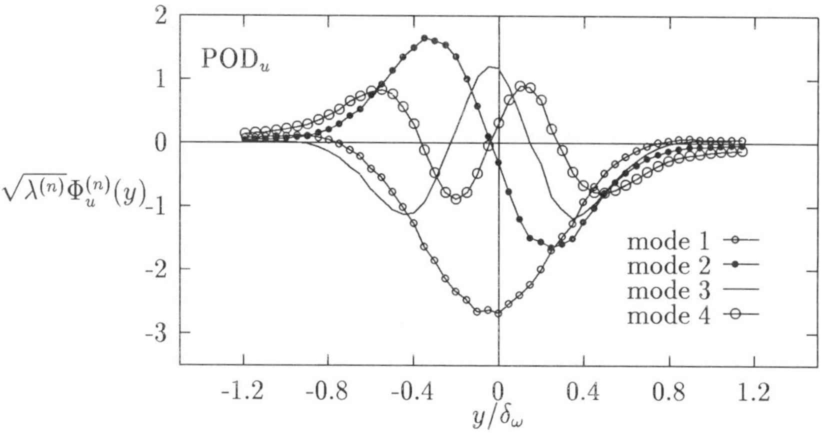

Figure 17: Shape of the first four eigenvectors of the POD base on u in the plane mixing layer.

course given by J. Lumley, we recall that the POD is optimal, that each flow field $u(y)$ can be reconstructed from the $\phi^{(n)}$ and that $R(y, y')$ can be retrieved from:

$$R(y, y') = \sum_{n=1}^{\infty} \lambda^{(n)} \phi^{(n)}(y)\phi^{(n)}(y'). \tag{14}$$

The energy contained in one mode can be defined as $\mathcal{E}^{(n)} = \int \lambda^{(n)} \phi^{(n)}(y)\phi^{(n)}(y)\, dy$.

Influence of the number of probes. Assuming N_y grid points, equation 13 can be approximated by the matrix eigenvalue problem:

$$R_{kl}\Phi_l = \Lambda\Phi_k \text{ ,where } k, l = 1, \cdots, N_y \text{ are the measurements points} \tag{15}$$

In this analysis we suppose that the points are equally spaced. The techniques for solving this problem are not discussed here but can be found elsewhere (v. [35]).

The number of points $N_y$used in the grid determines, in this case, directly the number of solutions or modes N_m that can be found when solving the eigenvalue problem. Hence the evaluation of the requested number of points can be compared to the search of the number M of modes necessary to reconstruct "properly"[6] the field. Lumley [48] argues that this number is proportional to the ratio $\mathcal{L}/l$ (spatial extent of the inhomogeneity divided by the integral scale of the field). For mixing layers, in plane [21] as well as in axisymmetric configurations [35] M is found to be of the order of 3-5. However, in bounded flows, such as boundary layers or channel flows,

[6] ie. capturing most of the energy.

the integral scale is a function of the distance to the wall and a sub-domain approach (inner-outer regions) provides a better description for a given mode number than a global description (v. Moin & Moser [50] or Chambers et al. [15]).
If N_y is chosen to be smaller than M, the corresponding eigenvalues present additional information from higher modes (leakage, aliasing). This is illustrated by figure 16-a where, in the case of POD applied to the plane mixing layer, the relative energy contained in the first n modes $(\sum_{i=1}^{n} \mathcal{E}^{(i)} / \sum_{i=1}^{N_m} \mathcal{E}^{(i)})$ is plotted for various numbers of points N_m. The energy within the first mode remains globally constant regardless of the mesh (42%). However, the coarser the mesh is, the higher the energy is in the lower modes. Another illustration of this effect can be found in Fig.16-b where the ratio $\mathcal{E}^{(n)} / \sum_{i=1}^{N_m} \mathcal{E}^{(i)}$ is plotted for the same grids in log-log representation. Here, a characteristic mode number related to half the number of modes is evident, similar to the Nyquist phenomenon.

One of the problems related to inhomogeneous directions is to provide a means for removing the higher modes from the lowest one in order to avoid any kind of aliasing effect. This cannot be performed in a simple manner unless a spatial filtering procedure, similar to the one used for homogeneous directions, is used. Some results however seem to indicate that for the highest modes, the eigenfunctions behave similarly to Fourier modes [50, 15]. This behavior is translated by an increase of the zero crossings of the eigenfunctions, following the mode number. This feature is shown in figure 17 where the shapes of the first 4 eigenfunctions obtained from a POD in a plane mixing layer are depicted.

In all of these aspects, it may be important to point out that a relationship exists between the spatial organization and the temporal scales in this configuration, (similar to the one of equation 6). The convection of the perturbation by the mean flow can limit the spatial resolution obtained. An example of the dependence of the "highest significant frequency[7]" on the mesh spacing for various POD is shown in figure 18. The solid line in the figure demonstrates a (monotonic) relationship between the resolvable frequency and the mesh spacing[8].

On the location of probes. Even if the abovementioned features are verified, there is no direct way to predict the optimum distribution of the measurement locations within the field. However it may be reasonable to concentrate the measurement locations where the kernel $R(y, y')$ (or the energy) varies the most. Such an approach is shown in figure 19. Here, results from 3 different PODs are shown. In two of them, the mesh points (48 and 24 points) are equally spaced while in the third one (23 points) the mesh spacing is proportional to the turbulent energy gradient. It is obvious that the third mesh provides the best representation of the flow, as the Nyquist cut off point is

[7] These results are obtained in the plane mixing layer, by using a slightly different form of the POD where the spectral-correlation is used as the kernel of the Fredholm problem [21].

[8] Surprisingly, the constant 0.1 found in this relation can be directly linked to the aspect ratio of the "large eddies" in the mixing layer.

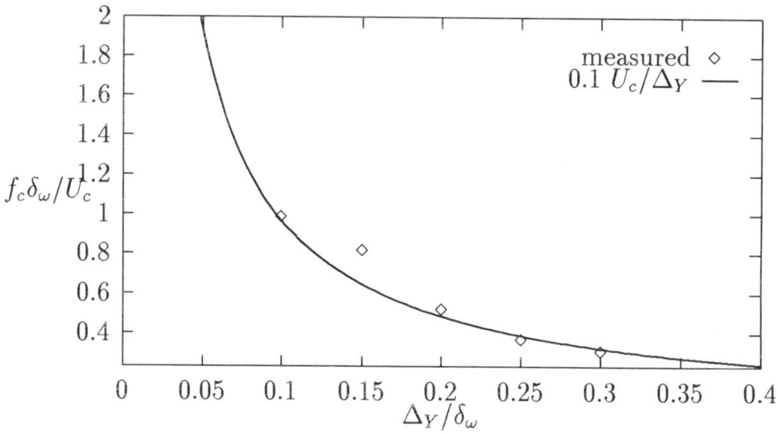

Figure 18: Evolution of the highest relevant frequency vs the mesh size for the first mode of the POD.

shifted towards the highest modes (as compared to figure 16).

In closing, it may be useful to recall that these analyses are valuable as far as energy is concerned but may not be easy to extrapolate to the case of conditional or topological approaches.

3 THE VARIABLE INTEGRATION TIME AVERAGE (VITA) AND PATTERN RECOGNITION ANALYSIS (PRA) METHODS

3.1 Introduction

Several attempts have been made to distinguish the passage of a turbulent structure embedded in spurious signals. In boundary layers, the different structures can be classified into several categories, each of them having their own signature [67]. This has led some researchers to split instantaneous signals into different categories and then to build detection criteria associated with each kind of structure. A typical example is the Four Quadrant analysis method where the relative signs of the velocity components are used for the classification of the detected events. This concept has been used mainly in boundary layer-type flows (see for example [61, 62]). More recently this concept has been used in free flows in association with Stochastic Estimation. Gieseke and Guezennec [70] have applied this combination to the study of a turbulent wake. We

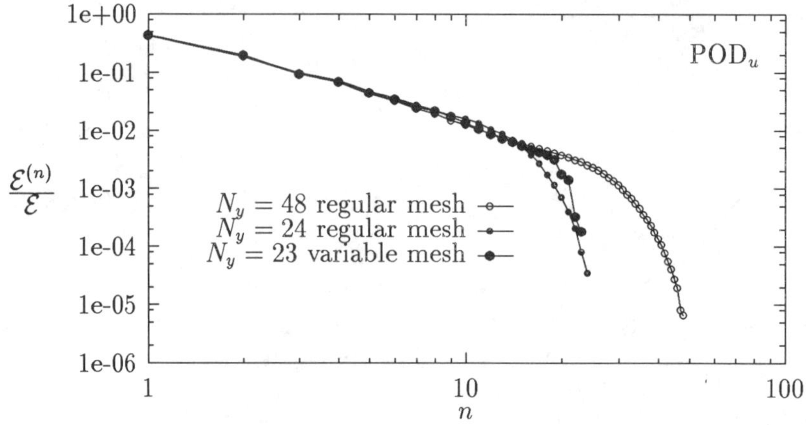

Figure 19: Influence of the spacing of the measurement points on the POD modal distribution.

will provide examples of these applications in the review of CS analysis in turbulent wakes.

3.2 The Variable Integration Time Average VITA technique

Another way to perform this selection is to introduce a time scale in the procedure. The Variable Integration Time Average (VITA) approach is now a popular method applied mainly to boundary layer flows, but easily applicable to free flows. The principle was introduced by Blackwelder and Kaplan [10] and is summarized in Fig. 20.

The raw velocity signal $u(\vec{x},t)$ (with a variance σ_u) is integrated over a short time scaled on the characteristic time scale of the structure. A new signal is then generated

$$\hat{u}_T(\vec{x},t) = \frac{1}{T}\int_{t-\frac{1}{2}T}^{t+\frac{1}{2}T} u(\vec{x},t)dt \tag{16}$$

The variance of the new signal, $\sigma_{\hat{u}}$ is then thresholded and a detector function is built.

$$D(t) = \begin{cases} 1 & \text{if } \sigma_{\hat{u}}^2 > K.\sigma_u^2 \\ 0 & \text{otherwise} \end{cases} \tag{17}$$

K is an arbitrary parameter defining the threshold.

The interpretation of the results is well established in bounded flows but less clear in free shear flows. As quoted by Ferré [26], this method is based on a local energy criterion. This criterion is not relevant for every CS, particularly in free shear flows.

Wallace *et al.* [61] have improved this approach by introducing a *pattern recognition*

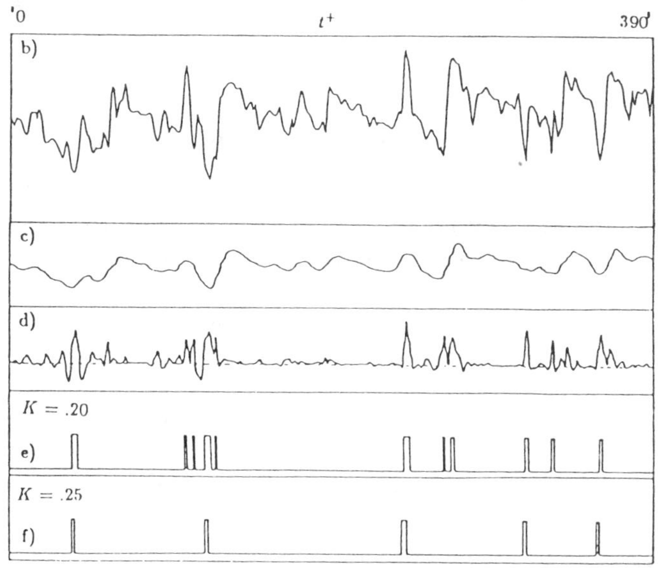

Figure 20: Example of VITA processing from Liandrat and Morlet-Bailly [46]. a) Raw signal; b) VITA integrated signal; c) Variance of the integrated signal $\sigma_{\hat{u}}$; d) Detector function with K= 0.2; e) Detector function with K= 0.25

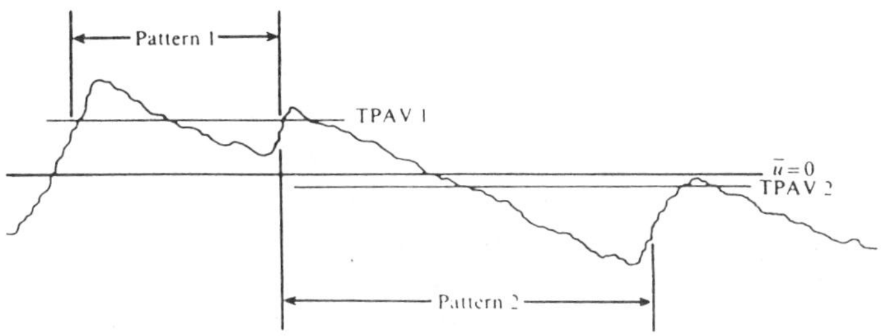

Figure 21: Typical pattern recognition method from Wallace *et al.* [61]

criteria for the velocity signals. The signals are examined for pre-defined typical behaviors such as a strong acceleration followed by a slow velocity decrease (see Fig 21). Here also, this approach is well suited to wall flows but is not convenient when the overall flow characteristics are unknown.

3.2.1 The Enhanced Variable Integration Time Average: EVITA

A new method, called EVITA for Enhanced Variable Integration Time Averaging , is proposed by Sullivan *et al.* [56] in order to combine VITA and PRA approaches. The idea is to retain the computational simplicity of VITA and the minimal subjectivity of the PRA.

The principle follows the abovementioned VITA approach, except that equation 16 is replaced by the convolution product:

$$\hat{u}_E(\vec{x},t) = \int_{-\infty}^{+\infty} g(\tau)\, u(t+\tau) d\tau \tag{18}$$

The measure of turbulent energy is used as the threshold quantity instead of the usual variances involved in eq. 17 and reads:

$$\hat{u_E^2}(\vec{x},t) = \int_{-\infty}^{+\infty} g(\tau)\, u^2(t+\tau) d\tau \tag{19}$$

The correlation function is the second derivative of the Gaussian function:

$$g\,(\frac{t-b}{a}) = [1-(\frac{t-b}{a})^2].\exp[-\frac{((t-b)/a)^2}{2}] \tag{20}$$

where t is the time, a is a dilatation coefficient of the function and b is a translation of the function.

Unlike PRA, the EVITA method does not require any actual pattern.

Another improvement is also introduced by Sullivan *et al.*. Instead of using a predetermined theshold as in the usual VITA method, relative thresholds are used: the detection occurs when the localized variance exceeds some percentage of its maximum. Fig. 22 shows examples of comparisons between VITA and EVITA applied to 10 ersatz structures arbitrarily seeded throughout a one-dimensional random signal. The scale numbers correspond to various normalized signal lengths, number 3 corresponding to the approximate scale of the ersatz structure. It appears that both techniques extract the approximate position of the structure for scale 3 (the scale of the structure). For other scales, EVITA is able to discriminate more completely than VITA between the structure and the background signal. EVITA is able to educe regions of structures regardless of their scale. In the abovementioned paper, EVITA is successfully applied to PIV results in a channel flow. Bursting events are effectively identified.

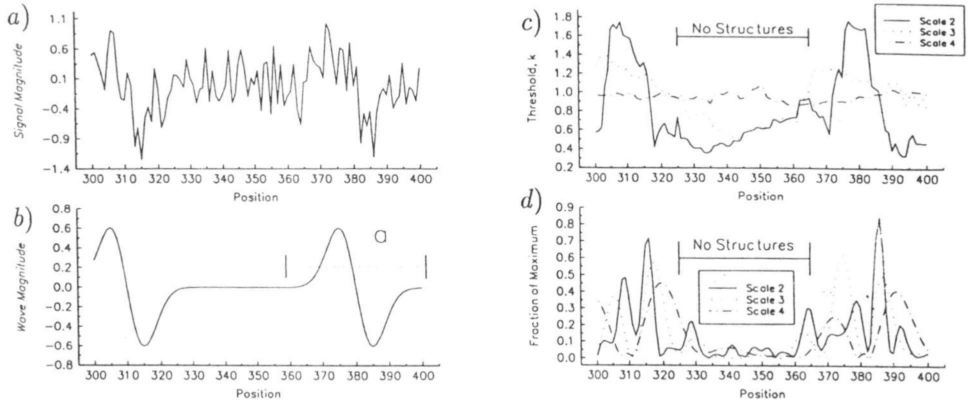

Figure 22: Comparison between VITA and EVITA methods on a synthetic signal from Sullivan *et al.* [56]. a) Raw signal; b) Embedded pattern; c) Detector function from VITA ; d) Detector function from EVITA

3.3 The Pattern Recognition Analysis technique

3.3.1 Principle of the PRA

The Pattern Recognition Analysis (PRA) technique was introduced by Mumford [51] and refined by Ferré and Giralt [26, 27] (for a review, see Ferré in [70]). This technique is an iterative process that is used to determine the ensemble average of the CS and to detect the patterns embedded in the time series. The process starts from an initial template of a conjectured trace of the CS, and, from a correlation analysis, converges towards the final template.

The PRA program initially proposed by Mumford and later refined by Ferré and Giralt corresponds to several successive steps. The raw data correspond to velocity time series (with I_m samples) obtained at J_m locations, $u(i,j), i = 1, \cdots, I_m; j = 1, \cdots, J_m$. It is necessary to start with a guess of the trace of the structure, called the "test pattern" (or later "template"), $T(k,l), k = 1, \cdots, K_m; l = 1, \cdots, L_m$. The dimension L_m of the template should be greater than that of the number J_m of sensors. Fig. 23 shows a sketch of the measurement and template grids.

The steps of the procedure are:

1. Form sets of convolutions:

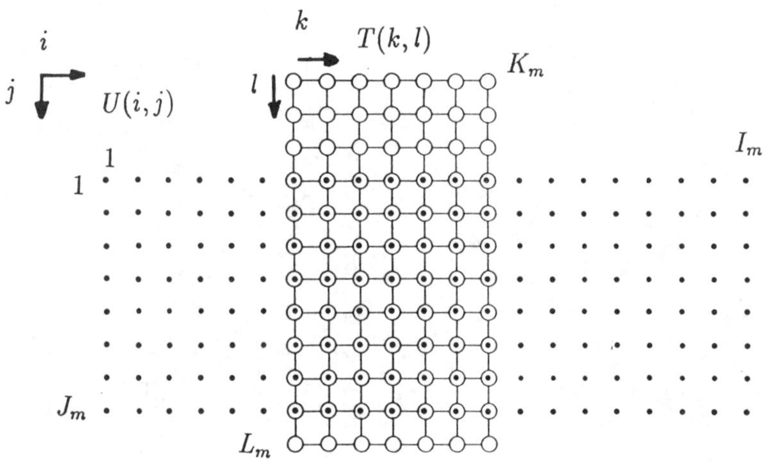

Figure 23: Sketch of the measurement and of the template grids.

$$C_{mn} = \sum_{i=1}^{I_m} \sum_{l=1}^{L_m} u(i, l+n-1).T(i+m-1, l) \tag{21}$$

With $m = 1, \cdots, K_m - I_m + 1$ and $n = 1, \cdots, J_m - L_m + 1$.

2. For this set, calculate the rms value σ_C.

3. For each value of m ($\equiv$ time steps), find the value of $n = N_{max}$ corresponding to the largest C_{mn} (or $\mid C_{mn} \mid$ depending on symmetries).

4. Select the values of m giving $C_{mn,\max}$ (or $\mid C_{mn,\max} \mid$) greater than $K.\sigma_C$, where K is a threshold number.
   Mumford defines a "burst" to be set of one or more successive occasions for which this condition is satisfied. The last averaging process follows the next steps:

5. Select bursts with lengths in a specified range. N.B.: this step can be skipped.

6. For each burst, find the value of m corresponding to the centre (or the maximum) of $C_{mn,\max}$ (or $\mid C_{mn,\max} \mid$). This gives the value of the lateral position $n_{\max}$.

7. Form the ensemble average from these values by adding the corresponding velocity pattern $u(i,j)$ with $i = 1, \cdots, I_m, j = m, \cdots, m + L_m - 1$ at the transverse position $n_{\max}$ with the convenient normalization.

Ferré and Giralt [26] introduce, instead of the correlation 21, the following two-dimensional correlation function:

$$R(m,n) = \frac{\frac{1}{I_m.L_m}\sum_{p=1}^{I_m}\sum_{q=1}^{L_m} u(p,q+n-1).T(p-m+K-I+1,q)}{\left[\left(\frac{1}{I_m.L_m}\sum_{i=1}^{I_m}\sum_{j=1}^{J_m} u^2(i,j)\right)\left(\frac{1}{K_m.L_m}\sum_{k=1}^{K_m}\sum_{l=1}^{L_m} T^2(k,l)\right)\right]} \tag{22}$$

Ferré and Giralt reformulate in [70] a more mathematical representation of PRA. They outline the procedure in six steps.

1. An arbitrary axis (unit vector) is selected as the initial template.
2. All the point vectors are projected over the template, and those showing an angle smaller (a correlation larger) than some threshold value are ensemble averaged.
3. The ensemble average replaces the initial template selected at step (1) and step (2) is repeated until convergence is achieved (when the template and the ensemble average are almost parallel). The last ensemble average obtained after iteration is the first feature axis.
4. An axis perpendicular to the first feature axis is selected again as initial template.
5. All the point vectors are projected over the template, and those showing an angle smaller (a correlation larger) than some threshold value are ensemble averaged.
6. The ensemble average replaces the initial template selected at step (4) and step (5) is repeated until convergence is achieved (when the template and the ensemble average are almost parallel). The last ensemble average obtained after iteration is the second feature axis.

For a simple time series, the abovementioned authors give an example of the 1-dimensional application of PRA. Fig. 24 shows the PRA of a synthetic signal derived from a single pattern embedded in noise with a signal to noise ratio of 2.2.

In this case the convergence is quite rapid and two iterations are necessary to achieve convergence. In this case a rough estimate of the pattern was used. In case of real signals, with no *a priori* knowledge of the flow, Ferré and Giralt state that the observation of the shape of the autocorrelation function can guide the estimation of the initial template.

The application of PRA to a two-dimensional array of sensors also exhibits a rapid convergence, with typically 2 iterations. The method has been tested also on synthetic signals by Mumford [51]. Fig. 25 shows the patterns retrieved with PRA.

Fig. 26 shows the application of PRA to experiments performed with a rake of hot-wires in the far wake of a cylinder. In such a real flow, studied by several authors [51, 26, 34], the initial template is determined from combinations of basic eddies corresponding roughly to 1. This procedure has been shown to be practically independent of the initial template.

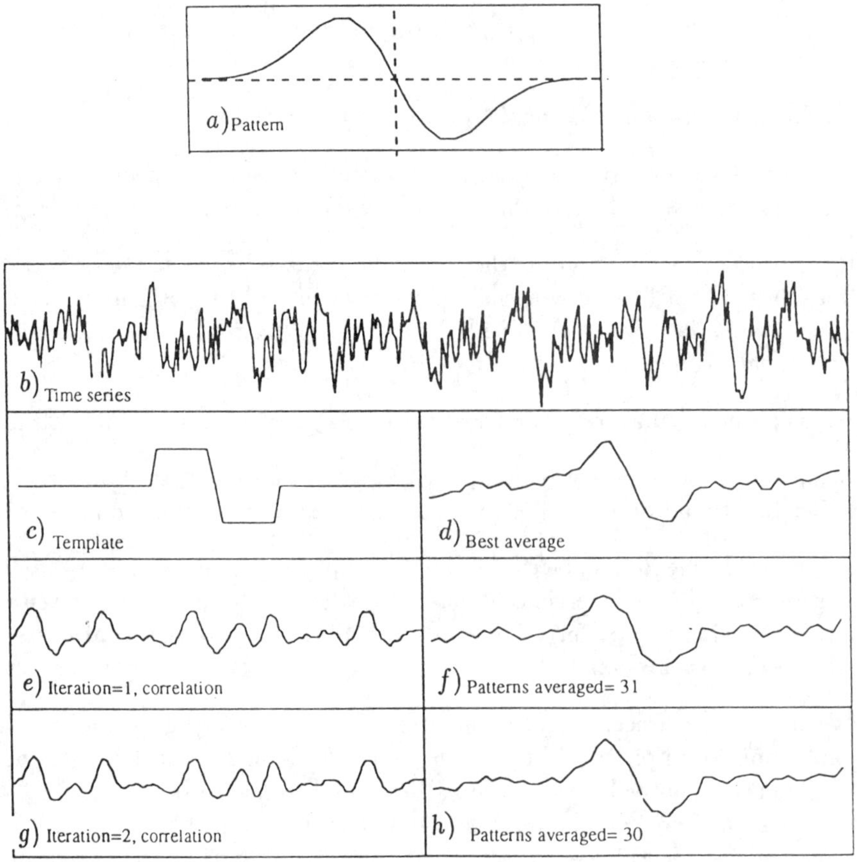

Figure 24: Example of PR processing on synthetic signal from Ferré and Giralt [70]. a) Basic pattern; b) Time series; c) Initial template ; d) Best average; e) Correlation coefficient at iteration 1; f) Pattern educed after iteration 1; g) Correlation coefficient at iteration 2; h) Pattern educed after iteration 2.

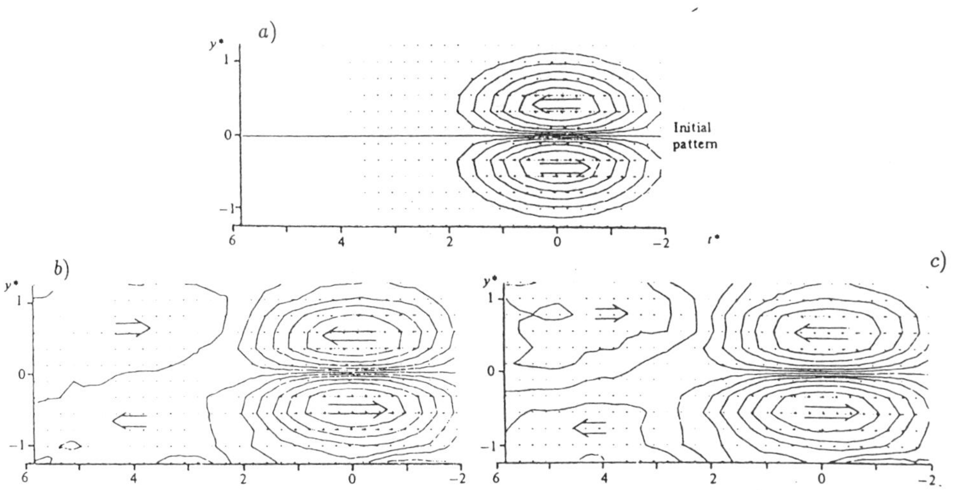

Figure 25: Example of PR processing on a synthetic 2D-array from Mumford [51]. a) Initial template; b) Pattern educed after iteration 1; c) Pattern educed after iteration 2.

PRA is also a useful tool for analyzing temperature patterns in turbulent flows [27]. The three-dimensional topology of the patterns can also be analyzed by scanning several slices of the flows [28]. In particular, a three-dimensional sketch of entraining motions in a far wake has been provided with this method. These results are recalled in the discussion on the different analyses of far wakes.

PRA can indeed be applied to bounded flows. Stretch *et al.* [55] have applied PRA to a simulated turbulent channel flow. In the paper of Savill *et al.* [70], PRA was applied to a manipulated channel flow. In this case, LES numerical results were used. Fig. 27 shows that the ensemble averaged pattern of the longitudinal velocity obtained from PRA (Fig. 27-a)) is remarkably similar to the instantaneous contours from an individual LES realization (Fig. 27-b)).

It has been proven that it is not necessary to use 3D PRA in order to extract the complete 3D structures from the 3D fields available from numerical computations. It is stated that a simple 2D PRA can be enough if the 3D field associated with the patterns detected is stored and ensemble averaged.

In addition to post-processing of experimental and/or numerical data in order to educe and analyze the CS, Savill *et al.* suggest that PRA can be used in real time in some computations: most of the generated data has to be overwritten for storage requirements. By detecting structures as they evolve, it can be possible to store only the most relevant data from successive time steps, allowing their dynamics to be studied

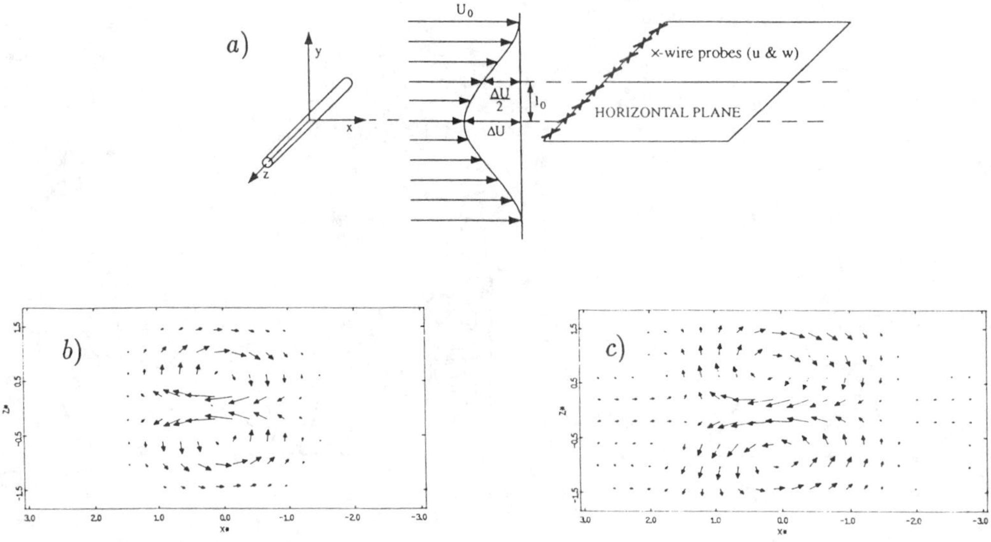

Figure 26: Example of PRA processing of hot-wire rake measurements in a far wake, from Ferré and Giralt [70]. a) Experimental arrangement; b) Initial template c) Ensemble average after 3 iterations.

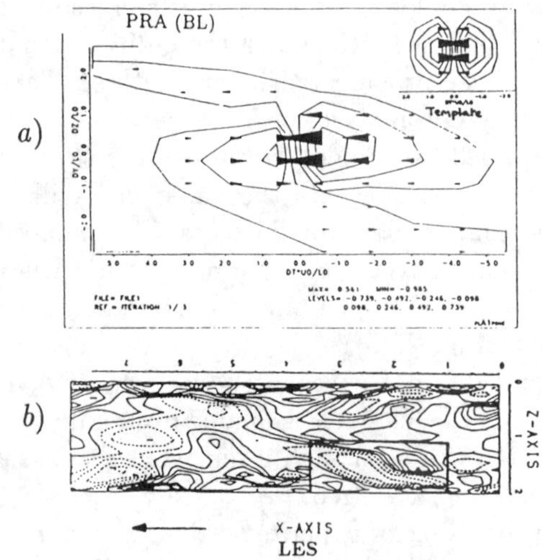

Figure 27: Example of PR processing of LES data in channel flow from Savill *et al.* [70]. a) PRA final pattern (upper right: initial template); b) Individual LES realization.

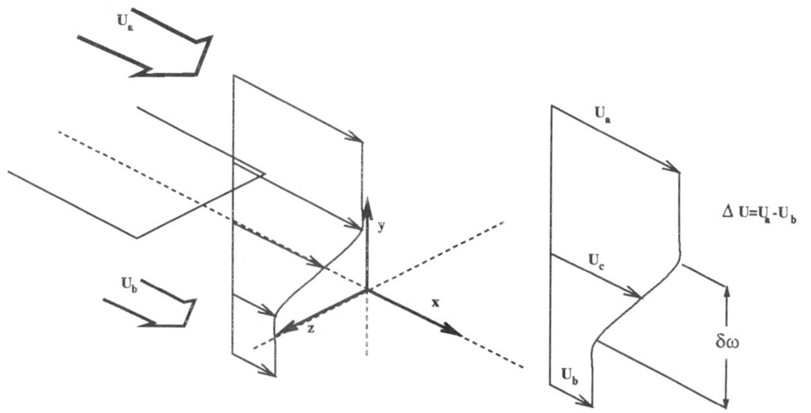

Figure 28: Experimental configuration for the mixing layer experiment.

in detail.

In comparing the similar objectives of PRA and Proper Orthogonal Decomposition, Berkooz *et al.* [8] suggest to interpret the final template with caution, since they have shown that any template with a suitable power spectrum could decompose the flow with an appropriate sprinkling function. This point is accentuated by the result of Stretch *et al.* [55] showing a median correlation, typically of the order of 0.3. Sullivan *et al.* in [56] also express some fear of the use of PRA in a very random flow field where there is equal likelihood for the existence of any presupposed candidate structure.

Despite some limitations in its use, PRA is a useful tool for extracting patterns embedded in turbulent flows. Ferré and Giralt state in [70] that:
"PRA is a technique that can overcome these drawbacks [of the procedures based on an eigenvalue problem] with a theoretical basis that is flow independent".

4 A COMPARATIVE ANALYSIS OF SOME CS CHARACTERIZATIONS IN THE PLANE MIXING LAYER.

In the following section, we will present two methods that can be used in view of CS characterization. The first one is a "pseudo-visualization" technique used to simply and visually analyze the data obtained from numerical simulations or multipoint measurements. The second one is a particular application of the conditional sampling technique specially designed for free shear flows. These two methods have been applied to a mixing layer flow. This flow is described in the next section and the pseudo-flow-visualization (PFV) technique is later presented and discussed. Finally, the Delocalized Conditional Sampling method is presented and some typical applications to the mixing layer are given.

4.1 Flow configuration for testing the detection methods.

The data are obtained from experiments performed in a two-dimensional subsonic plane mixing layer, in the fully developed region. The two streams, at U_a=42.2m/s and U_b=25.2m/s, merge at the trailing edge of a splitter plate with a velocity ratio $r=U_a/U_b$=0.59. The external velocity difference is $\Delta U = U_a - U_b$. The two boundary layers at the trailing edge are fully turbulent. A reference frame is defined as follows (Fig.28): X is the streamwise direction ; Y, the direction normal to the plate (y=0 lies on the mixing layer axis and $Y < 0$ in the low velocity side) ; Z, the direction parallel to the trailing edge of the plate. The corresponding velocity components are u, v and w respectively. A characteristic size of the typical large scale, quasi two-dimensional vortices can be defined from the mean velocity profile $\overline{u}(y)$ by the vorticity thickness:

$$\delta_\omega = \Delta U / \left(\frac{\partial \overline{u}}{\partial y} \right)_{y=0} .$$

Constant Temperature hot-wire Anemometry is used with multiple probes. Home-made rakes of hot-wires have been specially designed (v. Delville *et al* [20]). These rakes can be aligned either in the inhomogeneous direction (Y) or in the spanwise direction (Z). The actual bandwidths of the wires/anemometers are greater than 50kHz. The signals are simultaneously sampled at a maximum rate of f_s=100kHz per channel (12 bit conversions). The adjustment of gains and low-pass filtering frequency f_f of the signals are performed automatically under control of a micro-computer.

4.2 Pseudo Flow Visualization.

The use of dense rakes of probes combined with high speed simultaneous sampling is a way to determine accurately the space-time evolution of the flow field (George & Glauser [32]), at least for the large scale organization of the flow. The use of instantaneous reconstruction to analyze the behavior of the flow is not a new idea: Townsend [57], [12] or Keffer [42] refined this concept to the instantaneous velocity profiles or time histories to help to analyze the behavior of the flow. Delville *et al* [20] applied this approach to velocity signals in order to create "pictures" of the flow. The Pseudo-Flow-Visualisation (PFV) technique allows easy handling of a large amount of data, illuminating the large scale organization of the flow by simply considering the raw velocity signals as a picture of the flow. This approach can be compared with conventional experimental visualization techniques but avoids the problems due to seeding effects. Here the velocity field itself is considered as a "marker" of the flow. To get information on the spatial organization in the X direction, the Taylor hypothesis is used. Figure 29 gives some examples of such PFVs obtained. A more detailed description of the capabilities of this technique can be found in Bellin *et al.* [7].

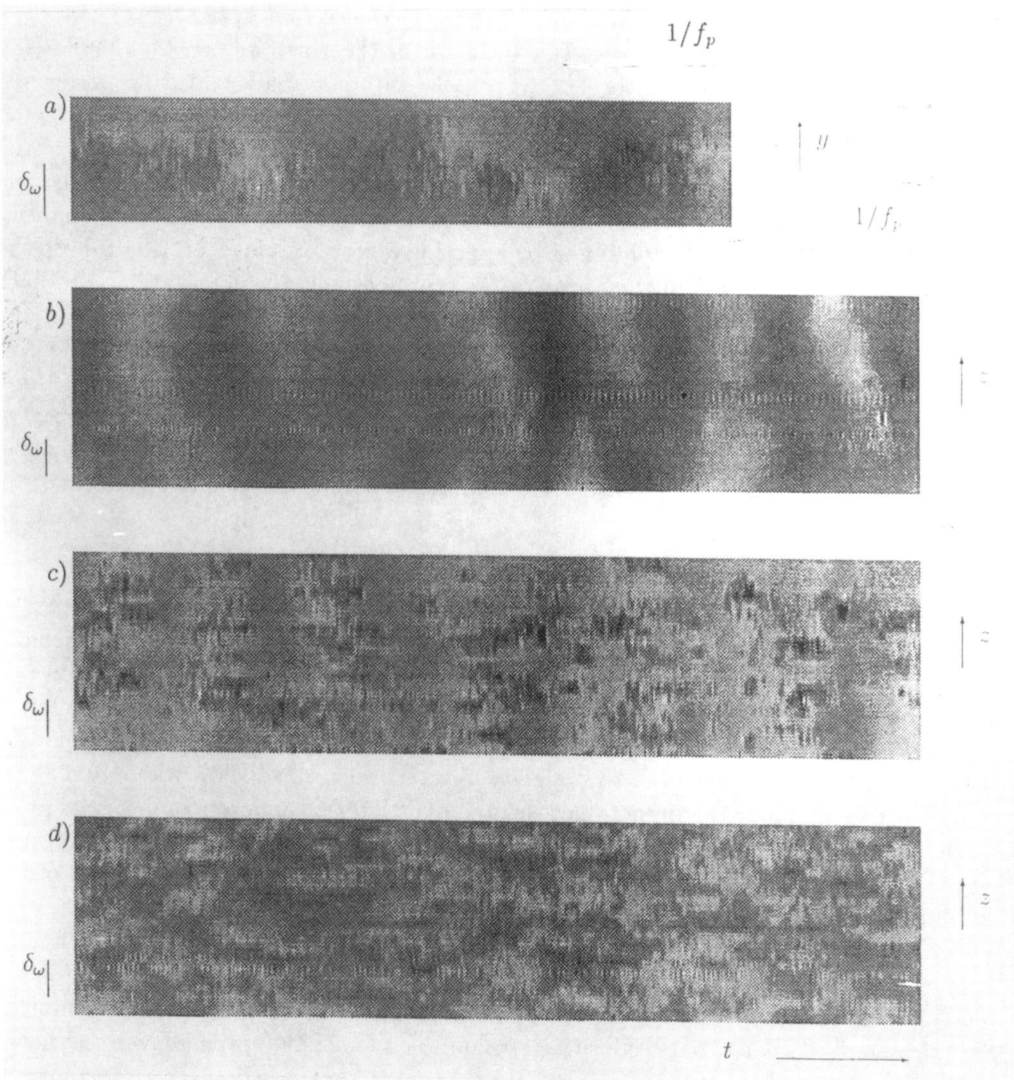

Figure 29: Examples of PFVs in the mixing layer. Sampling frequency $f_s = 100$kHz.

In these visualizations: dark gray indicates the negative fluctuations and light gray the positive ones.
— Vertical slice, raw visualization of $v'(y,t)$ *a)*
— Horizontal slices: visualization of $u'(z,t)$ *b)* at $y/\delta_\omega \sim 1$, *c)* at $y/\delta_\omega \sim 0.5$ and *d)* at $y \sim 0$.

4.2.1 Transversal (Y) slices.

The large scale organization of the flow is mainly characterized by the $\omega_z = -\partial u/\partial y + \partial v/\partial x$ component of the vorticity. The behavior of the vorticity can be analyzed by considering its distribution in the (x^*, y) plane[9]. Simpler analysis can be performed by looking only at the time histories of the u or v component of the velocity in the (x^*, y) plane. Then, considering variations in the y direction of the u component *or* the variations of the v component in the x^* direction provides information on the large scale organization. Such a typical visualization is shown in figure 29-a, where the quantity $v(y,t)$ is drawn. This PFV is obtained by applying a very simple coloration rule to the instantaneous velocities. The dark gray colors represent negative fluctuations and light gray represents positive fluctuations. The picture obtained is very similar to the one that could be obtained from conventional smoke visualizations. The passage of quasi-periodic large scale events are here clearly evident. The period of occurrence of these patterns agrees very well with the peak-frequency f_p measured from the velocity spectra which will be discussed in section 4.3. The presence of thin slanted links between the patches can also be observed and are very similar in shape and size to the braids.

4.2.2 Spanwise (Z) slices.

The secondary spanwise organization of the mixing layer appears clearly in the PFVs of figure 29-b-c-d. The fluctuating u-velocity time histories obtained from a horizontal rake experiment at high sampling frequency are plotted, using the same coloration rule as previously described. The use of the u component in such a horizontal PFV gives information on the flow in an indirect way. An interpretation of these PFVs could be to consider the u component of the velocity as a "marker" of the flow which undergoes a redistribution under the influence of the axial vorticity ω_x (pockets of low velocity or high velocity flow washed away by the axial vorticity). The sign of the fluctuation of u is then related to the penetrations of the low velocity or high velocity zones of the mixing layer in the horizontal slice observed.

When the rake is located in the free stream region of the upper velocity side $y/\delta_\omega \sim 1$ (figure 29-b), a large spanwise modulation of the velocity is observed. This modulation can be seen as the footprint of the passage of the large quasi two-dimensional structures. It corresponds to perturbations in the irrotational zone of the mixing layer associated with the passage of the large scale vortices (these structures are on average bounded to $y/\delta_\omega \sim 0.5$). When the rake is placed at the mixing layer edge ($y/\delta_\omega \sim 0.5$ figure 3-29-c), this two-dimensional feature is still observed. However, a secondary organization in the time direction (x^*), becomes noticeable. This organization is characterized by horizontal patterns of short duration, with small widths compared to the separations. This specific organization can be related to the two main organizations of the flow: the vertical band corresponds to the large scale quasi two-dimensional vortices and the small horizontal dark band corresponds to the upwash of low velocity fluid by the

[9]We will note x^* the spatial coordinate corresponding to the application of the Taylor hypothesis

x-aligned vortices. When the rake is located on the mixing layer axis (figure29-d), the large spanwise organization disappears and the secondary x organization becomes dominant.

4.3 Multiple spectral analysis.

4.3.1 Conventional spectra.

The large scale vortices occur quasi-periodically and their footprint can be to some extent retrieved from the temporal signals of the velocity. As seen previously, this footprint is strongly related to the variations of sign of the local velocity fluctuations and can be found on the frequency spectra. The evolution of the spectra of the u and v components across the mixing layer (Y direction) are plotted in figure 30-a, and important differences can be noticed. For the v component, a very well defined narrow band peak of energy appears regardless of the location within the flow: this peak can be observed in the mixing area as well as in the outer parts of the mixing layer. Contrarily, for the u component this peak can be found only in the external parts of the mixing layer; in the centre of the mixing zone, the dominant frequency is then located at lower frequencies and is much more broad band.

This behavior is illustrated in figure 30-b in terms of the Strouhal number evolution across the mixing layer. This Strouhal number is based on the vorticity thickness, on the convective velocity $U_c=(U_a+U_b)/2$ and on the dominant frequency f_p observed in each spectrum (corresponding to the maximum of energy).

These frequencies which are *a priori* related to the same phenomenon are strongly dependent on the quantity considered. Within the the mixing zone ($|y/\delta_\omega| < 0.5$) the Strouhal numbers remain stable but quite different: 0.21 for the u component and 0.33 for the v component. These Strouhal numbers attain an identical value of 0.33 at the edges of the mixing layer ($|y/\delta_\omega| \simeq 0.5$). In the external parts of the flow ($|y/\delta_\omega| > 0.5$), the typical frequencies for the u and v components evolve in the same way, slowly decreasing with growing distance from the axis of the mixing layer. This decrease has to be related to the fact that the farther the probe is from the axis, the more it is statistically influenced by vortices of larger size (smaller Strouhal number). These remarks illustrate the difficulty in providing a unique global definition of a characteristic time scale describing the large scale behavior particularly for the plane mixing layer.

4.3.2 Two-dimensional spectra.

Additional information on the dominant scales of this flow can be obtained by using spectral analysis involving simultaneously the spanwise and the temporal evolution of the velocity signals. With the rake of hot-wires aligned in the Z direction, 2D-spectra

a)

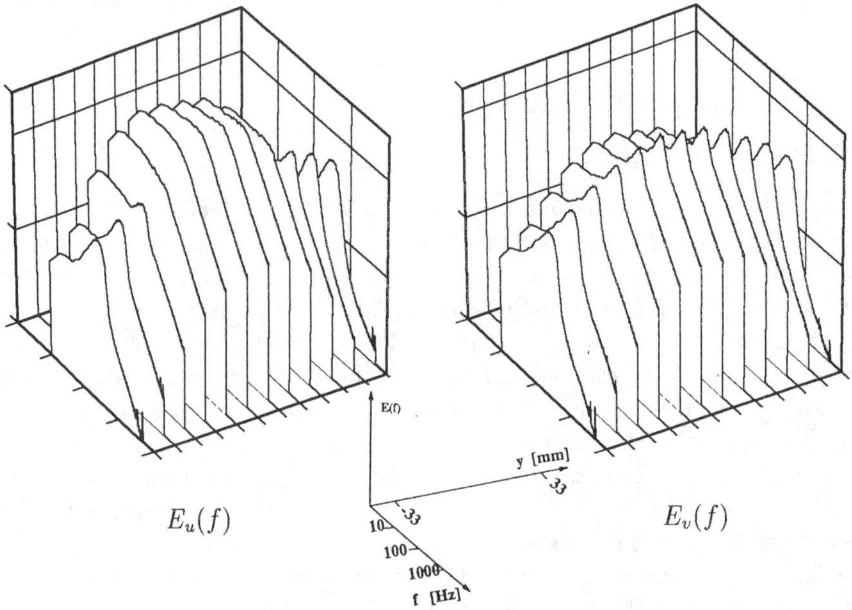

b)

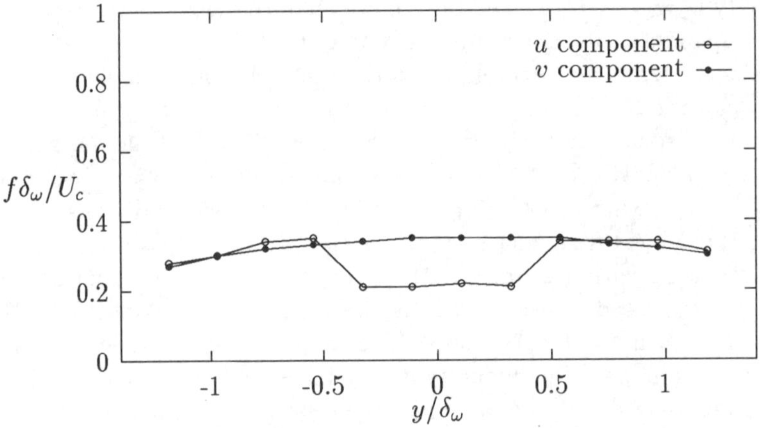

Figure 30: *a*) Distribution of the spectra of u and v across the mixing layer. *b*) Distribution of the most energetic scale.

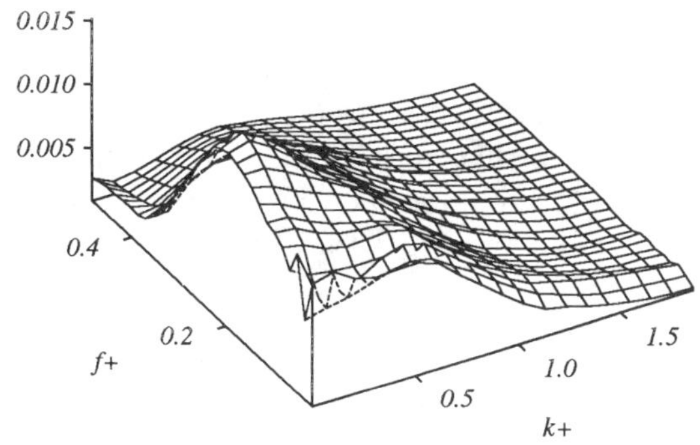

Figure 31: Example of two-dimensional spectra $\Psi_{uu}(y = y' \sim 0, f, k_z)$ measured on the mixing layer axis. ($f+ = f\delta_\omega/U_C$, $k+ = k_z\delta_\omega$)

can be defined as :

$$\Psi_{uu}(y; f, k_z) = \langle \hat{u}(y; f, k_z)\hat{u}^*(y; f, k_z)\rangle, \text{ where}$$
$$\hat{u}(y; f, k_z) = \tfrac{1}{TL_z}\int_{-T/2}^{T/2}\int_{-L_z/2}^{L_z/2} u(t, z)\exp(-2\pi j(k_z \cdot z + f \cdot t)\, dz\, dt,$$

As an example, the 2D spectrum measured on the mixing layer axis is plotted in figure 31. For each frequency/wave number (f or k_z), by considering only the small-wave numbers in the cross direction (k_z or f), two well defined wave-numbers can be sorted out, corresponding to the two main cross organizations of the flow. As for the previous conventional spectra, for small spanwise wave-numbers ($k_z \to 0$), the maximum energy is found for the typical frequency $f_p\delta_\omega/U_c$=0.21. For small frequencies ($f \to 0$) a maximum of energy can be found for a spanwise wave-number of $k_z \sim 0.66\delta_\omega^{-1}$.

These 2D spectra are particularly well designed for flows with 2D mean characteristics. They are more efficient tools than conventional 1D (temporal) spectra.

4.4 Delocalized Conditional Sampling.

The main purpose of this section is to illustrate how a self adapting Structure Education method can be used when the information on flows (velocity, pressure, passive contaminants, etc...) is simultaneously available in several locations. This situation particularly occurs when rakes of sensors are used, or for DNS or LES simulations.

This approach is an extension of the usual conditional detection technique that takes advantage of the simultaneous knowledge of information at several points in space: the detectors are automatically selected within the available data. Due to the small level of rejected events, this method, called Delocalized-Conditional-Sampling (DCS), is more efficient than the conventional method and can be applied when a single component of the velocity is available. This is not the case for the vorticity-based conditional methods (for example v. Hussain paper in this course). Moreover, it allows one to obtain new information on the structures in the flows: not only statistics on the size and position of the structures but also phase-shift within the structures may be retrieved.

4.4.1 Description of the method.

Conditional sampling based on one or two points as detectors has been extensively used in free shear flows by several authors, see for example [13, 4, 63]. This concept is extended to the case where the information is available from a rake of sensors, i.e. simultaneously in several locations in space. The principle is to automatically select the two optimal probes that will serve for the detection. With this technique each probe plays a role in the detection procedure and is used to compute the ensemble averages. The algorithm developed by Bellin [6] does not require crucial threshold conditions. One of the interests of this method lies in the possibility to detect structures of any size appearing at any transverse position and with any phase shift inside the structure. The DCS algorithm follows a two-step procedure: the first step determines a "pre-partition" of the flow allowing to split the time of observation into two parts, corresponding respectively to the instants where CS are present or not. The second step will specify the location of the CS within the pre-determined time periods and will compute the statistics of these events.

First step: Pre-partition The simplest and, in some sense, less ambiguous approach to select the different phases of the flow is to examine the instantaneous velocity field. This is performed according to the procedure illustrated in Fig. 32. First of all, low and high threshold values are applied to the velocity signals. This allows the determination of the instantaneous, external frontiers of the flow in, respectively the slow and the rapid side. These ragged limits are then smoothed. A correspondence between the two sides of the frontiers is established by fitting ellipses which link the local extrema. The choice of ellipses as fitting candidates is, at this stage, arbitrary but this choice is of secondary importance because these ellipses are only used to model a detector function. We arbitrarily define the detector function as 1 when an ellipse is detected and 0 when there is no link between the two frontiers. We will later demonstrate that for the present configuration this function is 0 when CS may be present. The conditional sampling itself can then be applied during these periods for better efficiency. This pre-partition procedure is rather insensitive to the threshold selected for the partition. It has been shown in [7] that a significant variation of this threshold (from 4 to 10 % of the external velocities) does not modify strongly the location of the

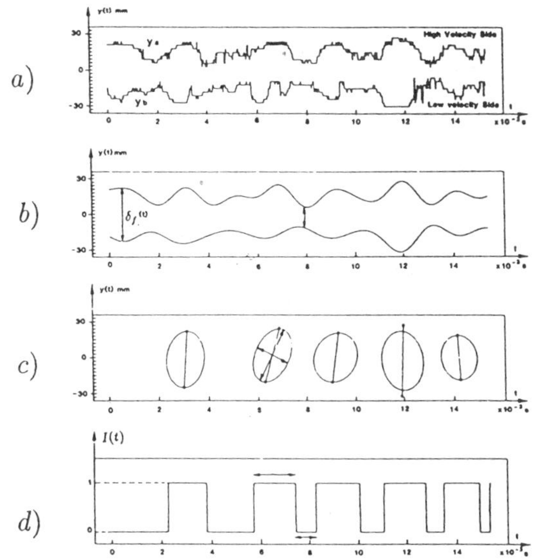

Figure 32: Events partition.

a) Edges extraction	*b*) Low pass filtering
c) Extrema linking and quadratic fitting of the edges	*d*) Introduction of a detection function $I(t)$

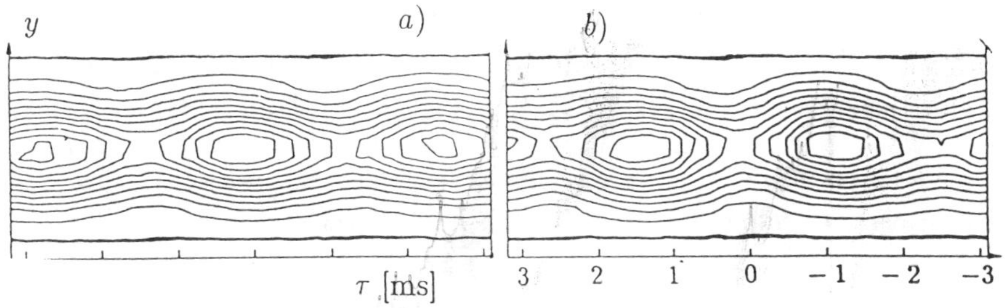

Figure 33: Signature of vorticity Ω_z based on the predetection function. *a*) centered on $I(t) = 0$, *b*) centered on $I(t) = 1$.

edges of the puffs and therefore the detector function $I(t)$. More precisely, the main consequence of a too low threshold is only to miss some $I = 1$ events. Thus the time period of the application of the DCS is increased without any benefit.
Indeed, conventional ensemble averages can be directly obtained from this detector function. Figure 33 shows the iso-contours of the vorticity obtained when the reference time is centered on $I = 0$ (figure 33-a) or on $I = 1$ (figure 33-b). From the topology of the iso-contours, it clearly appears that the results obtained when $I = 0$ correspond to the well accepted conventional characteristics of CS. The criterion I=0 can be associated with the more probable presence of CS, then justifying the abovementioned criterion.

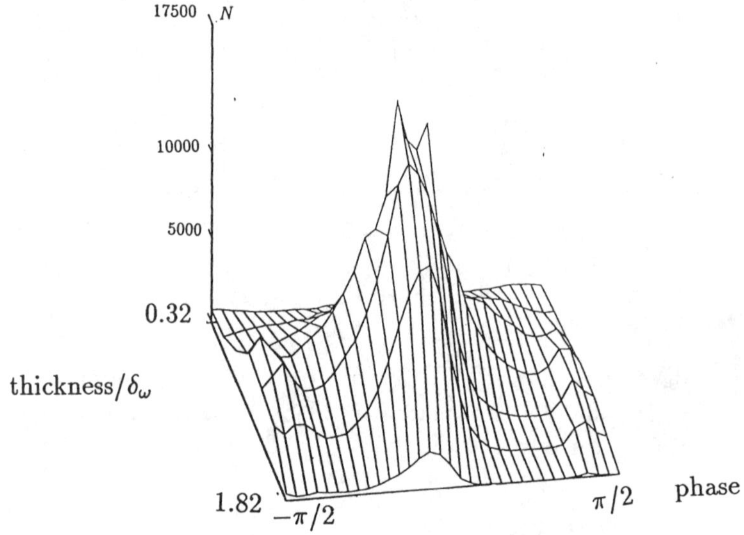

Figure 34: Joint PDF of thickness-phase from DCS.

Second step: DCS. After the pre-determined periods of time that correspond to high probable occurrence of CS have been selected by the first step, the conditional procedure itself can be applied. For each period selected, the algorithm can detect in space and time the two locations where the fluctuating streamwise component of velocity reaches its maximum value in the high speed side and where it reaches its minimum value in the low speed side. This is equivalent to automatically selecting, among the available measurement locations where the detectors should be optimally placed, instead of *a priori* positioning as is the case for conventional detector use [13]. This technique allows to detect CS of different scales, centered anywhere relative to the mixing layer axis. The DCS allows a lower rejection rate than conventional techniques and permits computations of new statistics for several characteristics of the CS.

In addition, the time shift between the two detections contains information on the stage of evolution of the CS. This can be related to the different phases of the life

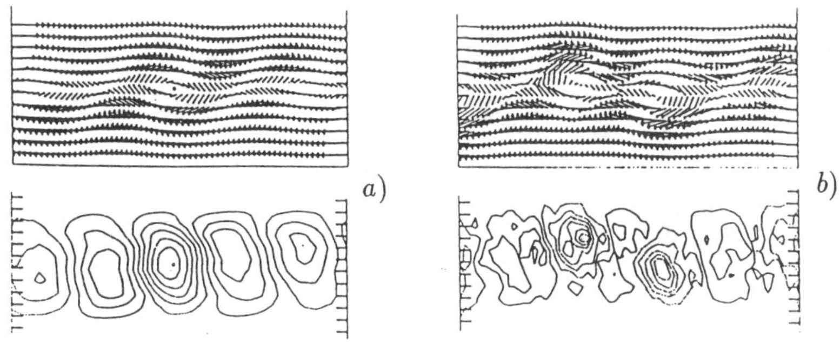

Figure 35: Ensemble average from DCS. *a*) average pattern, all events cumulated; *b*) average structure corresponding to a pairing stage. (lower: iso $< \omega >$).

of the structures. It is then possible to extract *conditional* ensemble averages taking into account 3 independent parameters or CS characteristics: lateral extent (size, or thickness), lateral position (relative to the mixing layer axis), and phase shift (between the two opposite sides of the CS).

Typical results. Figure 34 shows typical statistics which can be obtained from the data base. In this case, the joint probability of thickness and phase is plotted.

From this plot typical modes can clearly be detected. For the present data base, the most probable value of the thickness is 1.12 δ_ω and the most probable phase is 0 degrees.
It is also possible to analyze the time interval between two successive structures. With the Taylor hypothesis the most probable distance between structures can be evaluated to the order of $3\delta_\omega$. This value is in agreement with the results of other authors in comparable configurations [9]. A more detailed analysis of the statistics that can be derived from this DCS can be found in [7].

With this method, it is possible to select a class of size, position and phase shift and to determine the corresponding conditional averages. Figure 35 illustrates, using the velocity vector and the vorticity isolines, the influence of the time shifts on the resulting ensemble average. In this example it clearly appears that the selection of a given time shift allows to construct ensemble averages that correspond to given stages of CS dynamics. The resulting image then contains the average characteristics of the CS at different stages of its evolution. This procedure can be systematically used for any available turbulent quantity. Moreover, marginal statistics can be more convenient for the analysis of these characteristics.

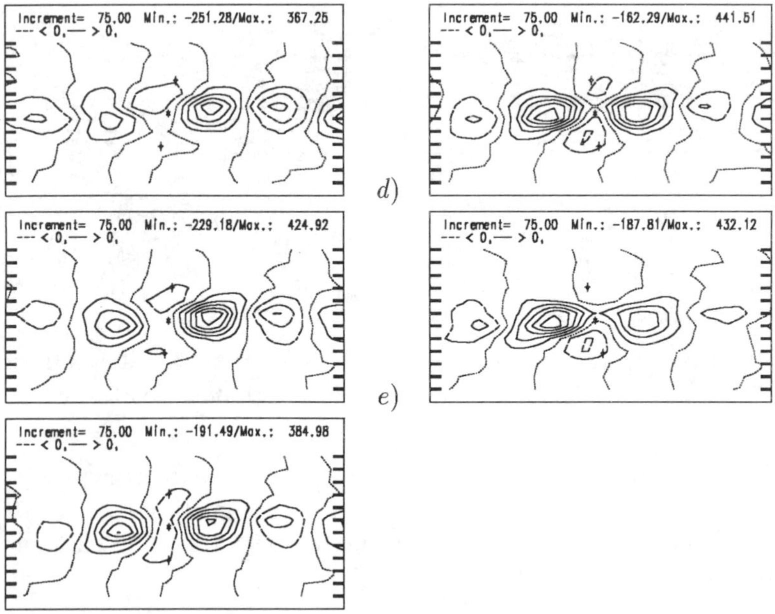

a) *d)* *b)* *e)* *c)*

Figure 36: Influence of the phase shift on the coherent production: $- < u_r v_r > (\frac{\partial u_c}{\partial y} + \frac{\partial v_c}{\partial x})$ Phase shifts: a) $-14°$; b) $-7°$; c) $0°$; d) $+7°$; e) $+14°$

Triple Decomposition As for any conditional sampling, double or triple decomposition as introduced in [54] can be applied. The efficiency of the DCS process and the possibility to select different characteristics of the CS (size and/or shift, ...) allow more detailed analysis than usual. Details on the results obtained in the data base can be found in [6]. A particular illustration is given in Figure 36. This figure shows the iso-contours of the production term $- < u_r v_r > (\frac{\partial u_c}{\partial y} + \frac{\partial v_c}{\partial x})$. In the same manner as we previously analyzed the influence of the phase shift on the ensemble average of velocity and vorticity, we can analyze the behavior of the production term during the different stages of the CS evolution.

Figure 36 shows that the increase of the phase induces a significant loss of coherence of the CS. As expected, the influence of the sign of the time shift (i.e. value of phase) has a non-symmetrical influence on the distribution of the production. The advance or delay in phase has a distinct influence according to the natural inclination of the CS.

4.5 POD in the Mixing Layer.

We present here briefly a few results obtained from the application of the POD in the flow under consideration. This POD corresponds to the resolution of the following

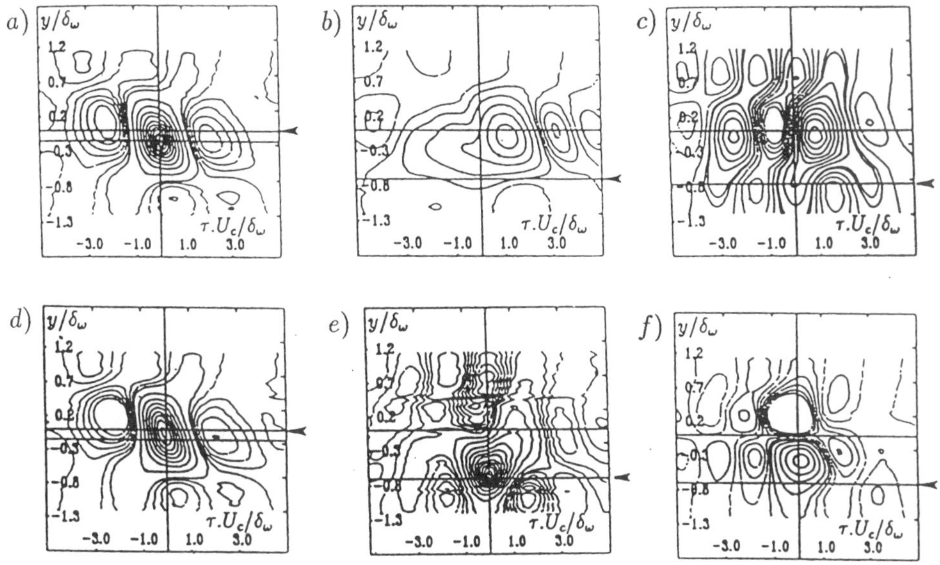

Figure 37: Contribution of the first two modes of the POD in the mixing layer. *a)* $n_c = 1$ mixing layer axis, *b)* n_c=1 edge of the mixing layer, *c)* $n_c = 2$ edge of the mixing layer, *d)* directly measured on the mixing layer axis, *e)* directly measured at the edge of the mixing layer, *f)* contribution of the second mode for n_c=2 at the edge of the mixing layer.

eigenvalue problem:

$$\sum_{j=1}^{n_c} \int_D \Psi_{ij}(y, y'; f) \widehat{\Phi}_j^{*(n)}(y'; f)\, dy' = \lambda^{(n)}(f) \widehat{\Phi}_i^{(n)}(y; f). \tag{23}$$

In this equation, $\Psi_{ij}(y, y'; f)$ is the cross-spectrum of u_i, u_j and corresponds to the temporal Fourier transform of the space time correlation

$$\Psi_{ij}(y, y'; f) = \int_\infty < u_i(y, t)\, u_j(y', t + \tau) > \exp(-2j\pi f \tau)\, d\tau \tag{24}$$

n_c is the number of components used for this POD. If n_c=1, only the u component is taken into account and if n_c=2, both u and v are simultaneously taken into account. The different consequences of the theorem of the POD for these specific applications is straightforward and is not discussed here. We consider here the contributions of the first modes of these PODs to the Space-Time-Correlations (STC), and to the instantaneous signals of velocity.

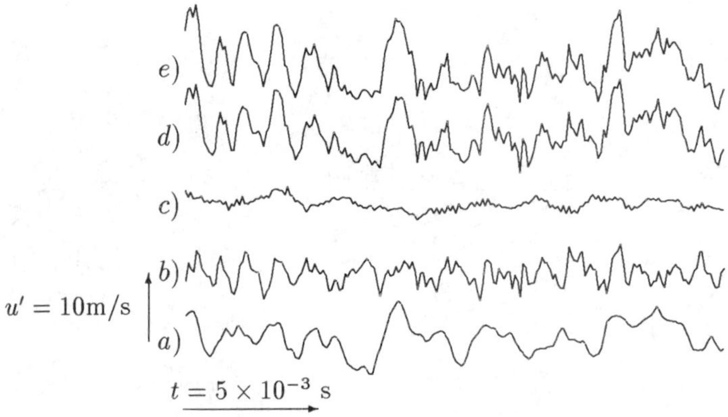

Figure 38: Contribution of the first modes of the POD ($n-c=1$) to the instantaneous velocity on the mixing layer axis: *a*) 1st mode *b*) 2nd mode *c*) 3rd mode *d*) summation over the first three modes *e*) measured velocity.

4.5.1 Contribution to the STC.

The STC $R_{ij}(y,y',\tau) =< u(y,t)u(y',t+\tau) >$ can be rebuilt from the eigenfunction of the POD by considering the inverse Fourier transform of $\Psi_{ij}(y,y';f)$:

$$R_{ij}(y,y',\tau) = \int \Psi ij(y,y';f)\exp(2j\pi f\tau)\,df \text{ where:} \tag{25}$$

$$\Psi_{ij}(y,y';f) = \sum_{n=1}^{N_m} \Psi_{ij}^{(n)}(y,y';f) \text{ and} \tag{26}$$

$$\Psi_{ij}^{(n)}(y,y';f) = \lambda^{(n)}(f)\Phi_i^{*(n)}(y;f)\Phi_j^{(n)}(y';f) \tag{27}$$

We compare in figure 37 the contribution of the first modes of both PODs to the STC. When the reference probe is chosen near the mixing layer axis, the reconstruction is acceptable for large scales for both PODs. However, when the reference probe is located at the edge of the mixing layer, the first mode of the POD for n_c=1 is poorly reconstructed and the second mode is more significant. Contrarily, for n_c=2 the first mode is much more representative.

The main features arising from these results are that when considering a scalar POD (involving u only) some kind of spatial partitioning of the field appears. Particularly, the first mode of this POD does not seem able to rebuild the phase opposition between the two edges of the mixing layer (this phase link is related to the passage of the large scale structures).

The use of a vectorial POD (n_c=2), where both u and v are taken into account, provides better results and the effect of partioning tends to disappear.

These results indicate that the more the POD takes various flow parameters into account, the more the decomposition captures the flow phenomena. One of the consequences is that care should be taken when interpreting results arising from scalar applications of the POD.

4.5.2 Contribution to the instantaneous signals.

The use of rakes of sensors allows one to access the instantaneous contribution of any given mode to the velocity field. For the PODs presented above, the way to retrieve this contribution is to calculate the following quantities:

$$u_i^{(n)}(y,t) = \int \Phi_i^{(n)}(y,t-\tau)a^{(n)}(\tau)\,d\tau, \tag{28}$$

or, in the Fourier space:

$$\widehat{u_i}^{(n)}(y,f) = \widehat{a}^{(n)}(f)\widehat{\Phi_i}^{(n)}(y,f), \text{ where} \tag{29}$$

$$\widehat{a}^{(n)}(f) = \sum_{i=1}^{n_c} \int \widehat{u_i}(y,f)\widehat{\Phi_i}^{*(n)}(y,f)\,dy. \tag{30}$$

Figure 38 shows an example of the contribution of the first POD modes to the velocity signal measured on the mixing layer axis when $n_c = 1$. Clearly the first mode is able to predict the behavior of the large scale evolution of the signal. When using three modes, the reconstruction is found to be very similar to the measured one.
An example of the reconstruction obtained when n_c=2 can be found in section 4.6.2 (fig.40).

4.6 LSE and Complementary technique.

4.6.1 Linear Stochastic Estimation

The Linear Stochastic Estimation (LSE) technique was first proposed by Adrian [1] in 1975. This method uses what Adrian terms as a "conditional eddy". This eddy is a *candidate* structure used to detect, given certain restrictions, other structures of similar type. Stochastic estimation uses the conditional information specified about the flow at one or more locations in conjunction with its statistical properties to estimate the information at the remaining locations. The reader can refer to the course of Adrian for further details and discussion on the LSE.

The LSE yields an estimate

$$\tilde{u}_i(y',t) = A_{ij}(y')u_j(y,t) \tag{31}$$

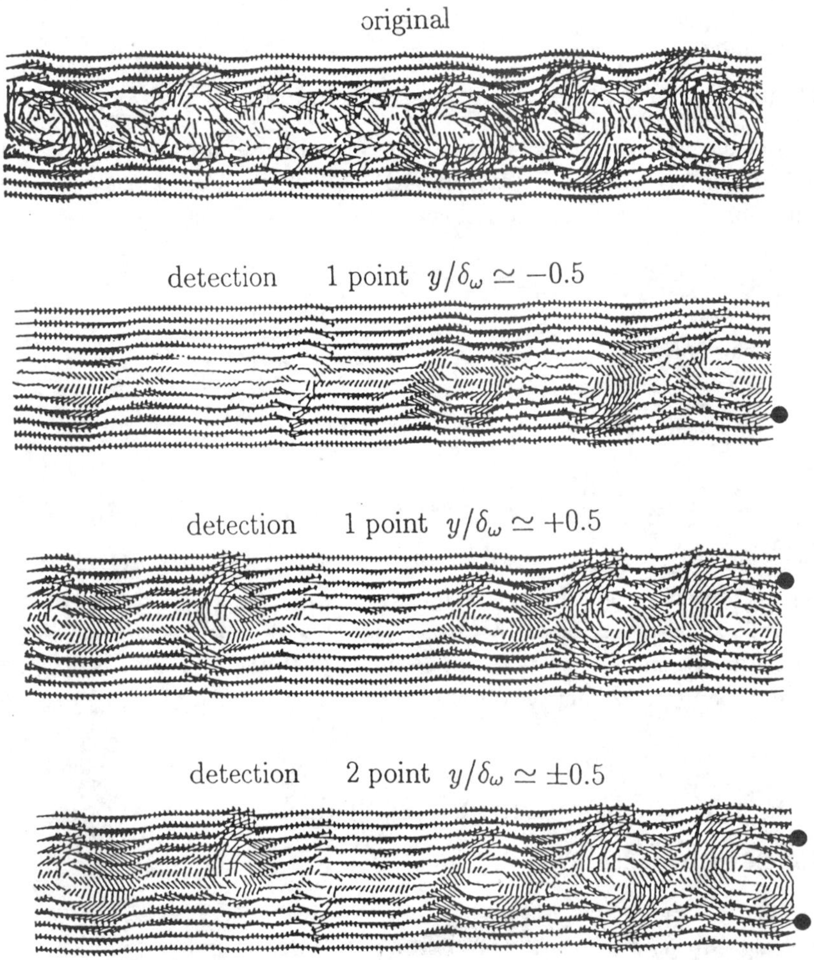

Figure 39: Estimation of the flow field via LSE (the "conditions" applied to the probe locations (•) correspond to the measured velocity).

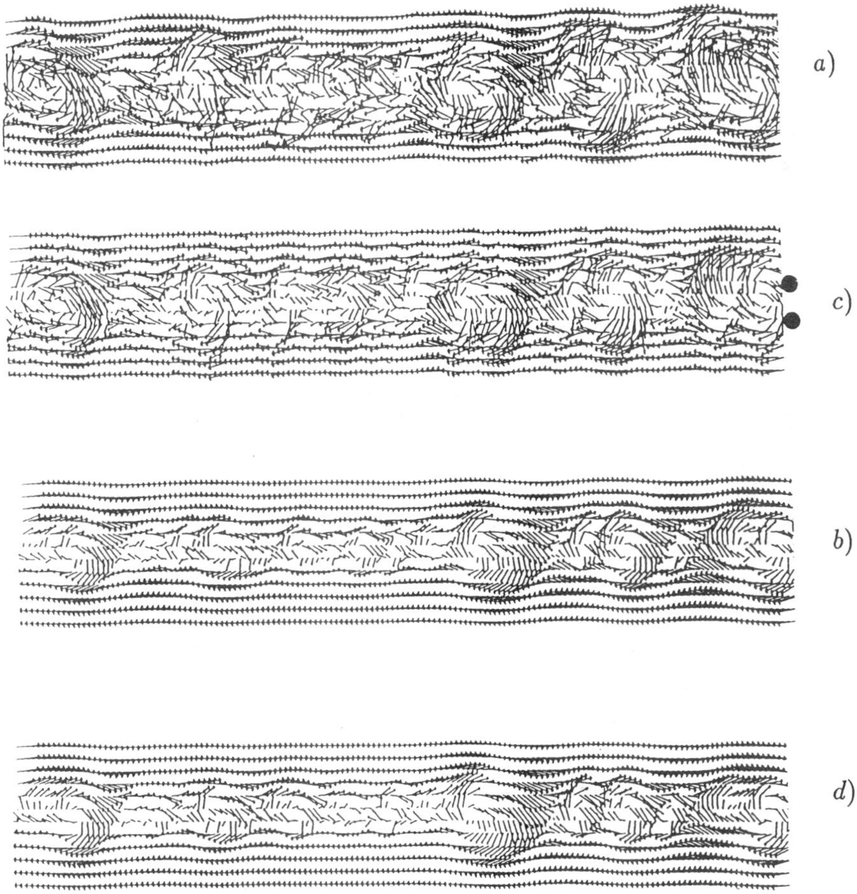

Figure 40: Estimation of the flow patterns via the complementary technique. *a*) measured *b*) direct application of POD (v. section 4.5 *c*) direct application of LSE *d*) application of the complementary technique.

with A_{ik} computed from,

$$\overline{u_j(y)u_k(y)}A_{ik}(y') = \overline{u_j(y)u_i(y')} \tag{32}$$

where $\overline{u_j(y)u_k(y)}$ and $\overline{u_j(y)u_i(y')}$ are the Reynolds-stress and two-point correlation tensors respectively. It should be noted that for these systems of equations, only the two-point space-time correlation data is utilized. These systems **are not** a function of the condition being investigated.

The estimated velocity components are found from the expansion of equation 31. It is in these estimated velocity equations that the condition selected plays a role. In this study the velocities measured at the "detection" locations are used as conditional events. We present in figure 39 several fields estimated from one point or two points. Vector plots are used to qualitatively examine the flow field. This figure presents the measured velocity field in a reference frame following the convective velocity (times goes from right to left), compared to the ones obtained via LSE. In these plots, the • correspond to the probe(s) location. The "detection" probes are here located at the edges of the mixing layer. Clearly this procedure is able to predict the large scale evolution of the flow even with only one probe. However using this technique, as opposed to POD, the turbulent energy within the field is greatly underestimated.

4.6.2 Complementary technique.

In order to gain both advantages of the two previous techniques (POD and LSE), we apply, in this section, a complementary method involving both approaches in order to identify the structures in the flow. This complementary technique consists of three main steps.
• First, the eigenfunctions and eigenvalues are obtained from direct application of POD to a two-point spectral tensor (v. section 4.5).
• Second, LSE is applied to the cross correlation tensor and multipoint estimates of the random vector field are computed as described by Cole *et al* [17].
• Third, the eigenfunctions obtained from step one are projected onto the estimated velocity field obtained from step two to obtain *estimated* random coefficients. These estimated random coefficients are then used in conjunction with the POD eigenfunctions to reconstruct the estimated random velocity field.

Mathematically the stochastic estimates of the random coefficients are calculated from:

$$a_n^{est}(f) = \int \hat{u}_i^{est}(y,f)\psi_i^{(n)*}(y,f)dy \tag{33}$$

where $\hat{u}_i^{est}(y,f)$ is either a single or multipoint linear stochastic estimate of the velocity field and $\psi_i^{(n)*}(y,f)$ is obtained from the original POD eigenvalue problem. The estimated streamwise velocity can be reproduced in Fourier space by

$$\hat{u}_i^{est}(y,f) = \sum_{n=1}^{\infty} a_n^{est}(f)\psi_i^{(n)}(y,f) \tag{34}$$

and then inversely transformed to obtain $u_i^{est}(y,t)$. For further details on this technique the reader is referred to Ukeiley *et al* [59].

The results corresponding to this application, presented in figure 40 show that with only a few percent of the initial information on the flow field (17%)[10], the first mode of the POD can be predicted accurately.

Acknowledgments The authors thank Prof. M. Glauser and Dr. L. Ukeiley for their constructive discussions. Several common studies have been used in several parts of this course. The bibliographical assistance provided by Prof. J. Ferré and Dr. M. Savill concerning the PRA was particularly appreciated. The authors also appreciate the fruitful exchanges during the sabbatical time of Prof. A. Pollard.

Finally the authors thank the many students having contributed to the studies presented herein. Among them Dr. E. Vincendeau and Dr. S. Bellin played a large role.

References

[1] Adrian R.J. (1975) "On the Role of Conditional Averages in Turbulence Theory." in *Proc. 4th Biennal symp. in Turbulence in Liquids.*

[2] Adrian R.J. (1979) "Conditional eddies in isotropic turbulence." *Phys. Fluids* 22(1), 2065.

[3] Adrian R.J. & Moin P. (1988) "Stochastic estimation of organized turbulent structure : homogeneous shear flow." *J. Fluid Mech.*,190, 531.

[4] Antonia R.A. (1981) "Conditional sampling in turbulence measurement". *Ann. Rev. of Fluid Mech.*, Vol. 13, pp. 131-156.

[5] Antonia R.A., Chambers A.J., Britz D. & Browne L.W.B. (1986) "Organized structures in a turbulent plane jet: topology and contribution to momentum and heat transport". *J. Fluid Mech.*, Vol. 172, pp. 211-229.

[6] Bellin, S. (1991) "Etude Experimentale des Structures Cohérentes d'une Couche de Melange Plane Turbulente de Fluide Incompressible". *Thèse Université de Poitiers.*

[7] Bellin S., Delville J., Vincendeau E., Garem J.-H. & Bonnet J.P. (1992) "Large scale structure characterization in a 2D mixing layer by Pseudo-Flow-Visualization and Delocalized Conditional Sampling." in *Eddy Structure Identification in Free Turbulent Shear Flows.* (J.P. Bonnet and M.N. Glauser eds.) Kluwer Academic Press.

[8] Berkooz G, Holmes P. & Lumley J. (1993) "The proper orthogonal decomposition in the analysis of turbulent flows", *Ann. Rev. Fluid Mech.*, 25, 1993.

[9] Bernal L.P. (1981) "The coherent structure of turbulent mixing layers". *PhD Thesis. University of Passadena Ca.*

[10] Let us recall that to access the first mode of the POD using equation 28 the distribution of the flow field needs to be known all over the domain.

[10] Blackwelder R.F. & Kaplan R.E. (1976) "On the wall structure of the turbulent boundary layer". *J. Fluid Mech.*, Vol. 76, part 1, pp. 89-112.

[11] Bousgarbies J.L., Geron M. & Khelif D. (1988) "Etude statistique de la structure tourbillonnaire de grande échelle au sein d'une couche de mélange plane" *17ème IUTAM*, Grenoble.

[12] Browand F.K. & Troutt T.R. (1980) "A note on spanwise structure in the two-dimensional mixing layer". *J. Fluid Mech.*, Vol. 917, part 4, pp. 771-781.

[13] Browand F.K. & Weidman P.D. (1976) "Large scales in the developing mixing layer". *J. Fluid Mech.*, Vol. 76, part 1, pp. 127-144.

[14] Brown G.L. & Roshko A (1974) "On Density Effects and Large Scale Structures in Turbulent Mixing Layers." *J. Fluid Mech.* Vol. 64, 775-816.

[15] Chambers D.H., Adrian R.J., Moin P., Stewart D.S. & Sung H.J. (1988) "Karhunen-Loeve Expansion of Burgers' Model of Turbulence". *Phys. Fluids* Vol. 31(9), 2573.

[16] Cimbala J.M., Nagib H.M. & Roshko A. (1988)"Large structure in the far wakes of two-dimensional bluff bodies" *J. Fluid Mech.*, 190, 265.

[17] Cole, D.R., Glauser, M.N. and Guezennec, Y.G. (1992) "An Application of Stochastic Estimation to the Jet Mixing Layer". *Physics of Fluids A.*,4(1) pp 192-194.

[18] Comte P., Fouillet Y., Normand X. & David E. (1990) "simulation numérique d'écoulements cisaillés compressibles" *25 ème Colloque AAAF*, Marseille, October.

[19] Debisschop J.R., Sapin S., Delville J. & Bonnet J.P. (1993) "Supersonic mixing layer analysis by laser planogram and hot-wires based P.O.D." in *Eddy Structure Identification in Free Turbulent Shear Flows.* (J.P. Bonnet and M.N. Glauser eds.) Kluwer Academic Press.

[20] Delville J., Bellin S., Garem J.H. & Bonnet J.P. (1988) "Analysis of Structures in a Turbulent Plane Mixing Layer by Use of a Pseudo Flow Visualization Method Based Hot-Wire Anemometry". *Advances in Turbulence II*, (eds. H.H. Fernholtz,H.E. Fielder), Springer Verlag, 251.

[21] Delville J., Bellin S. & Bonnet J.P. (1990) "Use of the Proper Orthogonal Decomposition a Plane Turbulent Mixing Layer". in *Turbulence and Coherent Structures*,(eds. O. Metais and M. Lesieur), Kluwer Academic Publishers, 75-90.

[22] Delville J. (1993) " Characterization of the organization in shear layers via the Proper Orthogonal Decomposition. " in *Eddy Structure Identification in Free Turbulent Shear Flows.* (J.P. Bonnet and M.N. Glauser eds.) Kluwer Academic Press.

[23] Delville J. (1994) "Experimental comparisons of PODs applied to different free shear flow regimes." In preparation.

[24] Favre A., Kovasznay L.S.G., Dumas R., Gaviglio J. & Coantic M. (1976) "La turbulence en mécanique des fluides." Gauthier Villars.

[25] Ferré J.A., Giralt F. & Antonia R.A. (1989) "Evidence for double-roller eddies in a turbulent wake from two-component velocity measurements." T.S.F.7 Stanford August 21-23.

[26] Ferré J.A. and Giralt F. (1989) "Pattern Recognition analysis of the velocity field in plane turbulent wakes" *J. Fluid Mech.*, vol. 198, pp.27-64.

[27] Ferré J.A. and Giralt F. (1989) "Some topological features of the entrainment process in a heated turbulent wake" *J. Fluid Mech.*, vol. 198, p. 65

[28] Ferré J.A., Mumford J.C., Savill A.M. and Giralt F. (1990) "Three-dimensional large-eddy motions and fine-scale activity in a plane turbulent wake" *J. Fluid Mech.*, vol. 210, pp. 371-414.

[29] Freymuth P. (1992) "Bibliography of Thermal Anemometry." TSI Incorporated. Fluid Mechanics Group.

[30] George W.K. (1978) "Processing of random signals.", in *Proceedings of the Dynamic Flow Conference.*

[31] George W.K. (1988) "Insight into the dynamics of coherent structures from a Proper Orthogonal Decomposition." in *Proceedings Symposium on Near Wall Turbulence* Dubrovnik, Yugoslavia, May 16-20

[32] George W.K. & Glauser M.N. (1991) "Flow Structure Identification from multi-point measurements." in *Second World Conference on Experimental Heat Transfer, Fluid Mechanic and Thermodynamic.* Dubrovnik, June 23-28.

[33] George W.K. & Taulbee D. (1990) "Designing experiments to test closure hypotheses.", in *Engineering turbulence modeling and experiments.* W. Rodi and E.N. Ganic Eds. pp. 383-397, Elsevier, New York.

[34] Giralt F. & Ferré J.A. (1993) "Structure and flow patterns in turbulent wakes.", *Phys. Fluids A* Vol.5, No 7

[35] Glauser M.N., Leib S. & George W.K. (1985) "Coherent structures in the axisymmetric turbulent jet mixing layer". *Fifth Symp. on Turb. Shear Flows.* Ithaca N.Y.

[36] Grant M.L. (1958) "The large eddies of turbulent motion". *J. Fluid Mech.* part 4.

[37] Herzog, S. (1986) "The Large Scale Structure in the Near-Wall of Turbulent Pipe Flow". *PhD Dissertation, Cornell University.*

[38] Hilberg D. & Fiedler H.E. (1989) "The spanwise confined mixing layer." in *Advances in Turbulence 2*, Fernholz-Fiedler eds.

[39] Hussain, A.K.M.F. (1981) "Role of Coherent Structures in Turbulent Shear Flows." *Surveys in Fluid Mechanics*, Narasimha and Deshpande (eds.).

[40] Hussain A.K.M.F. (1983) "Coherent structures - reality and myth". *Phys. Fluid*, Vol. 26, Nr. 10, pp. 2816-2850.

[41] Hayakawa M. & Hussain A.K.M.F. (1989) "Three-dimensionality of organized structures in a a plane turbulent wake". *J. Fluid Mech.*, Vol. 206, p.375.

[42] Tabatai M., Kawall J.G. & Keffer J.F. (1987) DANTEC Information Nr 4.

[43] Koochesfahni M.N., Catherasoo C.J., Dimotakis P.E., Gharib M. & Lang D.B. (1979) "Two-point LDV measurements in a plane mixing layer." *AIAA Journal*, vol. 17, No. 12, pp. 1347-1351

[44] Lehmann B. & Mante J. (1993) "Analysis of turbulent structure by mean of a two-component laser-Doppler scanning technique" in *Engineering Turbulence Modelling and Experiments 2* ed. W. Rodi & F. Martelli pp. 415-424 Elsevier, New York.

[45] Lesieur M. (1990) *Turbulence in Fluids*, 2nd edition, Kluwer.

[46] Liandrat J. & Moret-Bailly F. (1990) "Wavelet transform in fluid dynamics and turbulence." *European J. Mech. B* Vol. 9, No 1

[47] Lumley J.L. (1967) "The Structure of Inhomogeneous Turbulent Flows". *Atm. Turb. and Radio Wave Prop.*. Yaglom and Tatarsky eds. Nauka, Moscow, 166-178.

[48] Lumley J.L. (1970) *Stochastic Tools in Turbulence.* Academic Press, New York.

[49] Lumley J.L. (1981) "Coherent Structures in Turbulence". *Transition and Turbulence*, edited by R.E. Meyer, Academic Press, New York, p215-242.

[50] Moin P. & Moser R.D. (1989) "Characteristic-Eddy Decomposition of Turbulence in a Channel". *J. Fluid Mech.*, Vol. 200, 471.

[51] Mumford J.C. (1982) "The structure of large eddies in fully developed turbulent shear flows. Part 1. The plane jet.", *J. Fluid Mech.*, vol. 118, pp. 241-268.

[52] Mumford J.C. (1983) "The structure of large eddies in fully developed turbulent shear flows. Part 2. The plane wake." *J. Fluid Mech.*, vol. 137, pp. 447-456.

[53] Payne F.R. & Lumley J.L. (1967) "Large eddy structure of the turbulent wake behind a circular cylinder." *Phys. Fluids supplement*, vol. 10, S194-S196.

[54] Reynolds W.C. & Hussain A.K.M.F. (1972) "The mechanics of an organized wave in turbulent shear flow. Part3 : Theoretical models and comparisons with experiments." *J. Fluid Mech.*, 54, 263.

[55] Stretch D., Kim J & Britter R. (1990) "A conceptual model for the structure of turbulent channel flow" in *Notes for boundary layer structure workshop*, S. Robinson Ed. NASA Langley, Aug. 28-30.

[56] Sullivan P., Day M. & Pollard A. (1994) "Enhanced VITA techniques for turbulent structure identification." subm. to *Exp. in Fluids.*

[57] Townsend A.A. (1970) "Entrainment and the structure of turbulent flow." *J. Fluid Mech.*, vol. 41, 13-46.

[58] Townsend A.A. (1976) *The structure of turbulent shear flows* Cambridge University Press, Cambridge.

[59] Ukeiley L.S., Cole D.R. & Glauser M.N. (1993) "An Examination of the Axisymmetric Jet mixing Layer Using Coherent Structure Detection Techniques." *in Eddy Structure Identification in Free Turbulent Shear Flows.* (J.P. Bonnet and M.N. Glauser eds.) Kluwer Academic Press.

[60] Vandromme D. & Haminh H. (1990) "The compressible mixing layer." in *Turbulence and coherent structures*, Metais O. et Lesieur M. eds., Kluwer, 507-523.

[61] Wallace J.M., Broadkey R.S. & Eckelmann E. (1977) "Pattern-recognized structures in bounded turbulent shear flows." *J. Fluid Mech.*, 83, 673.

[62] Willmarth W. and Lu S. (1972) "Structure of the Reynolds stresses near the wall." *J. Fluid Mech.*, vol. 76.

[63] Zaman K.M.B.Q. & Hussain A.K.M.F. (1984) "Natural large-scale structures in the axisymmetric mixing layer". *J. Fluid Mech.*, Vol. 138, pp.325-351.

[64] *An album of fluid motion*, Van Dyke ed., The Parabolic Press-Stanford 1982.

[65] *Topological fluids dynamics* IUTAM Symposium Cambridge 1989.Moffatt H.K. & Tsinober A. edts.

[66] *Wither turbulence? Turbulence at the crossroads,* Proceeding, Ithaca, NY, 1989, Lumley J. ed., Springer 1990.

[67] *Boundary layer structure Workshop*, Langley Res. Center, Hampton, VA, Aug. 1990.

[68] *Structure of turbulence and drag reduction,* IUTAM symp. Zurich 1989, Gyr A. ed., Spinger 1990.

[69] *Turbulence and coherent structures,* selected papers from "Turbulence 89 : Organized structures and turbulence in fluid mechanics", Grenoble Sept. 1989 Metais O. & Lesieur M. eds., Kluwer Acad. Publ., 1990.

[70] *Eddy structure identification in free turbulent shear flows.* IUTAM Symposium. Poitiers - October 1992. (J.P. Bonnet and M.N. Glauser eds.) Kluwer Academic Press. 1993.

[71] *Special Issue* of *Theoretical and Computational Fluid Dynamics* Vol.5, 1993

[72] *Computational Fluid Dynamics Les Houches, Session LIX* M. Lesieur, P. Comte & J. Zinn-Justin, eds. Elsevier Sciences Publishers B.V., 1993

[73] from I.M.F.T. Toulouse

[74] from I.M.G. Grenoble

[75] from I.M.S.T. Marseille

NEW ASPECTS OF VORTEX DYNAMICS RELEVANT TO COHERENT STRUCTURES IN TURBULENT FLOWS

W. Schoppa and F. Hussain
University of Houston, Houston TX, USA

Abstract

Vortex dynamics offers a tractable avenue for explaining the dynamics of coherent structures, which appear to be the dominant features of turbulent shear flows and are the key to predicting and controlling turbulence phenomena such as mixing, entrainment, chemical reaction, drag, and aerodynamic noise. With this in mind, a series of investigations have been undertaken in our group; herein we provide a summary of these studies to date. In particular, we survey coherent structure eduction in free and wall-bounded shear flows, our new generic definition of a vortex, incompressible and compressible vortex reconnection, internal core dynamics within vortices, core dynamics-induced mixing layer transition, helical wave decomposition to extract polarized vorticity wavepackets from turbulent flows, and dynamical coupling of coherent structures with fine-scale turbulence. In addition to reviewing specific contributions in these topics, the results are discussed in context of their generic relevance to turbulence physics.

1. Introduction

Large-scale *coherent structures* (CS) have been the primary focus of turbulence research for the past three decades (*e.g.* see [1-10]). This intense research onslaught was motivated by the expectation that CS hold the key to understanding fundamental flow physics and to modeling turbulence as well as to developing control strategies for turbulence phenomena such as heat and mass transport, entrainment, mixing, combustion, and generation of drag and aerodynamic noise. In fact, we have long contended that turbulence control is possible only in the presence of CS; *i.e. no CS, no control*. While it is clear that CS must be incorporated in viable turbulence theories and models, it remains equally unclear as to how this is to be done. CS-based analytical approaches [11-13] as well as dynamical systems modeling [14-18] appear promising but fall far short of producing a viable turbulence theory. To reiterate: CS is central to turbulence in its understanding, theory, modeling, and applications.

Numerous CS studies have employed several extraction techniques: ranging from simple observation (*e.g.* flow visualization with passive markers) to rigorous mathematical treatment (*e.g.* proper orthogonal decomposition). There seems to be an evolving consensus that vorticity is the key defining property of CS [19-20], thus justifying application of vortex dynamics concepts to understand the evolutionary dynamics of CS, their interactions, their role in turbulent transport, and also their coupling with fine-scale turbulence (as shown herein, vortex dynamics provides physical explanations for this coupling as well as energy cascade and other turbulence processes – many of which were previously discussed in terms of k-space interactions). Unfortunately, the analytical foundation for the dynamics of inviscid thin-core vortex filaments cannot be directly exploited in a viscous turbulent environment. Therefore, the central thrust of our effort has been to further develop vortex dynamics to address CS dynamics and their roles in flows of practical interest.

Since CS are characterized by instantaneous space-correlated vorticity underlying the random vorticity field that we call fluid turbulence [21], their extraction (*i.e. CS eduction*) and analysis require quantitative vorticity field data. In this regard, the widely employed method of flow visualization suffers from the limitation of being frequently confusing or even misleading [22-23] and should be used only to supplement quantitative data. Although significant insight into CS has been obtained experimentally in transitional, nominally 2D flows, where CS either occur naturally in an almost periodic manner (due to feedback) or are induced to do so by controlled excitation, measurement of instantaneous 3D velocity and vorticity fields is currently beyond the capability of state-of-the-art experimental techniques. While multiple X-wire arrays, scanning LDA, and PIV can provide only planar flow field data (typically in planes of symmetry), there is a clear need for more ambitious, fully 3D techniques such as holographic particle velocimetry [24-26], scalar image velocimetry [27] and even electron spin resonance (ESR) and electron paramagnetic resonance (EPR) velocimetries. Currently, comprehensive studies of CS dynamics are possible only through analysis of well-resolved direct numerical simulations (DNS) of the Navier-Stokes equations, subject, however, to limitations of Reynolds number, flow domain geometry, size, complexity, etc.

Using idealized vortex configurations as suitable initial conditions for DNS, we can undertake a careful analysis that has three significant advantages: (i) tractable characterization of the influence of systematic initial condition changes on the flow physics, (ii) clear representation of the evolutionary vortex dynamics in terms of vortex line topology, and (iii) delineation of cause-and-effect physical-space mechanisms responsible for observed turbulence phenomena. As an example, consider DNS of a mixing layer, initialized with low wavenumber 2D and 3D linear instability modes [28]. Since these modes can also be forced experimentally, this study can be used to guide the design of effective control strategies for technological purposes, such as enhancement of mixing, heat transfer, and chemical reaction, reduction of drag and aerodynamic noise, jet vectoring, etc. We also obtain 'clean' evolutions free from the uncontrollable background noise and perturbations prevailing in experiments, thereby permitting detailed analysis of the vorticity field evolution. Furthermore, since the initialized instability modes are quite likely to be present in unforced, but noise-excited laboratory flows (due to blower, tunnel or laboratory acoustic modes), dynamical mechanisms uncovered in these controlled idealized simulated flows are certainly relevant to practical shear flows as well.

For interpretation of vortex evolutions, we rely on the concepts of vortex line geometry and topology – still a rare practice in turbulence research. Inviscid dynamics, which often (but not always) dominate large-scale CS evolution, are particularly amenable to analysis based on vortex lines since the advection and stretching of material vortex lines in this case completely govern the associated vortex dynamics. This way, physical mechanisms may be identified to explain CS evolution in terms of changes in vortex line patterns (*i.e.* geometry and topology). For instance, as shown in this paper, a rectilinear vortex with nonuniform core size exhibits a surprisingly complex vorticity evolution, whose understanding is greatly simplified once vortex line twisting and untwisting (and associated induced axial fluid transport) are recognized [29]. Noting that time history cannot be assigned to vortex lines in viscous flows, which admit topological changes (such as reconnection), the vortex line pattern in such flows, though useful, must be interpreted with extreme caution. We also emphasize the difference between a vorticity isosurface and a vortex surface, frequently confused; since vortex lines generally cross vorticity surfaces, this distinction is important to understand vortex evolutions.

Quantitative characterization of the vortex line geometry is possible via a new analytical tool – *helical wave decomposition* (HWD) – which involves decomposition of the velocity and vorticity fields into left- and right-handed (*i.e. left-* and *right- polarized*) components, using eigenmodes of the curl operator [30]. In flow regions where either the left- or the right-polarized component dominates, vortices have axial flow and vortex lines are locally twisted as left- or right-handed helices. Application of HWD concepts to transitional and fully turbulent free shear flows has revealed that CS generally contain strongly twisted vortex lines [28,31]. Thus, CS are not only vortical entities, but are also typically polarized with axial flow; the resulting Beltramization assures a relatively longer CS lifetime (before decay or breakdown, typically through interactions). This finding has opened a new avenue of study – *polarized vortex dynamics* – involving a virtually unexplored set of (nonlinear) interactions of like- and oppositely-polarized vorticity and velocity fields (*e.g.* §8.3). Research to date indicates that the development of strong local polarization is a key aspect of vortex reconnection, vortex breakdown, and shear flow transition, the last showing that transition can be more organized than previously presumed.

In the following, we present a survey of our recent efforts in extracting and analyzing evolutionary dynamics of CS, including some recent discoveries of new mechanisms unaccounted for in classical vortex dynamics theory. The remainder of this paper is organized as follows: our vorticity-based technique for CS eduction is reviewed in §2, followed by introduction of a general-purpose vortex definition (§3) and its application in §4 to CS eduction in a turbulent channel flow. In §5, we discuss the CS dynamics of vortex reconnection, for both incompressible and compressible flows. This is followed by an analysis of internal *core dynamics* within vortices in §6, shown in §7 to cause transition of spanwise vortices in a plane mixing layer. HWD is introduced and applied to analyze core dynamics in §8. The interaction of CS with fine-scale background turbulence along with the mechanism of the latter's organization and polarization are considered in §9.

I. CS eduction in turbulent shear flows

The logical first step in a CS approach to turbulence modeling and control is the development of techniques to extract organized, large-scale vortical motions from an apparently random field. While the CS concept has been used extensively in turbulence research, considerable confusion remains in developing appropriate eduction methodologies. Because the eduction procedure can influence the resulting CS features, the eduction process itself deserves careful analysis. We first define our stance on CS eduction in free shear flows, based on three-dimensional vorticity field data. Using our newly-developed vortex definition, we then develop a more general eduction strategy, applicable to boundary layers as well, whose effectiveness is demonstrated in a turbulent channel flow.

2. CS eduction in free shear flows

In this section, a scheme for vorticity-based eduction is outlined, along with comparisons with phase-locked eduction, and new insight into shear flow physics and flow control brought about by the CS concept. This scheme can be applied to laboratory or DNS data for any flow; for illustrative purposes, here we discuss eduction in a turbulent wake of a circular cylinder and also in a plane mixing layer.

2.1 CS and preferred modes

Although the CS concept is used quite freely in a qualitative sense, we utilize the following precise definition due to Hussain [19], supported by extensive research, as a basis for CS eduction: "a coherent structure is a connected turbulent fluid mass with instantaneously phase-correlated vorticity over its spatial extent". That is, underlying the apparently random, 3D vorticity field which characterizes turbulence, there is large-scale vorticity which is instantaneously coherent over the CS spatial extent. Note that the CS fluid mass does not remain constant as both entrainment and detrainment can occur. With this CS definition, turbulent flows may be decomposed into coherent and incoherent parts, shown in §2.3 to provide considerable insight into turbulence physics.

To implement this definition, an *eduction* scheme must be formulated to extract CS from instantaneous 3D vorticity fields. It is well known that CS vary greatly from flow to flow and from one flow region to another (as well as for different parameter regimes of the same flow) in their characteristic measures such as size, strength, and orientation. In general, our eduction approach involves the ensemble averaging of a large number of properly aligned realizations within particular parameter regimes. The occurrence of preferred large-scale vorticity patches characterized by distinct peaks of probability density function in parameter space – *preferred modes* – is implicit in the CS concept. The presence of preferred modes distinguishes the CS concept from statistical approaches, suitable for purely random processes. In fact, the absence of preferred modes would imply that the flow contains a numerous variety of CS, which could be linked only with an elaborate statistical approach, whose utility is doubtful [32].

It is important to note that even after proper CS eduction, considerable additional effort is needed to develop actual structure-based models. In particular, the governing equations which follow from CS-based flow decompositions (double: coherent and incoherent parts; triple: mean, coherent, and incoherent parts; see [33] for further discussion) require closure considerations analogous to those for the time averaged equations; this aspect of CS-based modeling is not addressed here. Instead, our focus is on understanding the geometry and dynamics of CS in turbulent flows to provide a physical foundation for future structure-based modeling and turbulence control, these being important subjects for future research.

2.2 Classical eduction strategies

Eduction using phase-locking. The most widely used CS eduction technique to date involves phase-locked sampling in flows containing a preferred mode, usually with controlled excitation [34]. However, such phase-locked measurements can be contaminated by phase jitter in both CS initiation and downstream evolution. Although jitter in initiation may be eliminated through periodic excitation, jitter in evolution becomes progressively larger downstream and is unavoidable due to inherently chaotic dynamics. Therefore, an ensemble average triggered by a remote signal will eventually be severely smeared sufficiently far downstream. These jitter effects can be alleviated by triggering sampling with a more local flow signal (typically u-fluctuations outside the shear region) to capture the CS footprint, with further refinement by streamwise and spanwise alignments [35-37].

Vorticity-based eduction. As an alternative approach to overcome the problem of jitter, we have developed a generic eduction technique which is triggered by the local vorticity signal itself (recall our vorticity-based CS definition). In this approach, CS are identified experimentally from relatively large-scale vorticity components, which are measurable with an array of X-wires. For example, in figure 1, rakes S1 and T1 are used to measure ω_z and ω_y respectively (using Taylor's hypothesis; see below), while rakes S2 and T2 provide simultaneous two-plane vorticity fields for spanwise and transverse coherence measurements. The more general eduction scheme utilized in §4, based upon our newly developed vortex definition (§3), requires instantaneous 3D velocity field data and is thus currently applicable only to DNS data and HPV measurements.

The technical procedure for vorticity-based eduction, adapted from Hussain [32] and Hussain & Hayakawa [38], is outlined below for ω_z eduction using rake S1 in the turbulent wake of a circular cylinder; this is easily adapted to ω_y (and rake T1) or even full-field 3D vorticity data. In the latter case, alignment of successive structures involves cross correlation of $\omega(x,y,z)$ in 3D space and time.

1. A spanwise vorticity (ω_z) map is constructed using velocity signals from rake S1 as $\omega_z = -\Delta v/(\Delta t\, U_c) - \Delta u/\Delta s$, where Δt is the time step between successive data points, Δs is spatial separation between adjacent probes (figure 1), U_c is the structure advection velocity, and u and v are the longitudinal and transverse velocity components. The inaccuracy in measuring vorticity using Taylor's hypothesis is not a serious constraint as we are interested primarily in recognizing the structure location and its geometry. Since the vorticity signal contains significant high-frequency fluctuations (small scales), smoothing is necessary, but only for recognition of the CS footprint and initial determination of the CS center; *i.e.* smoothing is used to determine the optimum alignment point, but not for the final ensemble-averaged data. For smoothing, a simple local averaging technique is adequate [38].

Examples of pseudo-instantaneous smoothed vorticity fields are shown in figure 2 [39], with ω_z in figure 2(a) and ω_y at two transverse locations in figure 2(b,c). These vorticity maps quantitatively reveal the organized events and guide the choice of strength and size thresholds for eduction. For instance, most vortex centers are located much closer to the wake centerline than the half-velocity defect (figure 2a), and pairs of strong, opposite-signed ω_y indicate significant kinking of the (originally spanwise uniform) Karman vortices.

2. Since vortices typically are widely dispersed across shear regions, it is useful to first select a transverse detection center y_c, which is the most probable y location of advecting structure centers. This is accomplished by first counting the number of ω_z peaks (higher than

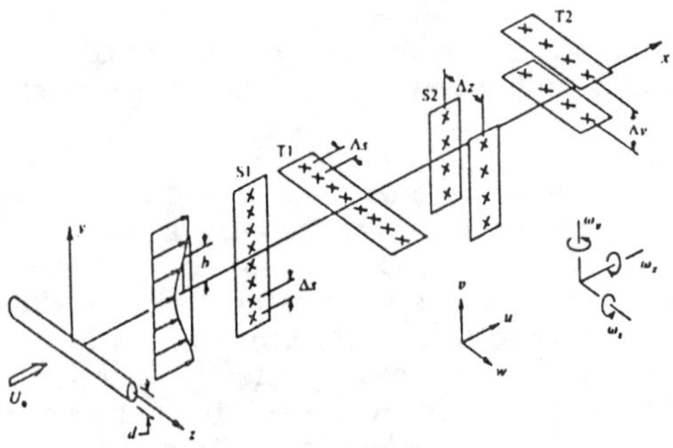

Figure 1. X-wire rake configurations and notation for a turbulent plane wake.

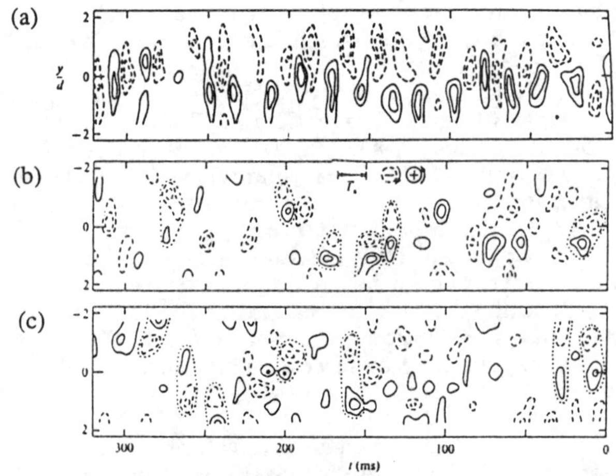

Figure 2. Experimentally measured pseudo-instantaneous vorticity maps in a turbulent plane wake at x/d=20: (a) ω_z; (b,c) ω_y at (b) y/b=0.94, (c) y/b=0.47. Negative values are dashed, and paired ω_y concentrations with opposite circulations are dotted in (b,c).

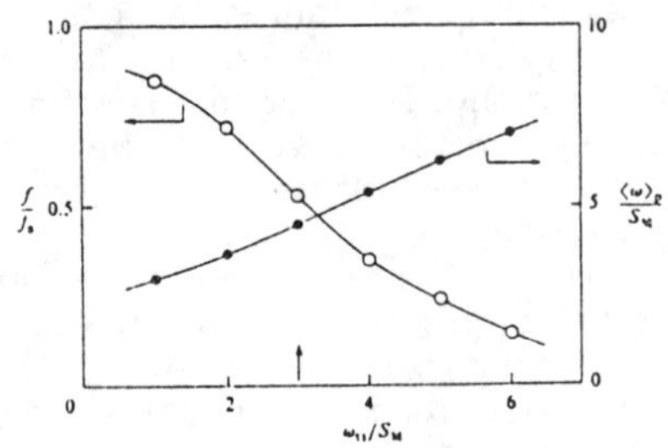

Figure 3. Number of accepted structures f and peak coherent vorticity $<\omega>_p$ as functions of the vorticity threshold at x/d=40 in a turbulent wake.

a certain level) as a function of y so that the transverse location with the largest structure count is denoted y_c. The X-wire rake is then centered at y_c (at each x), and all data are resampled. Note that this step is not necessary for full-field DNS data since it is possible to collect all identified CS in the simulation field independent of their y-location [40].

3. After centering the rake at y_c, only those structures are accepted which simultaneously: (i) are centered at y_c, (ii) have a peak value of (smoothed) vorticity of a given sign above a specified threshold, and (iii) have (smoothed) vorticity which is well-correlated over sufficiently large streamwise and transverse distances. With regard to (ii), the local maximum mean shear S_m is a logical choice to scale the vorticity threshold. Figure 3 shows the number f of detected structures at y_c and their ensemble-averaged peak vorticity $<\omega_z>$ as a function of the threshold ω_{t1}. From this data, the threshold level was somewhat subjectively chosen as $3S_m$, which results in acceptance of about half of the shed vortices, whose locations (y_c,t_c) are identified.

The transverse size criterion is enforced by accepting only structures whose vorticity at ($y_c \pm \Delta y$,t_c) is (i) of the same sign as that at (y_c,t_c), (ii) higher than a threshold ω_{t2}, and (iii) lower than the value at (y_c,t_c) – conditions which must be satisfied simultaneously at these three y locations, thus ensuring that the accepted structures are centered at y_c. The threshold ω_{t2} and the size Δy are chosen depending on their influence on the number of accepted structures. Virtually the same criterion is applied for the streamwise structure size Δx.

4. Large-scale structures are accepted only when the criteria in step 3 are satisfied simultaneously, so that weaker, smaller, fragmented, distorted, or transversely shifted advecting vorticity patches are discarded from the ensemble average. With the structure center as a phase (*i.e.* time) reference, signal segments around accepted structures are aligned with respect to their centers (y_c,t_c) and then ensemble-averaged, yielding the *zeroth-iteration ensemble average*.

5. Although CS centers are expected to be well identified by the location of smoothed vorticity peaks, the true center of the structure may be different from the location of peak vorticity or the centroid of the vorticity patch because of small-scales within the structure. To compute the true structure centers, we take the cross-correlation between the vorticity-signal segment ω_z of each realization and the initial (zeroth) ensemble average $<\omega_z>$. Each realization is then shifted by the time delay of the peak correlation. Structures requiring excessive time shifts or having a small cross-correlation value are discarded; these structures are considered to be freak. Then, we revise (y_c,t_c) to an improved estimate of the center and perform the ensemble average again, to obtain the *first-iteration ensemble average*. This step is repeated until convergence in both this shifting and the ensemble averaged vorticity is achieved. Typically, only a few iterations are required.

6. Unsmoothed signal fields are then time-aligned with respect to the final centers and ensemble averaged to educe CS velocity and vorticity fields. As mentioned earlier, the smoothing process is used only for recognizing organized events (*i.e.* large-scale vorticity patches satisfying selected criteria) and for determining the time shifts required for optimum relative alignment of the finally accepted events (before the final ensemble averaging). This elaborate eduction scheme does not create an artificial structure, but merely aids extraction of the dominant natural CS and sharpens its details. Note that if the vorticity field $\omega(x,y,z,t)$ is known as in HPV or DNS, the shifting can be performed in (x,y,z) at each time following the above procedure, before the CS is extracted from the 3D field. The deviation of the instantaneous signal from the ensemble average is totally accounted for as incoherent turbulence.

Advantages and limitations of vorticity-based eduction. The principal advantage of this procedure is that it is based on large-scale vorticity patches, an intrinsic CS property, as defined by Hussain [19]; thus, it is conceptually objective and self-consistent. In particular, the presence of large-scale vortical structures is recognized with filtered vorticity maps, which give a physical perception of the instantaneous fields. This allows more precise identification of vortical structures than a velocity-based approach, since vortical structures exhibit compact vorticity patches but a slowly decaying induced velocity. Additionally, the adverse effect of jitter is minimized – this is accomplished by maximizing cross-correlation of vorticity patches to locate the patch 'center', which is in turn used as a phase reference for alignment and ensemble averaging of all accepted patches. This feature is particularly important in order to clarify the relative spatial distribution of large- and small-scale (*i.e.* incoherent) motions. Finally, in contrast to phase-locked eduction, which requires the presence of regular preferred modes, vorticity-based eduction can be applied to any CS-containing flow provided that the local vorticity field can be measured, except in the near-wall region of boundary layers, where vorticity is an inappropriate CS identifier (discussed in §3).

The major limitations of experimental vorticity-based eduction are a direct consequence of the current inability to measure instantaneous 3D vorticity fields. For example, the use of Taylor's hypothesis and restriction of measurements to a single plane (thus only one vorticity component) are artifacts of measurement limitations. In fact, this eduction scheme can easily be adapted to accommodate space- and time-resolved 3D vorticity fields obtained via DNS [40]. Although the selection of detection criteria appears somewhat subjective, an appropriate choice of criteria is usually clear from the instantaneous vorticity patches, and the educed structures are typically insensitive to condition parameters.

Comparison of phase-locked and vorticity-based CS. As an illustrative example, CS educed using our vorticity-based technique are shown in figures 4(a,b) for the self-preserving region of a plane mixing layer with an initially turbulent boundary layer. The $\langle\omega_z\rangle$ distribution at both streamwise locations are quite similar in both geometry and peak vorticity values, in accordance with the self-similar evolution. Under nearly the same experimental conditions, CS educed with the phase-locking scheme (using u fluctuations outside the layer as a trigger signal) are shown in figures 4(c,d) [41]. In this case, the $\langle\omega_z\rangle$ contours are spiky on the high-speed side, unlike the symmetric distribution in figure 4(a). Apparently, phase-locking exaggerates ensemble-averaged flow properties near the trigger location, thereby distorting the actual CS shape. The superiority of vorticity-based eduction is also indicated by the close correspondence of figures 4(a,b) with the spanwise-averaged 3D DNS data in figures 4(e,f) for a turbulent mixing layer [40]. Despite the temporal and low Re simulation, the resemblance with the educed vorticity distribution is remarkable, both in structure shape and peak vorticity value.

2.3 CS features in free shear flows

Having illustrated the basic eduction procedure, we now discuss some fundamental dynamical features of CS in free shear flows. In particular, our extensive experimental data demonstrates that CS in free shear flows share some common features and dominate the Reynolds stress distribution, turbulence production, and mixing.

CS Topology. In figure 5, a spanwise cut through a typical shear layer CS is shown, with a low-level contour of coherent spanwise vorticity $\langle\omega_z\rangle$ denoting the structure boundary. Note that there are two critical points in the streamline pattern: the saddle (A) characterized by negligible spanwise vorticity and the center (B) characterized by the peak spanwise coherent vorticity. Also shown qualitatively are typical contours of coherent strain rate $\langle S\rangle$, incoherent turbulence intensities $\langle u_r^2\rangle^{1/2}$ and $\langle v_r^2\rangle^{1/2}$, incoherent Reynolds stress $-\langle u_r v_r\rangle$, coherent intermittency $\langle\gamma\rangle$, and coherent production $\langle P\rangle$. Note that these distributions are

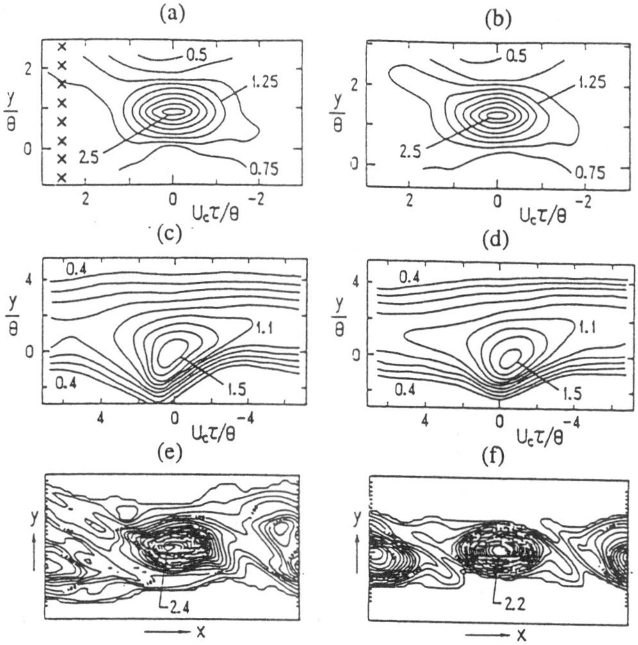

Figure 4. Educed $<\omega_z>$ in plane mixing layers, for (a,b) vorticity-based eduction at (a) $x/\theta=1000$ and (b) $x/\theta=2000$; (c,d) phase-locked eduction at (c) $x/\theta=600$ and (d) $x/\theta=4800$; (e,f) spanwise-averaged DNS data for (e) natural evolution (random noise perturbations) and (f) weak 2D forcing.

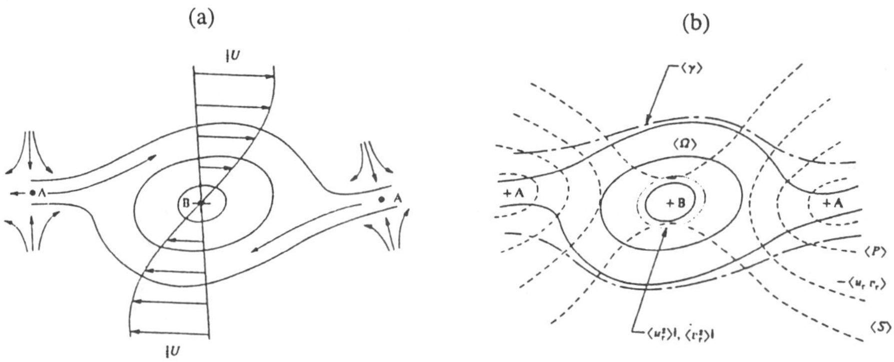

Figure 5. An example of a coherent structure in a plane mixing layer. (a) Coherent vorticity $<\Omega>$ contours and saddle A and center B; (b) approximate contours of other coherent structure properties: coherent intermittency $<\gamma>$; incoherent turbulence intensities $<u_r^2>^{1/2}$, $<v_r^2>^{1/2}$; incoherent Reynolds stress $-<u_r v_r>$; coherent strain rate $<S>$; coherent production $<P>$.

strongly linked to particular regions of the CS: $<u_r^2>^{1/2}$, $<v_r^2>^{1/2}$, and $<\gamma>$ have peaks at the center, while $<S>$, $-<u_r v_r>$, and $<P>$ have peaks in the saddle region.

The pseudo-instantaneous normal vorticity ω_y maps in figures 2(b,c) indicate that these 'roll' CS, while quasi-2D near their initiation point, typically become highly 3D upon downstream evolution. Using essentially the same eduction procedure as outlined above for ω_z, ensemble-averaged $<\omega_y>$ contours were obtained, as shown in figures 6(a-c). Each panel indicates opposite-signed vorticity concentrations located preferentially on one side of the detected structure, such that larger negative $<\omega_y>$ appears in negative z when triggered by positive ω_y (figures 6a,b), and vice-versa when triggered by negative ω_y (figure 6c). Two illustrations of possible roll deformation are shown in figures 6(d,e); frequent outward kinking (figure 6d) is indicated by the preferential locations of secondary, opposite-signed vorticity in figures 6(a-c) and also the instantaneous patterns (demarcated by dotted lines) in figures 2(b,c).

Another important 3D feature of free shear flows is longitudinal vortices ('ribs') between the primary roll CS, inferred from $<P>$ data [42] and a prominent feature of 3D DNS data [43] and flow visualization [44]. Although it is difficult to distinguish rib ω_y from that of roll contortions, Hayakawa & Hussain [39] demonstrate, by comparison of simultaneous ω_y and streamwise velocity maps, that many of the unpaired closed contours in figures 2(b,c) are manifestations of ribs.

CS-based conceptual model. Based on our extensive CS studies, we have developed a dynamical model of the turbulence physics of free shear flows, illustrated in figure 7 for the case of a turbulent plane wake. Due to its intense strain rate, the dynamically most significant region of the flow is the saddle, denoted by X-marks in figure 7(a). Since production is observed to be the highest in the 'braid' region connecting adjoining structures, the braid obviously cannot be a continuous vorticity sheet, but instead consists of intermediate-scale streamwise ribs separated in the spanwise direction (figure 7b). The predominantly spanwise roll induces motion of external fluid toward the saddle regions, as denoted by the solid arrows in figure 7(a). This fluid can percolate through the braid by rib induction, which also causes significant spanwise contortions of the rolls. Fluid advected to the saddles acquires rib vorticity (through diffusion), which is then subjected to vortex stretching along the diverging separatrix; this is the mechanism of smaller-scale (incoherent) turbulence production, the regions of which are shown as shaded areas in figure 7(a).

The turbulence produced is continuously advected into the rolls, causing a higher incoherent turbulence intensity near the structure center and also at the back of the rolls; these paths are shown as broken arrows. The strong outward motion, together with the turbulence intensity on the back side of the structure, constitute an important mechanism for the transport of small-scale turbulence to the edge of the wake.

We deduce that the most effective mixing occurs in the hatched areas A and B in figure 7(a) because of direct interaction of nearly orthogonal rib and roll vortices, which presumably produces strong 3D vorticity fluctuations. Furthermore, owing to the combined effects of this interaction and the presence of high turbulence intensity at the back of the structure, mixing in A should be greater than in B. Although our measurements cannot accurately quantify the actual extent of mixing, our numerical studies of chemically reacting mixing layers confirm the view that intense turbulent mixing occurs in such highly localized regions [45].

2.4 Examples of turbulence control using CS

As noted earlier, the dynamics of most shear flows are dominated by a few preferred modes (CS); in turn, CS generation, evolution, and interaction can be modified by controlling the instabilities responsible for their creation. Therefore, turbulence phenomena can be

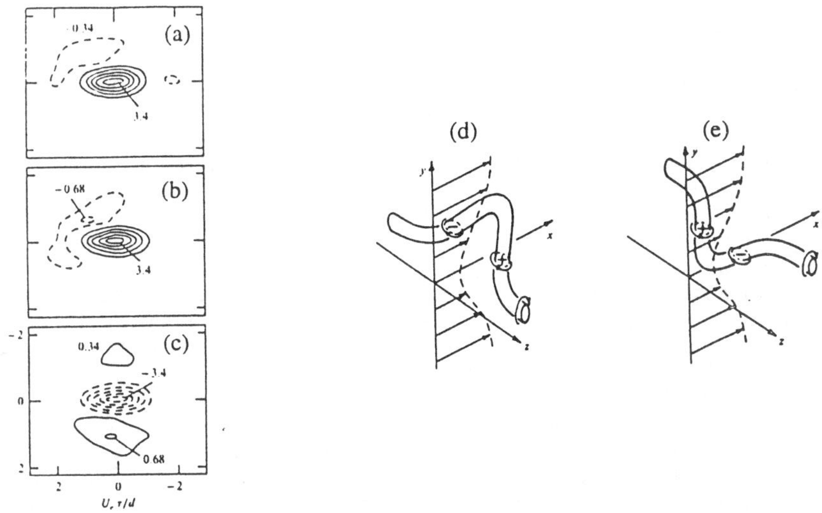

Figure 6. Roll three-dimensionality in a turbulent wake: (a-c) ensemble-averaged ω_y contours at x/d=20 and (a) y/b=0.94, (b) y/b=0.47, (c) y/b=0.47 with triggering by negative ω_y. Note that a stronger non-reference structure appears preferentially on one side of the reference structure. (d,e) Conceptual sketches of roll deformation; (d) outward kinking, (e) inward kinking.

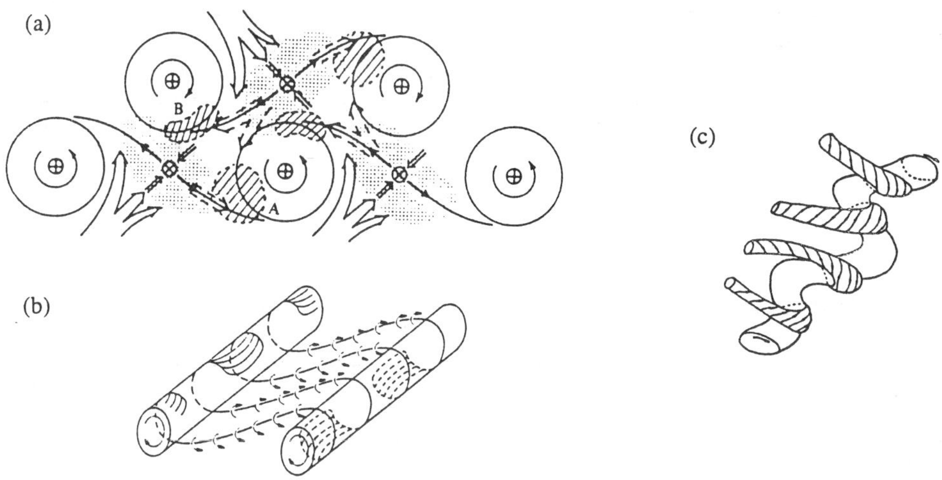

Figure 7. A conceptual picture of the CS topology in a turbulent plane wake; (a) a plane cut of the wake; (b) a simplified schematic of rib vortices; (c) a more realistic schematic of the rib-roll arrangement.

selectively augmented or suppressed through appropriate control of the underlying CS. We now briefly mention some techniques we have found to be useful in controlling CS initiation and interaction in free shear flows.

Boundary conditions. By changing the physical boundaries of the flow system, the CS geometry and evolution can be controlled to a certain degree. For example, in a jet, different nozzle shapes (*e.g.* circular, plane, elliptic, rectangular, etc) produce drastically different CS evolutions. This type of passive manipulation (*i.e.* requiring no external input) of CS utilizes the fact that self-induced vortex motion, produced by local curvature, can cause systematic deformation into a highly 3D structure, thereby enhancing transports of mass and momentum (*e.g.* high aspect-ratio elliptic jets). Some nozzle shapes with sharp corners also cause a CS to fold on itself, thus promoting reconnection and increased mixing.

Initial conditions. The inflow conditions (*e.g.* mean and fluctuating velocity profiles, spectral content, etc) of free shear flows are known to critically affect the downstream evolution, which can be controlled by adjusting the initial boundary layer characteristics. In this way, the instability modes responsible for CS generation are altered, which in turn affects the strength and spacing of the CS and their interactions farther downstream.

Excitation. One of the most popular means of CS control is to perturb the entrance flow (near its separating point) using excitation at selected frequencies. Examples of excitation devices include external sources such as acoustic drivers, oscillating flaps, and vibrating ribbons, and internal sources set up by flow resonance such as in whistler jets, pulsed combustors, and chamber cavity resonance. Controlled excitation can generate a particular CS mode, possibly one which is not dominant naturally, to either enhance or attenuate turbulence characteristics. For instance, since laminar shear layers generate intense aeroacoustic noise through laminar CS interaction, the suppression of structure interactions through active excitation can significantly reduce jet noise [46-47].

To summarize, we have outlined a general vorticity-based technique to extract CS, the tractable part of turbulence which dominates transports of mass, momentum, and heat. Although it is unclear as to how CS details can be incorporated into a viable turbulence theory, we emphasize that effective turbulence modeling necessitates cognizance of the geometry and dynamics of the underlying CS. In addition, the CS approach is not only of conceptual and academic interest, but also of significant practical importance in understanding and designing systems involving turbulent flows. Manipulation of turbulence via CS control has already shown great promise in a variety of technological applications, particularly mixing, combustion, aerodynamic noise, and drag.

3. Definition of a vortex

In the previous section, we discussed an effective eduction scheme for free shear flows, in which CS are identified as compact regions of large-scale vorticity. Nevertheless, vorticity magnitude is not necessarily an appropriate identifier of CS, especially in boundary layers. For example, vorticity is the largest at the wall, where no large-scale CS are present. Although the concept of and fascination with vortices are as old as the subject of hydrodynamics itself (*e.g.* many sketches of DaVinci), no universally accepted quantitative indicator of vortices currently exists. Intuitive perceptions of a vortex, mostly valid in 2D, frequently fail in 3D, particularly where vortex interactions are involved. This issue superficially appears academic, but is quite important to provide an analytical framework for CS dynamics and modeling in both wall-bounded and free shear flows, noting that CS are vortical events dominating turbulence phenomena: *i.e. vortices are the muscle and voice of turbulence.* In this section, we introduce a quantitative definition of a vortex and demonstrate its validity in a wide range of test flows; this new indicator is then applied in §4 to educe and analyze near-wall CS dynamics in a turbulent channel flow.

3.1 Intuitive quantities for vortex identification and their limitations

First of all, in a viscous flow, vorticity is ever present, and thus a threshold has to be selected to identify vorticity patches. The presence of a high vorticity level is necessary but not sufficient for a vortex to exist; for example, a laminar homogeneous shear flow has vorticity everywhere, but no vortex. As a sophistication of this intuitive idea, a vortex is often considered to be a vortex tube whose surface consists of vortex lines. However, the existence of a vortex tube does not imply the presence of a vortex; for instance, a vortex tube in a laminar pipe flow is not a vortex in any sense. Before we attempt to define a vortex, we will first clarify what is expected of an effective definition/identification scheme.

The first necessary condition imposed is that regions identified as vortex cores must have net vorticity, *i.e.* nonzero circulation. Thus, potential flow regions are excluded from vortex cores, and a potential vortex is a line vortex. In viscous flows, we obviously expect vortices to have non-zero cross-sections. Secondly, we require the indicator to be Galilean invariant (*i.e.* identical in all inertial reference frames).

In the following, we consider three commonly used indicators of vortices: (i) low-pressure regions; (ii) regions with closed or spiral streamlines and pathlines; and (iii) isovorticity surfaces; for each quantity, counter-examples are presented to illustrate their inadequacy in detecting/identifying vortices in unsteady 3D flows.

Local pressure minimum. The physical reasoning for associating vortices with low pressure regions is that pressure tends to have a local minimum within regions of swirling motion when the centrifugal force is balanced by the pressure force, the so-called *cyclostrophic balance*. For instance, Robinson [10] used a low pressure criterion to identify vortical structures in DNS data of a turbulent boundary layer. However, a cyclostrophic balance is guaranteed only when the flow is inviscid, 2D, and steady. As a counter-example, a well-defined pressure minimum can exist in an unsteady irrotational flow as well; obviously, this flow cannot contain a vortex because all contours enclose zero circulation. To illustrate, consider an *unsteady* irrotational, axisymmetric motion with a stagnation point: $u_r = -a(t)r$, $u_\theta = 0$, $u_z = 2a(t)z$, where a(t) is the unsteady strain rate and r,θ,z are the radial, azimuthal, and axial coordinates respectively. Upon integration of the Euler equation, we find that the pressure (P) is given by

$$P = (da/dt - a^2)\, r^2/2 + (-da/dt - a^2)\, z^2. \tag{3-1}$$

When $da/dt - a^2 > 0$, the pressure has a local minimum in any (r,θ) plane, even though the flow has no vorticity, and hence, no vortex core. Similar problems in identifying vortices by low pressure regions arise in several other cases, including irrotational sink and source flows and Karman's viscous pump (where centrifugal forces are balanced by viscous effects). An inverse example is a 2D Stokes flow, where there is no pressure minimum even within vortical motion.

In addition to the above difficulties, the length scale of pressure may be vastly different from that of the vortex core. For the case of Burgers vortex, this scale difference is clearly illustrated in figure 8(a), where the lengthscale of vorticity is fixed, while the lengthscale of the pressure hump (*i.e.* the region of $\partial p/\partial r \geq 0$) increases with the value of vorticity on the axis. This inherent scale difference between a vortex core and the associated low-pressure region (pressure, being a solution of the Poisson equation, is a nonlocal variable) may disallow a clear demarcation of weaker, yet dynamically significant vortices, especially when these are located near stronger vortices. In fact, in DNS data of a plane mixing layer, isopressure

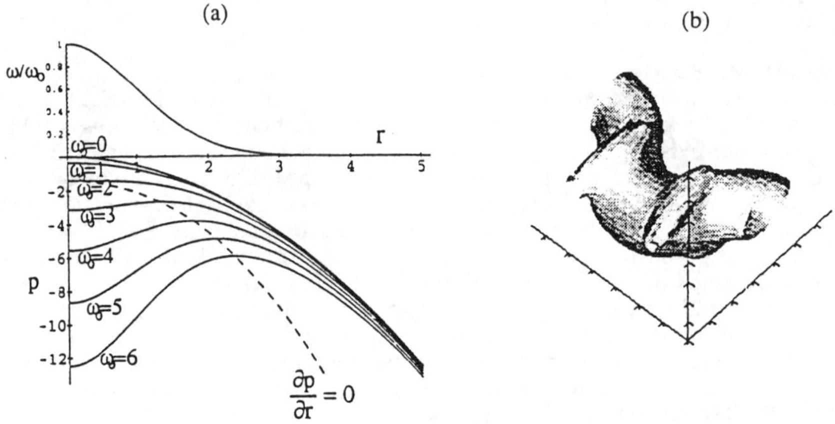

Figure 8. (a) Pressure and vorticity distributions in a Burgers vortex; ω_0 is the vorticity value at the vortex center; (b) isopressure surface in a transitional plane mixing layer.

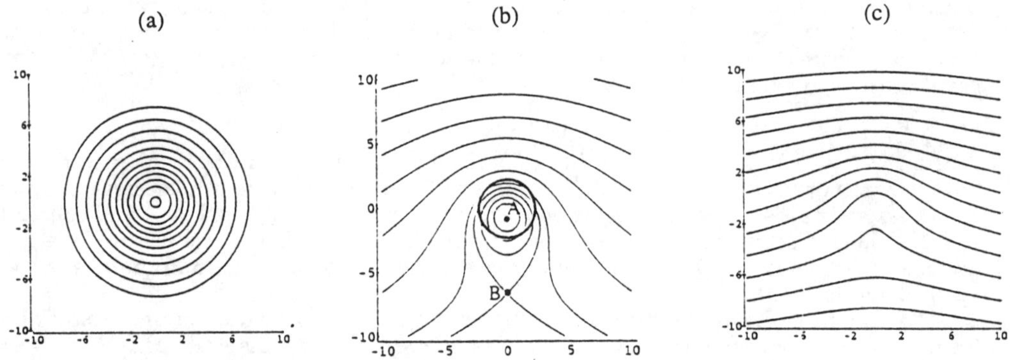

Figure 9. Streamlines of a Lamb vortex in reference frames moving (a) with the vortex center; (b) with a point within the vortex core; (c) faster than any point.

surfaces cannot capture rib and roll vortices simultaneously (figure 8b); while both are dynamically dominant, the pressure level is quite different within the two vortices.

Closed pathline and streamline. The use of pathlines and streamlines to qualitatively identify vortices [48] is fundamentally inappropriate as these are not Galilean invariant. Thus, in different reference frames, drastically different 'vortices' are observed from the streamline pattern (figure 9). Another critical inadequacy of the closed pathline criterion is that a particle may not complete a full revolution around the vortex center (hence no closed pathline) during the lifetime of a vortex. This can occur if vortices undergo transition to turbulence due to nonlinear processes such as pairing, tearing, core dynamics, or fine-scale transition, even though these vortices may have been dynamically significant during their lifetime. Thus, even in a swirling motion with closed streamlines, pathlines are not necessarily closed.

Vorticity magnitude surface. Vorticity magnitude is undoubtedly the most widely used quantity to represent vortices in experimental eduction and DNS data [38,43,49]. While typically adequate for wall-free flows, this can pose a problem in wall-bounded flows. In fact, in a 2D wall-bounded flow, the maxima and minima of vorticity magnitude must occur at the wall [48], where a vortex obviously does not exist. In a turbulent boundary layer, the maximum vorticity magnitude occurs at the wall directly below the high-speed streaks [50]. In both cases, the near-wall flows do not exhibit any swirling motion, characteristic of vortices. In addition, the use of vorticity magnitude or isovorticity surfaces to identify vortices can be inappropriate even in free shear flows. For instance, vorticity sheets or layers with locally parallel flow are not vortices. Additional counter-examples include misrepresentation of a vortex embedded in a homogeneous shear flow (figure 10b) and an axisymmetric vortex with axial variation of core size and vorticity (figure 10a). Generally speaking, nonzero vorticity is a necessary, but not sufficient condition for vortical motion. In fact, vortices with strong (*i.e.* internal) core dynamics, including those involving vortex breakdown, have regions of negligible vorticity on the axis.

3.2 Previously developed vortex indicators

In this section, we briefly introduce two recently proposed, rather sophisticated measures for vortex identification (which satisfy the two requirements in §3.1), based on local characteristics of the velocity gradient tensor $\partial u_i/\partial x_j$.

Complex eigenvalues of velocity gradient tensor: $\Delta>0$. Chong *et al.* [51] proposed to identify vortices as flow regions where the velocity gradient tensor exhibits complex eigenvalues. The physical motivation of this criterion is that in a reference frame moving with a particular fluid particle, the existence of complex eigenvalues of the velocity gradient implies a local spiral/swirling motion. Computationally, the boundary of the region with complex eigenvalues is easily obtained from the discriminant of the cubic characteristic equation

$$x^3 - Px^2 + Qx - R = 0, \qquad (3\text{-}2)$$

where P, Q and R are the first, second, and third invariants of the velocity gradient tensor. Complex eigenvalues occur when the discriminant $\Delta = (Q/3)^3 + (R/2)^2 > 0$, so that the vortex core boundary is defined in this case by the surface $\Delta = 0$. Note that since this definition is derived from $\partial u_i/\partial x_j$, it is Galilean invariant.

Second invariant of velocity gradient tensor: $Q>0$. Hunt *et al.* [52] defined a vortex core (eddy) as a region with positive $Q = (\Omega_{ij}\Omega_{ij} - S_{ij}S_{ij})/2$, where Ω_{ij} and S_{ij} are the antisymmetric and symmetric parts of $\partial u_i/\partial x_j$ respectively. Physically, positive Q identifies regions where the vorticity magnitude is disproportionately larger than the strain rate, thus excluding nearly parallel vorticity layers. This condition can also be expressed as $-(\lambda_1+\lambda_2+\lambda_3)/2 > 0$, where $\lambda_1>\lambda_2>\lambda_3$ are eigenvalues of the tensor $\Omega_{ik}\Omega_{kj} + S_{ik}S_{kj}$.

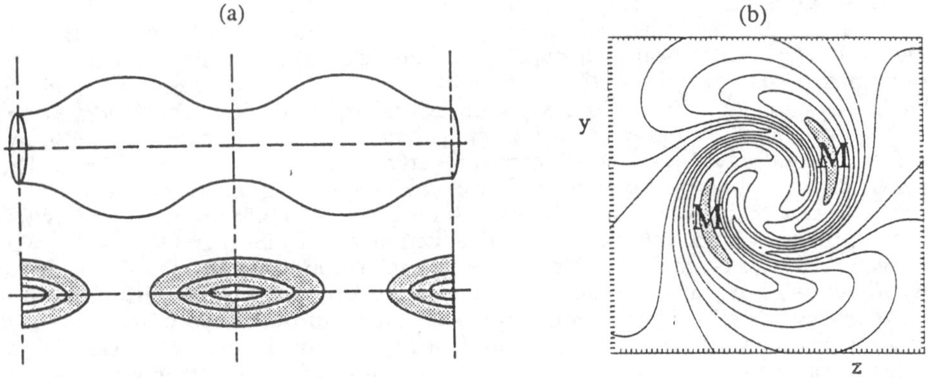

Figure 10. (a) Axisymmetric vortex with an axial variation in vorticity: top panel -vortex surface, bottom panel - vorticity contours; (b) contours of vorticity magnitude for an inviscid streamwise vortex embedded in a homogeneous shear flow.

In a similar vein, Truesdell [53] introduced the 'kinematic vorticity number', defined as

$$N_k=(1+2Q/S_{ij}S_{ij})^{1/2}, \tag{3-3}$$

which measures the 'quality' of rotation, distinguished from the local rotation rate given by $|\omega|$. In their study of vortex core dynamics, Melander & Hussain [29] identified the core of axisymmetric vortices as connected spatial regions with $N_k>1$. Obviously, in terms of vortex core demarcation, regions with $N_k>1$ have $Q>0$ and vice-versa; thus, these two definitions are operationally equivalent. Since N_k is based on the ratio of vorticity to strain rate, it has the problem that regions with extremely low vorticity and negligible strain rate are indicated as vortices even though they may not be dynamically significant.

It can be shown that both Q and Δ vanish identically at a no-slip wall, unlike vorticity magnitude, and thus may be suitable for vortex identification in boundary layer flows. Nevertheless, we will demonstrate in §3.4 that these definitions fail as well under certain circumstances, thus motivating our development of a new definition.

3.3 New vortex definition: $\lambda_2<0$

Although low pressure regions often poorly represent vortices as shown earlier, this concept serves as a starting point for our new definition [54]. In particular, we seek to isolate and eliminate consideration of non-vortical contributions to the pressure minimum, in particular, unsteady straining and viscous effects (§3.1). Since information on local pressure extrema is contained in the Hessian of pressure ($P_{,ij}$), we seek an equation involving $P_{,ij}$, derived as the gradient of the Navier-Stokes equations, where the symmetric part is the tensor equation

$$DS_{ij}/Dt - \nu S_{ij,kk} + \Omega_{ik}\Omega_{kj} + S_{ik}S_{kj} = -P_{,ij}/\rho. \tag{3-4}$$

The first term on the LHS represents unsteady straining effects, which can cause a pressure minimum without any associated vortical motion. In highly viscous flows, the second term tends to mask pressure minima even in the presence of strong vortices (*e.g.* 2D Stokes flow). In this sense, $\Omega_{ik}\Omega_{kj}$ and $S_{ik}S_{kj}$ can be isolated as the true vortical contributions to pressure minima. Elementary variational calculus shows that the sum of these terms attains a local maximum (hence contributing to a pressure minimum) if and only if two eigenvalues (guaranteed to be real) of $\Omega_{ik}\Omega_{kj} + S_{ik}S_{kj}$ are negative. Equivalently, we *define a vortex core as a connected spatial region in which the tensor $\Omega_{ik}\Omega_{kj} + S_{ik}S_{kj}$ has a negative second eigenvalue λ_2.*

3.4 Comparison of definitions

In this section, we compare this λ_2-definition with the Δ- and Q-definitions discussed in §3.2, using illustrative DNS and analytical solutions, in which the vortex geometries are intuitively clear. We justify λ_2 as a more effective vortex indicator by demonstrating its success in flows where the other definitions are clearly inadequate.

Interestingly, these three definitions become identical in planar 2D flow, despite their varied physical foundations and the fact that they are not explicitly related. Thus, in simple planar flows, these three definitions are all equally effective in identifying compact vortex core regions. Furthermore, in axisymmetric flows with $v_\theta = V(r)$, $v_r = v_z = 0$, all three definitions require $dV^2/dr > 0$, thereby excluding the surrounding, essentially potential region with decreasing azimuthal velocity in r. Nevertheless, as shown by the following counter-examples, in 3D flows (or even axisymmetric flows with swirl), our λ_2-definition is effective over a wider range of vortical flows.

Limitations of $\Delta>0$ definition. A particularly revealing test flow is our analytical model for a tornado [55], consisting of a conically symmetric, swirling jet emerging into a half-space with momentum and circulation sources located at the origin. Since this flow has an axial velocity in addition to an azimuthal velocity around the vortex axis (the z axis), it exhibits helical vortical motion. In addition, due to the conical symmetry of this flow, the vortex core boundary should also be conical in shape. The Δ-definition shows two vortex cores marked by grey shading in figure 11(a): one on the axis (region 1) and another (region 2) detached from the axis. Because there is significant vortical motion only immediately near the axis, the presence of region 2 indicates clear misrepresentation of the vortex geometry by the Δ-definition. In contrast, $-\lambda_2$ is maximum and positive on the axis, without a spurious detached vortex core (figure 11b). The effectiveness of our λ_2-definition for this idealized vortex with axial flow is especially important, noting our finding (discussed in §9) that CS generally contain helically twisted vortex lines and hence strong axial flow (*i.e.* partially Beltramized).

Additionally, in DNS data, the Δ-definition tends to produce very noisy vortex core boundaries; for example see the rib and roll vortices of a transitional plane mixing layer in figure 12. We have ensured that the small scales in figure 12(a) for $\Delta>0$ are an artifact of the definition itself and *not* numerical noise. In contrast, the λ_2-definition reveals both the rib and roll vortices clearly, with smooth vortex core boundaries (figure 12b). Vorticity surfaces and contours are similarly smooth, confirming that the small scales in figure 12(a) for Δ are in fact unphysical.

Limitations of Q>0 definition. In the conically symmetric, swirling jet mentioned above, the vortex core based on the $Q>0$ definition shows a narrow hollow core along the axis (figure 11c). This exclusion of the near-axis region is unphysical because fluid near the axis has the highest vorticity, undergoes nearly solid body rotation, and is clearly an integral part of the vortex. Thus, the Q-definition misrepresents the vortex geometry in this case, unlike our λ_2-definition (figure 11b).

As an additional test case of vortical motion with strong axial flow, we consider an inviscid, steady axisymmetric axial vortex embedded within a thin vortex ring, as illustrated in figure 13(a). In this flow, the axial vortex maintains its local expansion due to the induced motion of the ring. The Q-definition incorrectly shows disconnected regions (figure 13b) along the continuous *axisymmetric* vortex column, implying that the Q-definition becomes inadequate whenever a vortex is expanding locally due to an imposed non-uniform strain field. In contrast, the λ_2-definition correctly reveals a continuous axial vortex core (figure 13c).

In summary, among these potential definitions, only our λ_2-definition was found to correctly represent the topology and geometry of vortex cores for the large variety of flows considered by Jeong & Hussain [54]. Thus, connected regions with $\lambda_2<0$ accurately demarcate vortex cores for studies of CS dynamics and related flow physics. Based on these results, we refine our vorticity-based definition by suggesting that a coherent structure is defined as a *connected region with phase-correlated negative λ_2 over its spatial extent*. This λ_2-definition has been used to study CS in a number of turbulent flows, of which the following discussion of near-wall turbulence is an example.

4. λ_2-based CS eduction in near-wall turbulence

In this section, we develop a general-purpose CS eduction scheme applicable to both free- and wall-bounded shear flows, which incorporates our λ_2 definition into the eduction technique outlined in §2.2. For illustrative purposes, we utilize this approach to educe quasi-streamwise vortices near the wall from DNS of a turbulent channel flow. Based on the educed CS, we explain near-wall dynamics and provide a conceptual model which captures virtually all experimentally observed events such as VITA, VISA, ejection (Q2), sweep (Q4), kinked low- and high-speed streaks, etc.

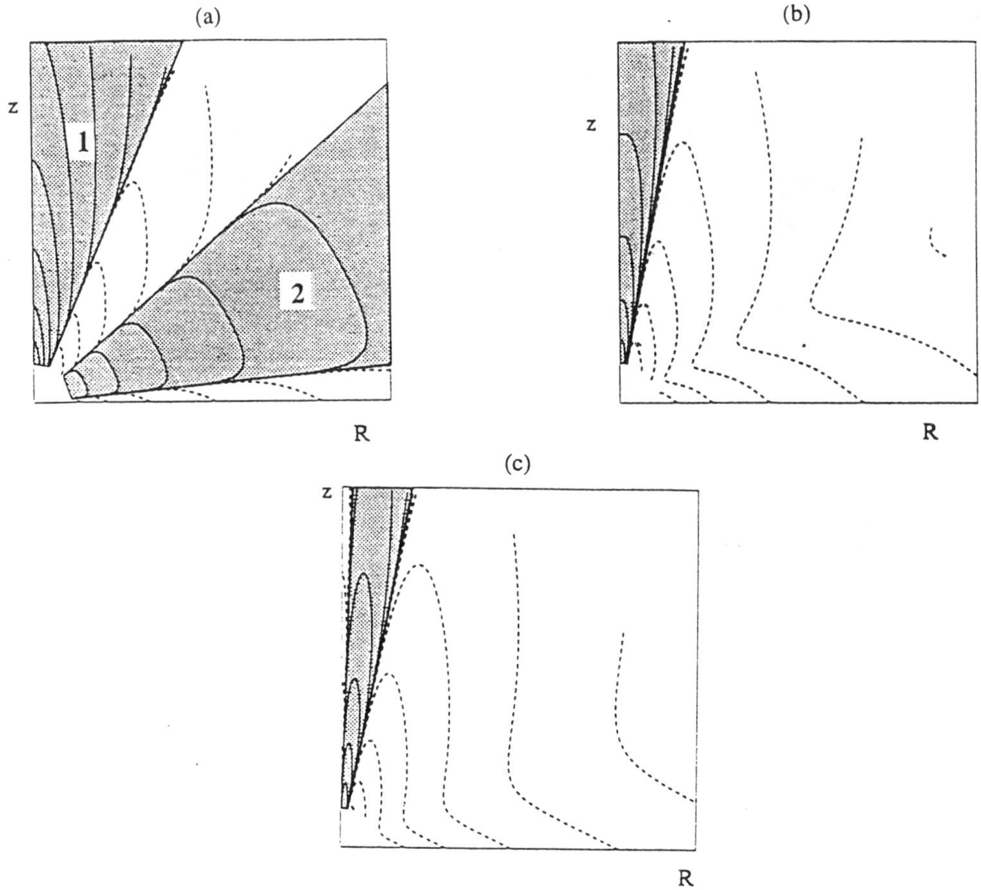

Figure 11. Vortex cores in a swirling jet with conical symmetry, depicted by shading for the (a) $\Delta>0$; (b) $\lambda_2<0$; (c) $Q>0$ vortex core identification schemes.

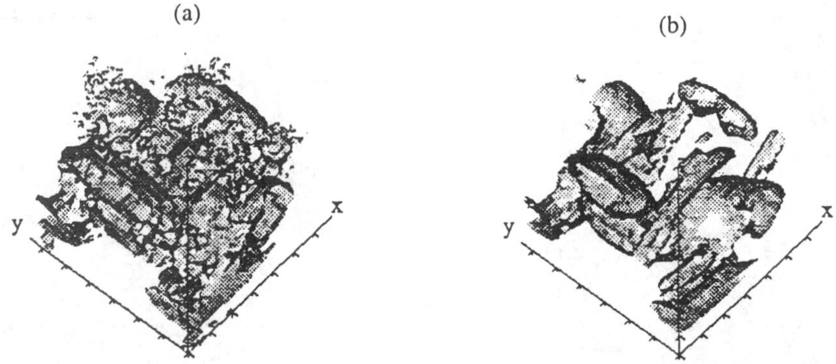

Figure 12. Vortex cores in DNS data of a transitional mixing layer, depicted as isosurfaces defined by the (a) $\Delta>0$; (b) $\lambda_2<0$ vortex identifiers.

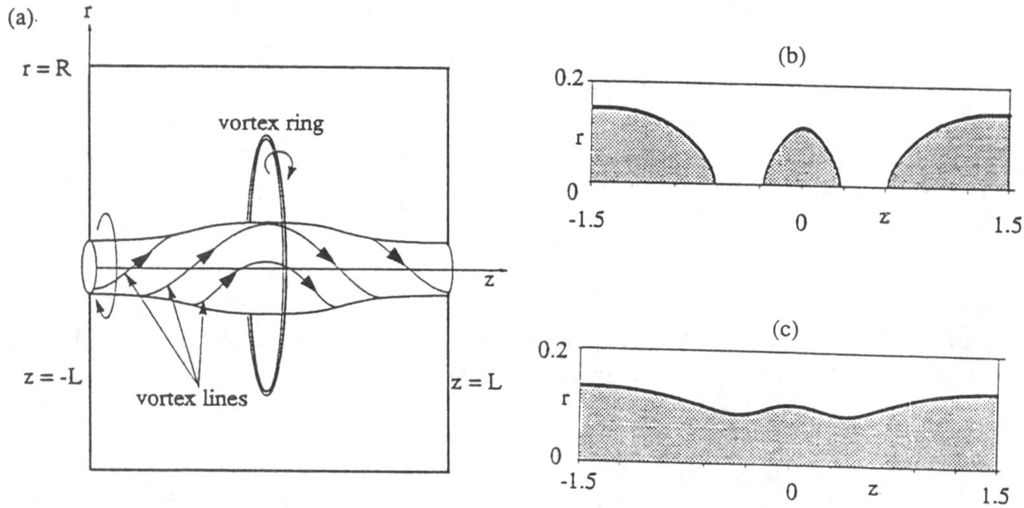

Figure 13. An axial vortex passing through the center of a vortex ring: (a) schematic; shaded vortex cores based on (b) $Q>0$; (c) $\lambda_2<0$ vortex definitions.

4.1 Eduction technique

In near-wall boundary layer turbulence, CS are buried within a layer of strong spanwise vorticity, which is present everywhere near the wall. As explained, CS identification using vorticity magnitude is inappropriate in this case. In this regard, our λ_2-based vortex definition is powerful enough to extract organized events from the background shear characteristic of boundary layers, to permit CS eduction.

For this study, the vorticity-based eduction scheme (for free shear flows) discussed in §2.2 has been modified in two ways: (i) $\lambda_2<0$ is used as a sampling criterion instead of vorticity, and (ii) the entire CS extent is detected using full-field DNS data, in contrast to vortex cross-sections (unavoidable in experimental data with a rake of X-wires). For the purpose of (ii), conditions regarding CS length and maximum permissible deviation from the streamwise direction are imposed, and detected structures are aligned at the center of their streamwise extent. Using cross-correlation between the ensemble averaged and instantaneous events, the latter are then shifted to further enhance the ensemble average.

4.2 Educed CS and conceptual dynamical model

To distinguish between left- and right-handed longitudinal vortices, we educed two sets of quasi-streamwise vortices: those with *positive* ω_x (called SP) and *negative* ω_x (called SN). The λ_2 surfaces for the educed CS, shown in figure 14, indicate that SP and SN are symmetric counterparts, both inclined 9° to the wall and tilted from the streamwise (x) direction on each side by 4° in the top view. The relative spatial orientation of SP and SN, shown schematically in figure 14, can be inferred from contours of ensemble averaged ω_x near the ends of these structures' streamwise extent. This data clearly indicates that these CS have a general tendency to arrange in an array of opposite-signed streamwise vortices which overlap in x. It is important to note that we find no tendency for neighboring CS in the spanwise direction to be opposite-signed, contrary to the previously popular hairpin vortex model, by now well-recognized as a rarity [56]. Although vortex lines can attain hairpin shapes (discussed below), elongated hairpin vortical structures do not occur in general.

From the fundamental spatial orientation of SP and SN in figure 14, we have developed a conceptual model to explain important experimentally observed features, as shown schematically in figure 15 [57]. In particular, the spatial locations of an internal shear layer detected by an asymmetric VITA event, a kinked low-speed streak, and Q2/Q4 events are shown relative to the SP and SN structures. In this scenario, the celebrated 'bursting' process is simply the passage of several SP and SN past a fixed location to create a series of ejection events, consistent with the multiple ejection bursting process reported by Bogard & Tiederman [58].

The tilting of SP and SN in the top view is seen to be a consequence of mutual vortical induction. In figure 15, the mutual induction between SP and SN advects the regions of SN marked A and B in the negative and positive z directions respectively. Thus, the effect of this induced motion is to *precess* the SN structure clockwise with an angular velocity of approximately W/L, where W is the velocity induced by SP at A and L is the length of SP. In a frame moving with the x centerplane of SN, its downstream end moves forward while its upstream end moves backward (because of the background shear), both with a speed U. Therefore, the structure does not precess because $W/U = \tan\theta$ is constant, but is instead stretched along a line inclined at the angle θ with the x-axis. Hence, the observed CS tilting angle results from a balance between mutual induction (in z) and the velocity difference (in x) between the two ends of the structure (due to the boundary layer shear). Although the tilting and inclination angles are small, this three-dimensionality of CS (relative to the mean flow) is responsible for several significant effects, which we discuss briefly below.

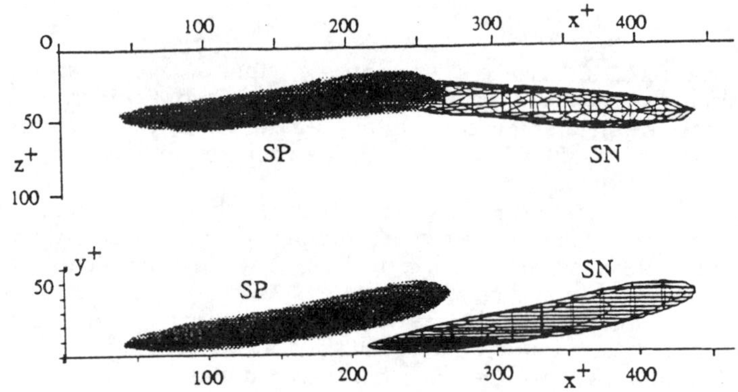

Figure 14. Isosurface plot of educed λ_2 for SP and SN structures, superimposed to show their relative spatial locations: top panel - top view, bottom panel - side view.

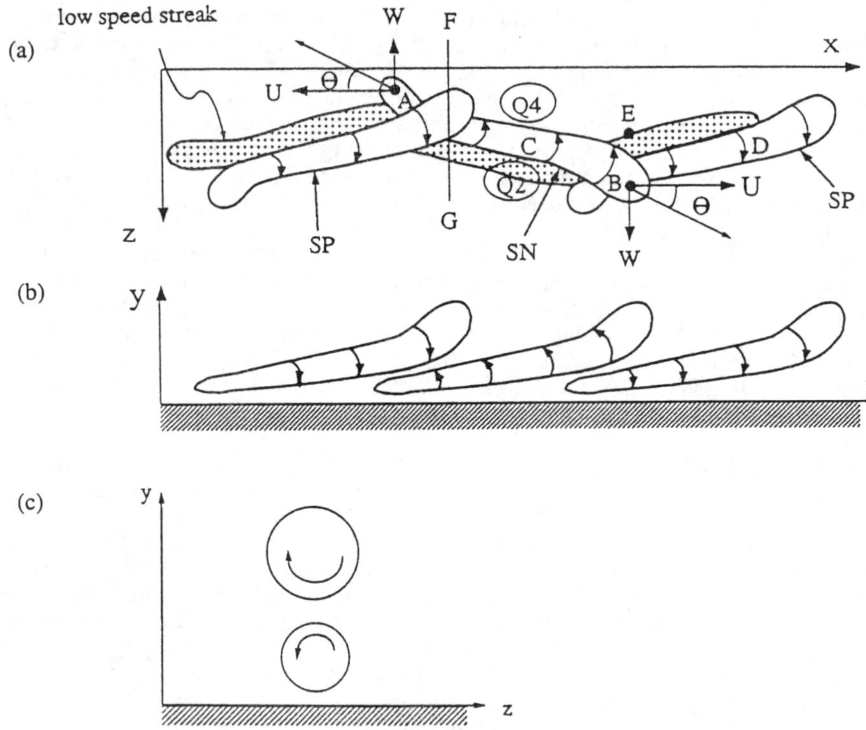

Figure 15. Conceptual model of the spatial relationship between several events corresponding to the near-wall CS: (a) top view; (b) side view; (c) cross-section of SP and SN at F-G in (a).

Near-wall shear layer. Experimentally measured internal shear layers with positive and negative $\partial u/\partial x$ also occur for this SP-SN configuration. We have found that an asymmetric internal shear layer occurs at point E in figure 15, between the structures C and D. This shear layer is created as the structure C induces motion toward the wall upstream of E, while the structure D induces velocity away from the wall downstream of E. The associated u-velocity fluctuations are positive and negative on the upstream and downstream sides of point E respectively, resulting in negative $\partial u/\partial x$ (figure 16a). Positive $\partial u/\partial x$ events can also occur within the structure due to its tilting (figure 16b), since ejection (negative u) appears upstream and sweep (positive u) occurs downstream of the SP; the vice-versa is true for SN. This scenario is quite consistent with experimental observations (*e.g.* see [59]).

Vortex lines around CS. Although SP and SN are not legs of a hairpin vortical structure, hairpin-like vortex line patterns do appear in the educed vorticity field, as shown in figure 17(b). Note the predominance of the spanwise vorticity component everywhere outside the CS core. The inclination angle of the vorticity vector at the structure center is about 17°; this is significantly higher than the 9° structure inclination angle. Therefore, vortex lines beginning at the vortex center immediately deviate from the vortex axis and cross the boundary of the vortex (figure 17a). Thus, vortex line tracing is not generally effective in ascertaining vortical structures within turbulent boundary layers [56].

Effect of phase difference between velocity components. The induced velocities of SP and SN cause a phase difference between ensemble-averaged u and v velocity components in cross-stream (y-z) planes (figure 18a,b), which is crucial to Reynolds stress generation. In particular, this phase difference causes the peak Q2 to be located farther away from the wall than the peak Q4 within a given structure (figure 18c). This is consistent with the observation that Q4 events are more dominant than Q2 events below $y^+ = 12$ [56], provided that the dominant structures are located in the buffer region. The phase difference also creates negative Reynolds stress above and below the CS; on the top, $\langle u \rangle$ is negative (because fluid there has been advected from a lower-momentum region) while $\langle v \rangle$ is negative, thus causing Q3 events. In the same way, Q1 events occur below the structure. Nevertheless, figure 18(c) demonstrates that Q1 and Q3 events are both significantly weaker than Q2 and Q4 events (also observed experimentally).

The origin of the underlying phase difference is easily understood by considering an inviscid streamwise vortex with initial azimuthal velocity profile f(r) embedded in a homogeneous shear flow with $dU/dy = S$. The analytical solution for this class of flows is

$$u(r,\theta,t) = Sr[\sin(f(r)t + \theta) - \sin\theta];\ v(r,\theta,t) = f(r)\cos\theta;\ w(r,\theta,t) = f(r)\sin\theta, \qquad (4\text{-}1)$$

where r is the distance from the center of the vortex and θ is the angle with respect to the x-axis. When t is small, $u \approx Srf(r)t\cos\theta$, so that u and v are in phase. Upon time evolution, the phase difference $\Delta\theta$ between locations of peak u and v is given by $\Delta\theta = -f(r)t/r$, showing linear growth in time for fixed r. Thus, the phase difference in the educed fields is due primarily to cross-stream CS advection within a background shear U(y) (see [60]).

Effects of CS on intercomponent energy transfer. In turbulent shear flows, it is well-known that the time-averaged pressure strain $p\partial u/\partial x$ is negative [61]; for near-wall turbulence, we find that pressure strain events are connected with CS. The tilting of SP and SN in the x-z plane is responsible for producing positive $\langle \partial u/\partial x \rangle$ within the structure. Since $\langle p \rangle$ is negative within the structure, $\langle p \rangle\langle \partial u/\partial x \rangle$ is also negative (figure 19a). Note that if SP were tilted in the opposite direction, $\langle p \rangle\langle \partial u/\partial x \rangle$ would be positive. Negative $\langle p \rangle\langle \partial u/\partial x \rangle$ also occurs in the internal shear layer (point E in figure 15); in this region (denoted A in figure 19b), $\langle \partial u/\partial x \rangle$ is negative and $\langle p \rangle$ is positive (due to flow deceleration), so that $\langle p \rangle\langle \partial u/\partial x \rangle$ is again negative. Thus, redistribution of turbulent kinetic energy due to

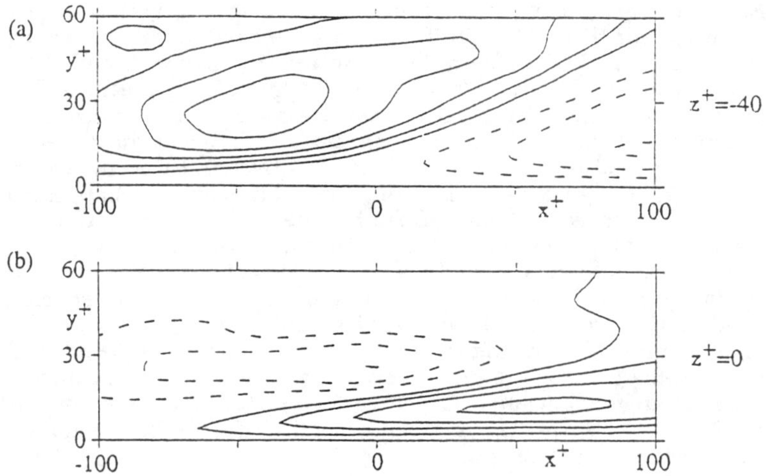

Figure 16. Contours of <u-U> at (a) z^+=-40, across an internal shear layer; (b) z^+=0, cutting through SP.

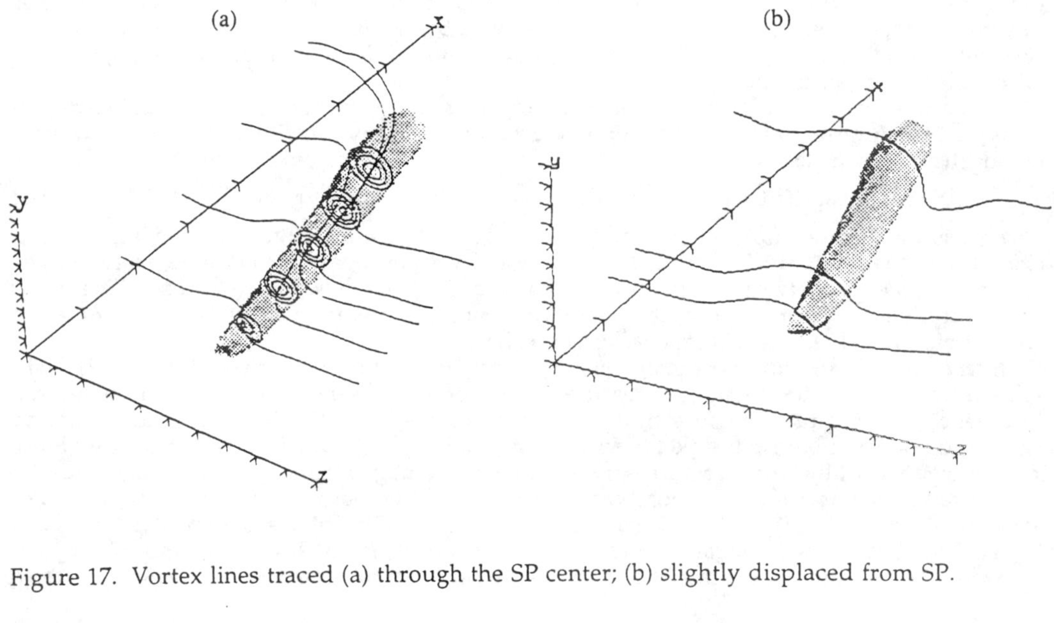

Figure 17. Vortex lines traced (a) through the SP center; (b) slightly displaced from SP.

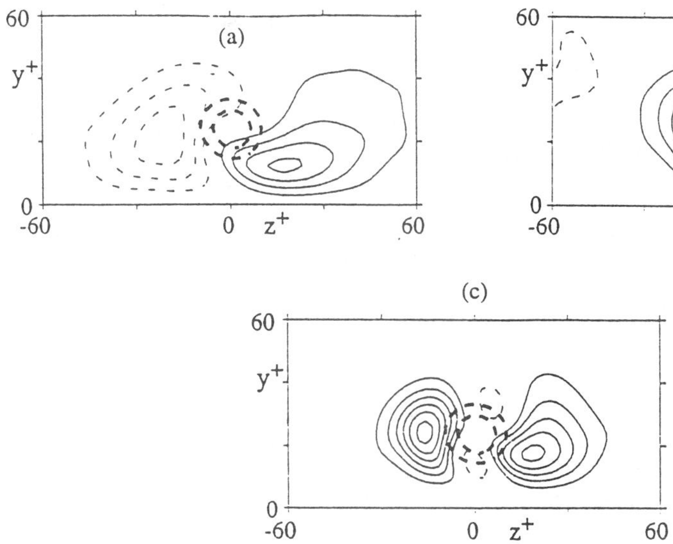

Figure 18. Ensemble-averaged quantities at $x^+=0$ (the middle of SP's x extent). (a) <u-U>; (b) <v>; (c) -<u-U><v>.

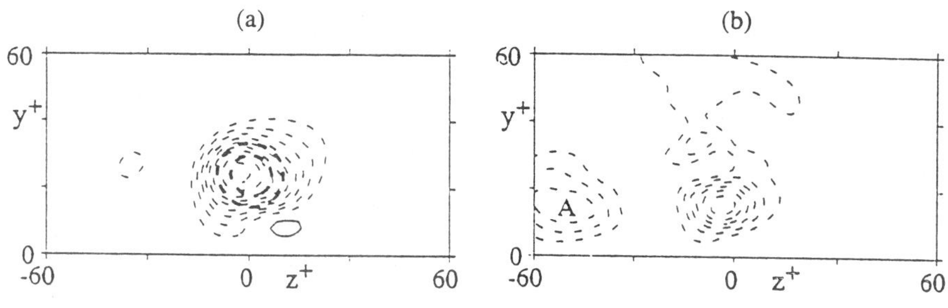

Figure 19. The coherent intercomponent energy transfer term <p><∂u/∂x> at (a) $x^+=0$; (b) $x^+=-150$ (the upstream end of SP).

pressure-strain effects is active not only within the CS core, but also in the internal shear layer.

To summarize, these results serve as clear evidence that fully 3D CS can in fact be accurately extracted from turbulent shear flows using our general-purpose λ_2-based eduction scheme. This robust eduction scheme enables us to obtain an ensemble averaged flow field which not only retains the footprints of instantaneous vortices, but also exhibits their spatial arrangement with respect to well-known dynamical events. Thus, the near-wall turbulent flow is reduced to an ensemble-averaged flow unit, permitting vortex dynamics-based analysis which reveals the fundamental turbulence physics. This illustrates that the combination of a robust eduction scheme with interpretations based on vortex dynamics is indeed a powerful approach to study fully turbulent flows, enabling development of conceptual dynamical models which can, in turn, be utilized in turbulence modeling and control.

II. Evolutionary dynamics of CS

Having discussed in detail effective techniques to extract CS from turbulent flows, the next crucial step toward structure-based turbulence modeling and control is to consider evolutionary CS dynamics. Since CS are widely accepted as vortical entities, classical vortex dynamics theory – centered around inviscid dynamics of slender vortices – emerges as a potentially effective tool in understanding and predicting CS evolution. Nevertheless, in the following, we uncover some fundamental dynamics which operate only for vortices with *finite core size* in a *viscous* flow, distinct from and unaccounted for by classical vortex dynamics.

Since our focus is on 3D vorticity field evolutions, which currently cannot be obtained experimentally, we must rely exclusively on DNS pending further development of modern velocimetry techniques.

5. Viscous vortex reconnection

As will be described in detail in this section, vortex reconnection is responsible for changes in vortex line topology, which occur during evolution of viscous, vortical flows. Note that even though such topology changes can occur only in the presence of nonzero viscosity, vortex reconnection occurs even at high, albeit finite, Re. Additionally, the results below suggest that vortex reconnection occurs continuously at all scales in turbulent (viscous) flows and serves as a fundamental, physical-space cascade mechanism.

Since we have extensively studied reconnection dynamics in incompressible flows, this nontrivial 3D interaction also serves as a useful ground to extend the rather limited current understanding of compressible vortex dynamics.

5.1 Reconnection phenomena

In general, viscous vortex interactions can be classified into two categories: augmentation and cross-diffusion of circulation. Typically in 2D, axisymmetric, or otherwise locally constrained flows, merger of like-signed vortices causes circulation accumulation, in which viscosity acts as a smoothing agent. A special case of cross-diffusion – circulation annihilation – occurs when perfectly aligned antiparallel vorticity comes into contact; head-on collision of vortex rings [62,63] is one example. In general 3D vortex geometries, viscous cross-diffusion of circulation is accompanied by reconnection, as is clearly demonstrated by: aircraft trailing vortices (figure 20d; [64]), pinched-off of hairpin tips in boundary layers [65], colliding vortex rings at arbitrary orientations [23,66], elliptic jet bifurcation into two vortex rings (figures 20a-c; [67]), and collision of two antiparallel vortex tubes (discussed below).

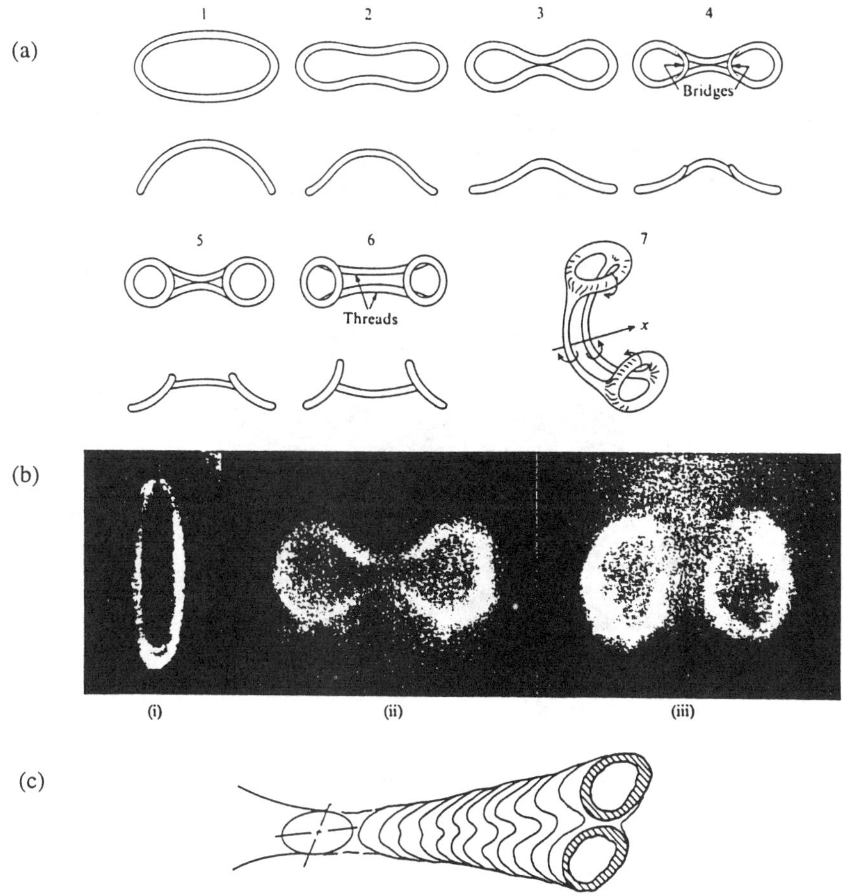

Figure 20. For caption see following page.

(d)

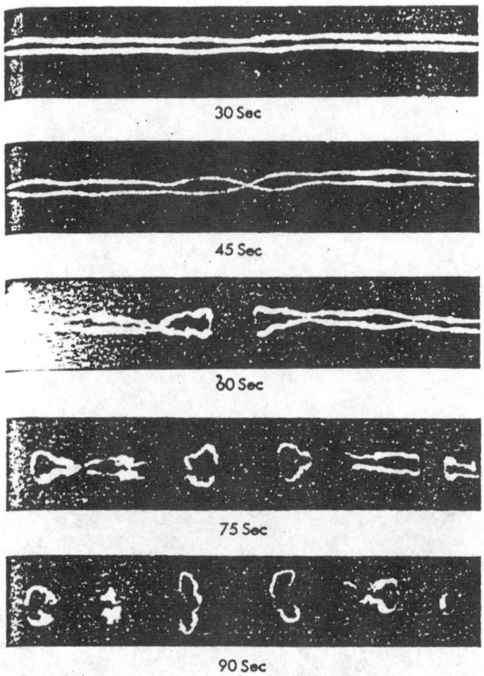

Figure 20. Examples of vortex reconnection, during evolutions of (a) elliptic vortex ring; (b,c) elliptic jet; (d) Crow instability of aircraft contrails.

Vortex reconnection, a mechanism of topology change in viscous vortical flows, is of fundamental interest in many turbulence phenomena. It has also been proposed as an agent for aerodynamic noise generation [33], mixing, turbulence production, and helicity generation [32]. For some time, the role of viscosity in reconnection remained controversial, as also did the fundamental reconnection mechanism. Previously studied initial vortex configurations such as two orthogonal vortex tubes [68] or the so-called trefoil vortex [69] exhibit a very complex evolution, for which the fundamental physics remained largely hidden. In this paper, we focus on the reconnection of two sinusoidally perturbed antiparallel vortex tubes, enabling us to reveal the essential physics. The sinusoidal perturbation is merely to assure collision of the two tubes. Since two vortex filaments in close proximity tend to become antiparallel, this study is of generic importance.

5.2 Incompressible reconnection of antiparallel vortex tubes

Through careful observation of the flows in figure 20, one can see that in each case, the local CS geometry immediately near the reconnection zone is that of two antiparallel vortices with locally curved axes. Using the carefully designed DNS initial conditions shown in figure 21, we are able to mimic this essential precursor of reconnection in a simplified way, thus permitting in-depth interpretation. Two symmetry planes are evident for this configuration: a symmetric plane XY (π_s)and a dividing plane YZ (π_d). In both planes, the normal velocity and the tangential vorticity vanish at all times. Consequently, it can be shown that the circulations Γ_s and Γ_d in the half planes π_s^+ (x>0) and π_d^- (z<0) can change only by viscous effects, *i.e.* reconnection. Also, $d\Gamma_d/dt = -d\Gamma_s/dt$, since an arbitrary vortex line must intersect either π_s or π_d. The initial sinusoidal axis perturbation models the vortex axis curvature observed experimentally and causes self-induced collision of the vortex tubes to initiate reconnection. The symmetries in the initial conditions are very useful in simplifying the analysis, and it was ensured that the resulting reconnection is not noticeably altered by introducing small asymmetries [70].

Three phases of reconnection. The evolution of the vorticity norm $|\omega|$ during vortex reconnection is shown in figure 22 for $Re=\Gamma/\nu=1000$, where Γ is the initial circulation of each vortex tube. A comparison of the frames of figure 22 with the evolution of circulation Γ_s (figure 23) reveals three distinct phases: (i) inviscid advection, (ii) 'bridge' formation, and (iii) 'thread' formation.

The *first* phase is governed essentially by inviscid vortex advection, as nearly no circulation is transferred from π_s to π_d. During this phase, while the two vortices move upward by mutual induction, self-induction presses them together, thus flattening their cross-section and forming a contact zone with steep vorticity gradients. In the *second* phase, a dramatic change in topology takes place as a rapid transfer of circulation from π_s to π_d (figure 23) occurs due to viscous vorticity cross-diffusion in the contact zone. This phase is marked by 'bridging', *i.e.* the linking of originally distinct vortices (figures 22c,d), which is the heart of the reconnection process. The growing bridges acquire a curved horseshoe shape, causing them to pull apart by self-induction. Simultaneously, the flow induced by the bridges stretches the dipole in π_s, thus sustaining annihilation. The *third* phase is the evolution of the remnants of the original dipole after the initial rapid circulation transfer. These slender vortices, called 'threads' in the following, are well defined and intense due to their stretching by the bridges. These threads are the consequence of incomplete circulation annihilation and are an inherent feature of reconnection.

The bridging mechanism. The rapid circulation transfer from π_s to π_d begins when the vortices come into contact near $\pi_s \cap \pi_d$ (figures 22b, 24b) to form a contact zone C. Since the flow is highly viscous around C and nearly inviscid elsewhere, we may consider the vortex

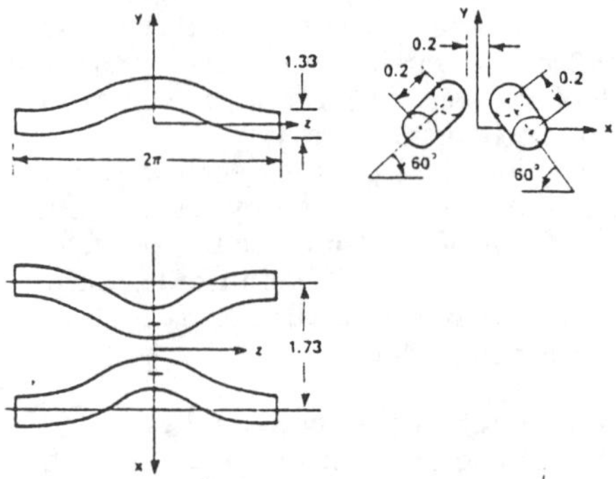

Figure 21. Initial antiparallel vortex configuration for reconnection DNS.

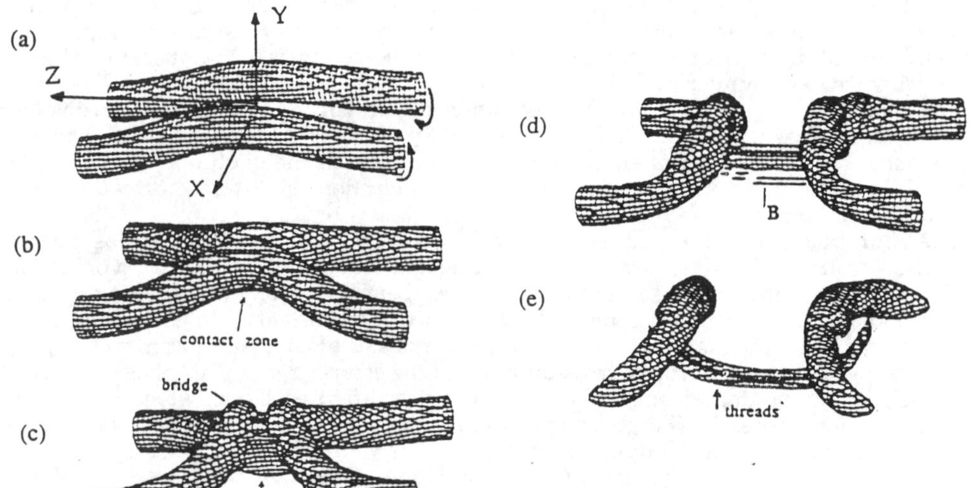

Figure 22. Evolution of 30% $|\omega|$ isosurface during reconnection. (a) t=0; (b) t=2.25; (c) t=3.5; (d) t=4.75; (e) t=6.0.

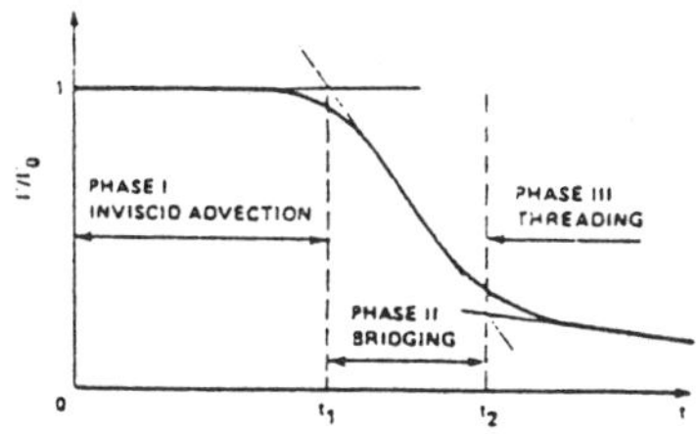

Figure 23. Circulation in $\pi_s{}^+$ (x>0), defining phases of reconnection.

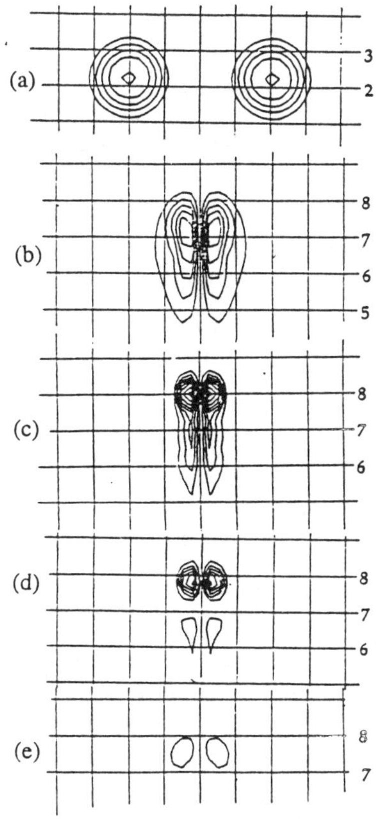

Figure 24. Evolution of $|\omega|$ in π_s at (a) t=0; (b)t=3.0; (c) t=3.75; (d) t=4.5; (e) t=6.0.

lines far away from C as material lines for the sake of discussion. We use this idea schematically in figures 25(a-c), where each vortex is represented by three vortex lines; one line from each intercepts C (figure 25a). Shortly thereafter, these vortex lines annihilate each other at x=z=0 by cross diffusion, producing two reconnected lines intercepting π_d away from π_s (figure 25b) at a location denoted z_o in the following.

The cross-section AB (figure 25c) at the edge of the contact zone C reveals the two-dimensional dipole structure of the original vortices. In a co-translating frame, the streamline pattern is the well-known recirculation bubble of a vortex dipole (figure 25d). Although this frame is not precisely defined for an unsteady dipole, the existence of two stagnation points S_F and S_B — at the front and back of the dipole — can be inferred. The large velocity between S_F and S_B advects z_o towards S_F, ahead of the dipole (figure 25d). The relative upward advection of z_o stops at S_F, and a reconnected vortex line unfolds along the diverging separatrix a-a (figure 25d). This mechanism is illustrated schematically in a perspective view of one half of the vortex pair in figure 25(e), which shows a vortex line at five different instants (1-5). Far away from C, a vortex line is simply swirled around the vortex tube. However, at C, vortex lines are reconnected as soon as they come into contact between instants 1 and 2 and then swept upward by the pumping motion of the dipole. At instant 3, the upward motion of the reconnected vortex line stops at S_F, where it is stretched and begins to wrap around the dipolar vortex pair (instants 4 and 5). Meanwhile, the circulatory motion in the dipole continuously pumps fresh non-reconnected lines into C near S_B (figure 25d), which in turn follow the same sequence of reconnection. Due to this bridging mechanism, reconnected vortex lines accumulate above the antiparallel vortices and align orthogonal to them.

Dynamics of bridges and threads. Although helicity density (h) is a frame-dependent quantity and hence not a suitable measure of topology (unlike volume-integrated helicity), in the present idealized configuration it does reveal some important features. Helicity is very small initially and during the first phase of reconnection, since vortex lines inside the vortex tubes are nearly parallel. During reconnection, significant helicity develops in a particularly interesting pattern, consisting of intertwined positive and negative regions (figure 26a,b). Helicity is most prominent in the threads (figure 26b), where the strong swirling velocity of the bridges is nearly aligned with the thread vorticity. The alternating-sign helicity distribution surrounding the bridges is enigmatic until the vortex line geometry is examined. Figure 26(c) shows that vortex lines emerging from the threads do not wrap all the way around the bridges as one would expect from the $|\omega|$ surfaces. Instead, they attain asymmetric hairpin-like shapes (see A and B, figure 26c), where the vorticity is larger in one hairpin leg than in the other. Unfortunately, this aspect is not faithfully reflected in surface plots of $|\omega|$. The intense leg (A) is the extension of the thread (C), and its high vorticity causes a large value of helicity. The diffuse leg is difficult to identify in the helicity distribution since $|\omega|$ there is small; also, the angle between the vorticity and velocity vectors is larger here. The intense leg induces a flow away from π_d outside the bridges, thus causing stretching and vorticity intensification of this portion of the bridges (point A in figure 27c).

Inside the bridge, the vortex lines have a slow helical twist near π_d (barely discernible in figure 26c). Contrary to the intense thread hairpin leg, twisting within the bridge produces an axial flow toward π_d at the center of the bridge (figure 26d). The direction of the axial flow is determined by the orientation of the twist, which in turn is determined by the location of fastest swirl (around the vortex). Since in this case, the axial flow is directed toward π_d from both sides, it causes negative enstrophy production due to vortex compression, as seen in a cross-section of bridges parallel to π_d (dotted lines in figure 27b). These twisted vortex lines result from the bridges' skewed (nonconcentric) vorticity distribution (figure 26c). Specifically, the peak vorticity in the bridge is not at the location of the geometric centroid of low level

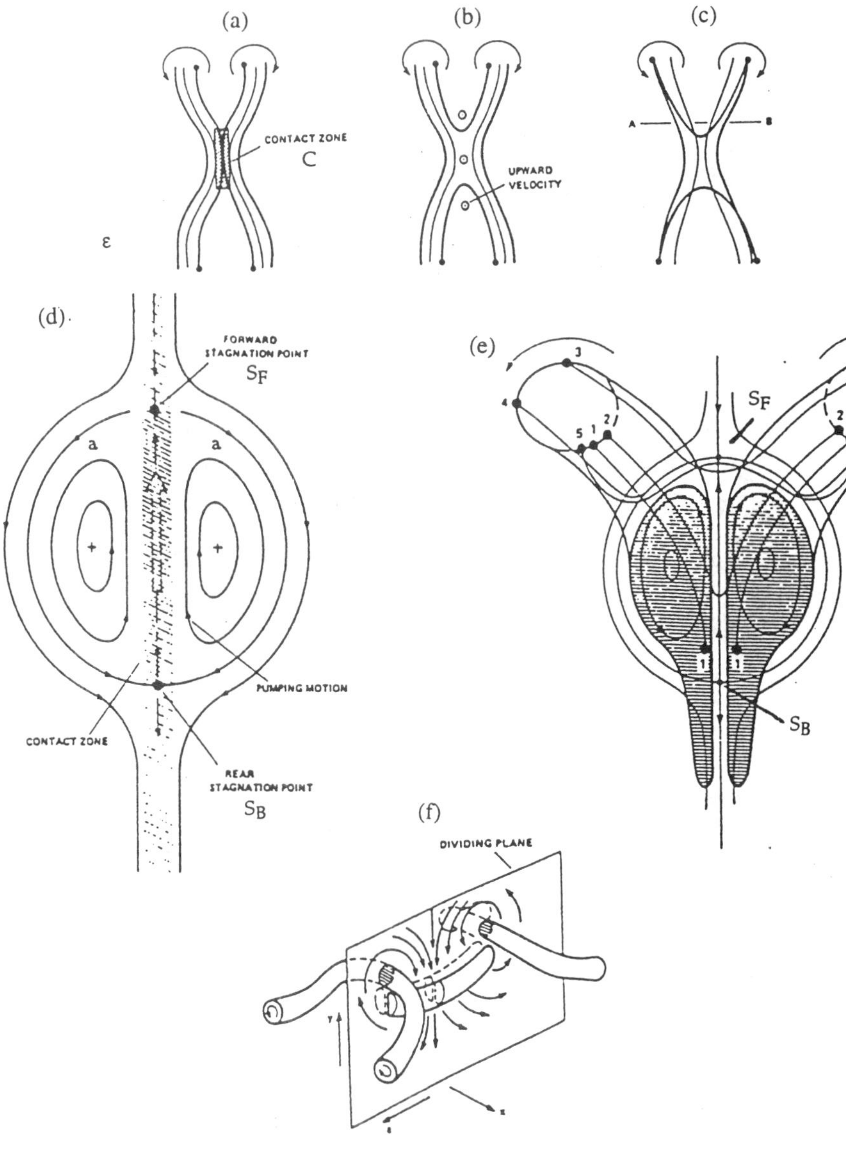

Figure 25. (a-c) Schematic of annihilation and reconnection; (d) streamline pattern in section A-B of (c) in a co-translating frame; (e) advection and accumulation of reconnected vortex lines forming bridges, with successive positions of a vortex line marked 1-5; (f) schematic of flow induced by bridges.

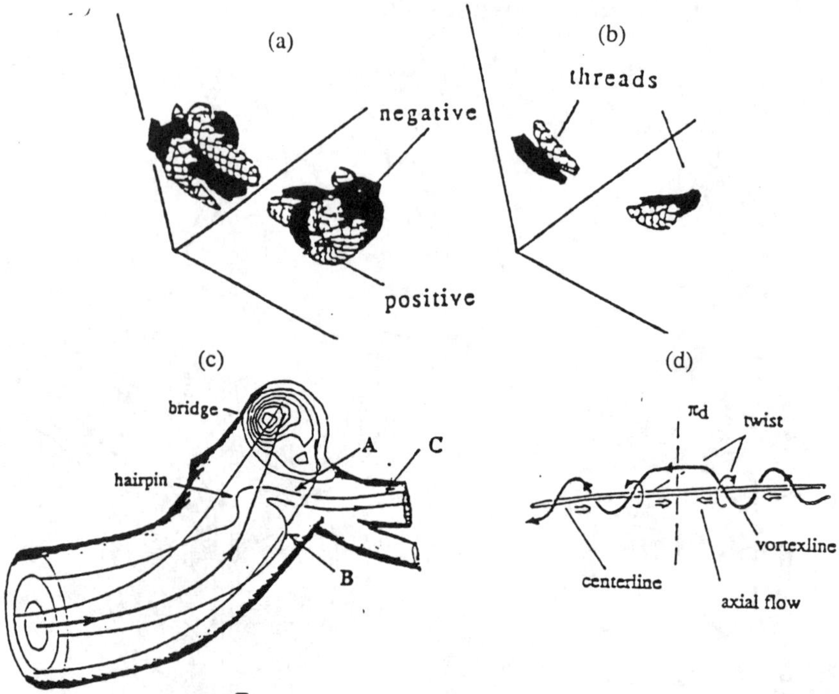

Figure 26. (a,b) Surfaces of helicity density after reconnection (t=4.5) at (a) 20% level; (b) 50% level; (c) vortex lines through the bridge and thread; (d) schematic of axial flow induced by twisted bridge vortex lines. In (c), plane cross-sections of $|\omega|$ in π_s, π_d, and the domain boundary have been overlaid with a sketch of the vortex to orient the reader.

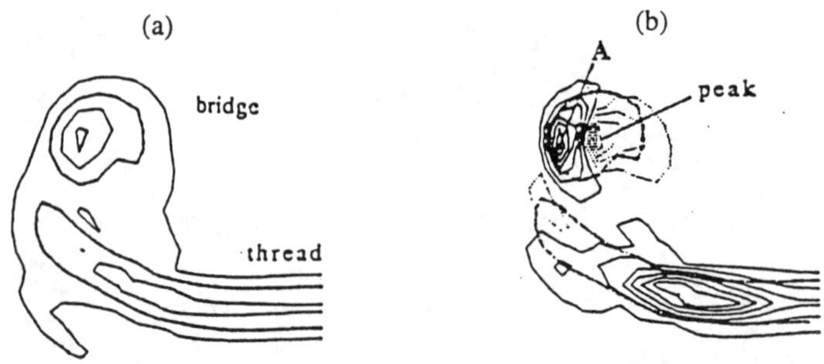

Figure 27. Plane cross-section parallel to, but slightly displaced from π_d at t=4.75, showing (a) $|\omega|$; (b) enstrophy production P_ω, where the dotted lines represent negative values and chaindotted lines are $|\omega|$ contours.

vorticity contours. Furthermore, the peak occurs away from the contact zone as a consequence of earlier stretching by the dipole during bridging. This local vorticity reduction due to vortex line twisting constitutes an interesting inviscid mechanism, clearly distinct from diffusion, which smooths the vorticity distribution of the reconnected vortices. As will be discussed in §6-7, such behavior of vortices with core nonuniformity, which we term *core dynamics*, constitutes a significant, virtually unexplored aspect of vortex dynamics.

Curvature reversal and reconnection cascade. Curvature reversal of the original vortex tubes, responsible for incomplete reconnection, is a characteristic and crucial feature. In essence, it occurs because the upward motion of the dipole in π_s slows relative to that in ε, outside the contact region (figure 24; note that the y-position is denoted in each panel); the question remains, why does the dipole slow down near π_s?

This slowing is caused by a downward velocity field in C — downwash — which opposes the dipole's own upward motion in π_s (figure 25f) and is induced by the growing bridges as well as by non-reconnected vortex lines in ε. Furthermore, the dipole loses circulation through vorticity annihilation, and its upward motion weakens as it relaxes into a head-tail shape. This relaxation into a head-tail shape, an inviscid process, decreases the dipole propagation speed, which is determined almost exclusively by the head. Although this inviscid effect is weak at Re=1000, it may in fact be the primary mechanism for curvature reversal at higher Re. An indication of this would be that the area A of the nearly self-similar head decreases with increasing Re, such that $\omega_m A^{1/2}$ (propagation speed of the head to leading order) is bounded as $Re \rightarrow \infty$. This indicates the possibility that the amount of circulation transferred from π_s to π_d tends to zero as Re tends to infinity, with a convective (as opposed to viscous) transfer rate.

These results have some interesting new implications with regard to energy cascade. Foremost, the vortex reconnection evidently cannot be completed in a single step. As the bridges move apart by self-induction, the thread-dipole advects forward in the presence of the weakening 'downwash' caused by the bridge-dipole (orthogonal to the original vortex dipole). In this way, the threads reverse their curvature again and then collide at the midpoint by self-induction, thus starting the next burst of reconnection, which occurs at a smaller scale - a mechanism of *reconnection cascade*. In this physical-space cascade scenario, the frequency of reconnection bursts will increase and circulation transfer will decrease as the Reynolds number increases.

5.3 Compressible vortex reconnection

Having obtained a fairly detailed understanding of incompressible reconnection, we now study the evolutionary changes brought about by compressibility effects. This is accomplished using the same antiparallel vortex configuration as above, with perfect gas thermodynamics. Interestingly, we have found that the selection of initial pressure (P) and density (ρ) fields, whose relation is nonunique in compressible flows, can drastically influence the subsequent evolution. Since this has broad implications for all compressible DNS, we carefully demonstrate that polytropic initial conditions correctly capture compressibility physics and are most appropriate.

Possible initial conditions. For compressible simulations, the initial P and ρ fields must be specified, in addition to vorticity. Ideally, such simulations should be initialized with an analytical solution of a compressible vortex. Since such solutions are not currently available, we will compare the initial conditions used to date for compressible turbulence simulations. The fact that flow evolutions typically depend strongly on the initial P and ρ distributions is clear from numerical simulations [71-74]. Using vortex reconnection as a test case, we will compare the following three P and ρ initial conditions:

(i) *Constant density initial conditions (CDIC):* This initialization, first used by Feiereisen *et al.* [71], assumes $V=1/\rho$=constant, with P obtained by imposing
$\partial(\nabla.\mathbf{u})/\partial t = 0$ from

$$\nabla^2 P = \frac{-\gamma M_o^2}{V}\ \nabla\cdot(\mathbf{u}\cdot\nabla\mathbf{u}), \tag{5-1}$$

where γ is the ratio of specific heats at constant pressure and constant volume, and M_o is the reference Mach number represented by the ratio of velocity scale and sound speed at infinity. This is equivalent to specifying the pressure gradient required to balance the centrifugal force in a 2D axisymmetric flow.
(ii) *Constant initial conditions (CIC):* Used by Lee *et al.* [74], this initialization specifies $V=1/\rho$=constant and P=constant.
(iii) *Polytropic initial conditions (PIC):* We define a new polytropic initial condition, found to be the most suitable among those considered here. We simultaneously impose

$$P = aV^{-b} \text{ and } \nabla\cdot(V\cdot\nabla P) = -\gamma M_o^2\ \nabla\cdot(\mathbf{u}\cdot\nabla\mathbf{u}). \tag{5-2}$$

The proportionality constant a was chosen to obtain the desired pointwise maximum Mach number M, and b was chosen to be γ (homentropic initial conditions) for all simulations reported here. It should be noted that the above polytropic relation is used only for initialization; the evolution for $t>0$ is governed by the Navier-Stokes equations for a perfect gas.

Initial condition influence on reconnection. In the following, results are reported for Re=1000 (as for the incompressible case above) and two M: subsonic with M=0.5 (denoted M_1), and supersonic with M=1.45 (denoted M_2). The evolution of reconnection for M_1 was found to be essentially the same as that of the incompressible case discussed above for all initial conditions. At M_2, substantial differences among the three initial conditions are observed, shown in figure 28 by iso-vorticity surfaces at a late stage of reconnection. With CDIC, the compressible flow evolves very similarly to the incompressible case, in terms of bridge and thread enstrophy magnitude. In contrast, the bridge vorticity for CIC at M_2 is absent at this surface level as for PIC, which also produces very little bridging. A consequence of the strong bridges for CDIC and incompressible reconnection is curvature reversal and hence self-induced separation of the threads at later times. This mechanism is clearly suppressed in the simulations with PIC and CIC, where the net compressibility effects are obviously significant.

The rate of reconnection for each initial condition at M_2 is represented in figure 29 by the circulation evolution in π_s. With CDIC, the evolution is similar to that at M_1 except for a slight increase followed by a drop in circulation at early times. We have found that this initial transient for CDIC is due to baroclinic vorticity generation. With PIC and CIC, the circulation quickly decreases (at $t\sim0.5$) due to the appearance of a shocklet between the initial dipolar vortices. The PIC case shows a slower rate of circulation decay until t=3.0, after which the decay rate becomes faster than CIC, but still slower than CDIC. With CIC, we see an oscillatory decay of circulation; plots of P and ρ reveal that the rapid decreases correspond to periods of interaction of acoustic waves (from adjacent boxes) with the dipole vortices.

The enhanced circulation transfer for M_2 at early times is due to shocklet formation, which causes an increase in Γ_d and a decrease in Γ_s via both viscous cross-diffusion and baroclinic vorticity production. In figure 30, we show the relative location of this shocklet for the three initial conditions as an overlay plot of iso-vorticity and iso-dilatation surfaces at t=0.5. The shocklet is characterized by a thin region of large negative velocity divergence (implying compression of fluid), across which P, ρ, $\mathbf{u}$ and M change sharply. We have verified that the jumps in these quantities are consistent with the Rankine-Hugoniot relations across a shock.

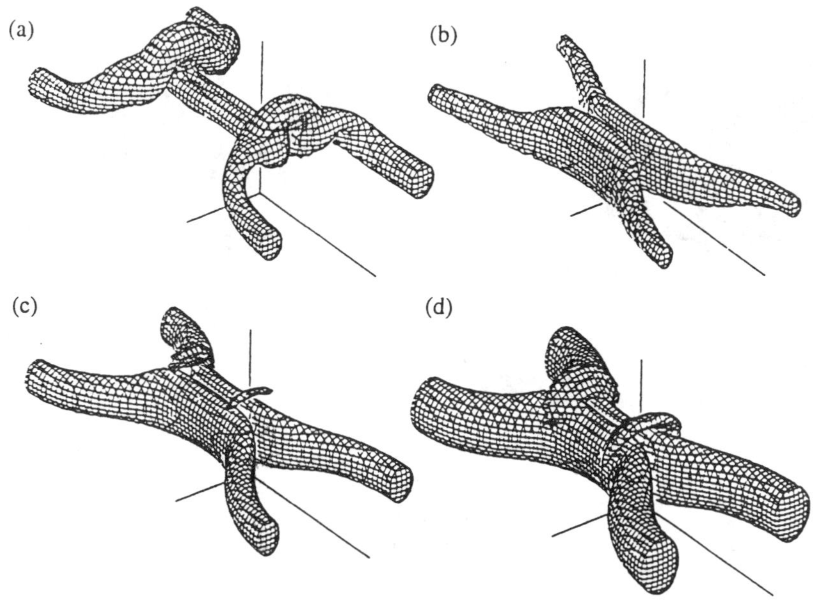

Figure 28. (a-c) 30% $|\omega|$ surfaces at t=6.0 for compressible reconnection at M_2 with various initial conditions: (a) CDIC; (b) CIC; (c) PIC. The 25% surface for PIC is shown in (d).

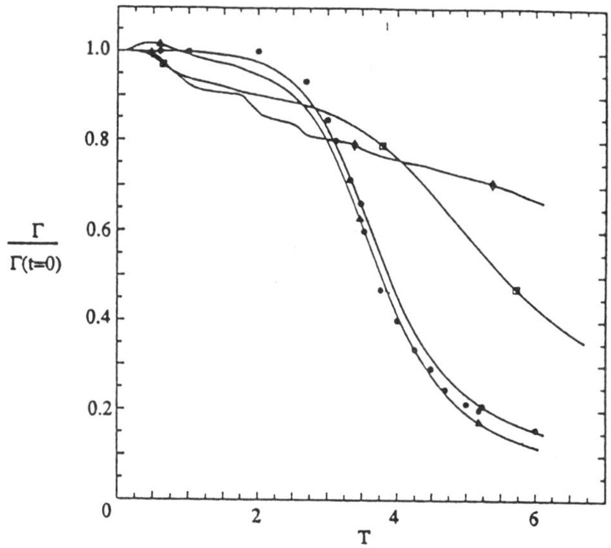

Figure 29. Evolution of circulation in π_S for PIC at M_1 (O), and M_2 cases with PIC ([]), CDIC (Δ) and CIC (◊). Solid circles denote data for incompressible reconnection.

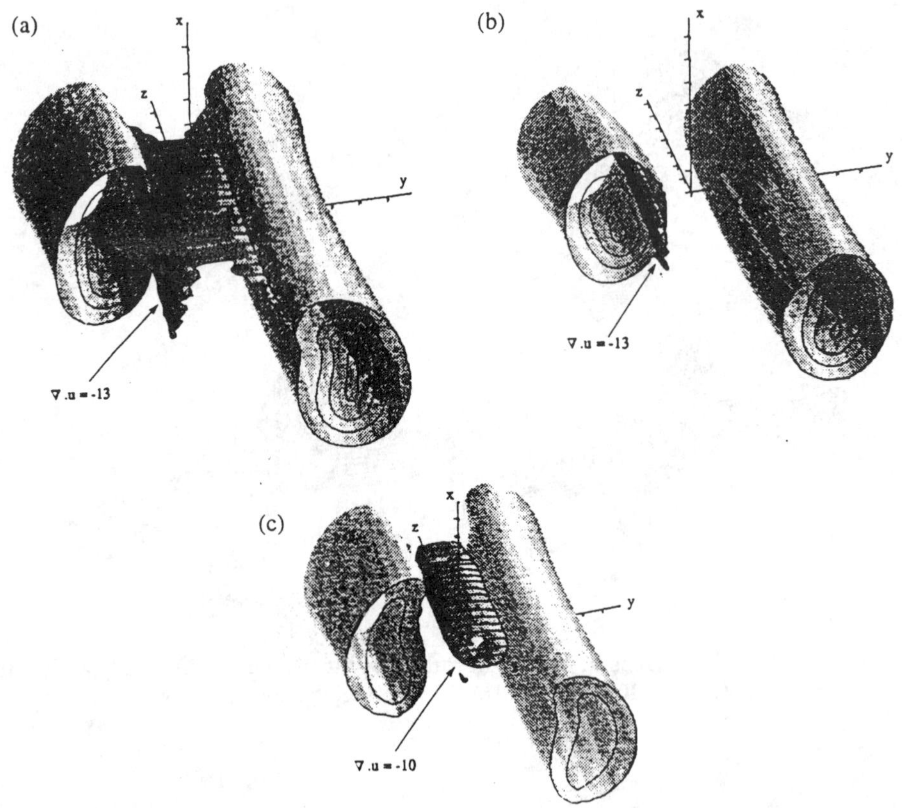

Figure 30. Overlay of isovorticity and iso-dilatation (hatched) surfaces at t=0.5 for M_2 cases with (a) PIC; (b) CDIC; (c) CIC.

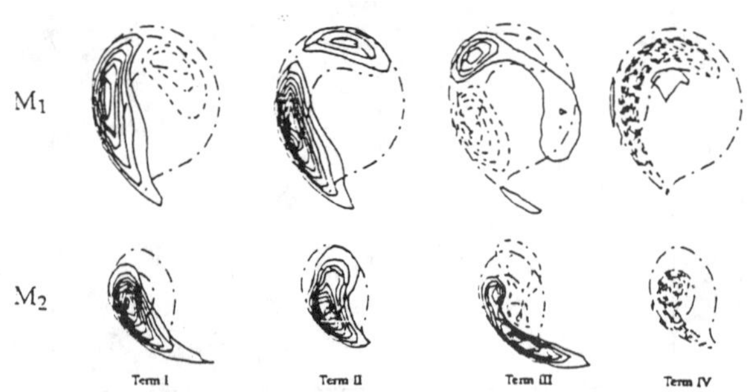

Figure 31. Enstrophy production terms in equation (5-4) in π_d for M_1 and M_2 with PIC at t=6. The chaindotted lines are vorticity contours.

Note that π_s is a material plane since the fluid velocity normal to it is zero (due to periodic boundary conditions). For this plane, the transfer of circulation can be expressed simply as

$$\frac{d\Gamma}{dt} = \int \frac{-1}{\gamma M_o^2} V\nabla P \cdot dc + \int \mu \nabla^2 \mathbf{u} \cdot dc + \int \frac{\mu V}{3} \nabla(\nabla \cdot \mathbf{u}) \cdot dc \tag{5-3}$$

where dc is a line element along the curve demarcating the boundary of the material plane. On the RHS of (5-3), the first term is due to baroclinic production, the third term represents compressible viscous effects, and the second term is the viscous cross-diffusion term (the only term present in the incompressible case). In the following, we denote the first term as R_B and the sum of the second and third terms as R_v. With PIC at M_2, both R_B and R_v are negligible at early times (t<0.3), and after shocklet formation (t>0.4), they are both negative and two orders of magnitude larger than at M_1, causing a rapid decrease of Γ_s.

In contrast, for CDIC at very early times (t<0.3), R_B is positive (as gradients of ρ caused by dipole motion are inclined to the initial P gradients) and an order of magnitude larger than negative R_v, causing Γ_s to increase. After shocklet formation (t~0.4), both R_B and R_v become negative, and Γ_s is reduced to nearly that at M_1. We find that after this time, R_v for CDIC and PIC are comparable, but R_B with CDIC is much smaller, perhaps due to baroclinic generation away from the shocklet region arising from misaligned P and ρ. For all M_2 cases, the early appearance of a shocklet leads to faster transfer of Γ from π_s; for CDIC, this effect is offset by baroclinic generation, which causes an early increase in Γ_s.

After shocklet dissipation, the larger Γ_s of CDIC leads to faster growth of the initial vortex perturbation (by mutual induction of the dipole) compared to PIC. A larger perturbation corresponds to higher curvature in the interaction region and, consequently, faster self-induced motion of the interacting vortices toward π_d; this is why reconnection is faster for CDIC. At late times (t>5), the P and ρ distributions inside the core in π_s are similar for both CDIC and PIC, as are the compressible enstrophy production terms' distributions. In both cases, the late-time Γ transfer is predominantly due to viscous cross-diffusion. However, the vorticity gradients in the interaction region are more diffuse with PIC due to slower self-induced motion of the vortices towards π_d, resulting in slower circulation transfer.

To summarize the initial condition dependence, we find that an initial transient occurs with CDIC, in which the baroclinic production is opposite to that observed at late times (or with PIC). As a result, additional Γ_s is generated, which critically affects the subsequent evolution of vorticity. Nevertheless, the evolutionary dynamics for CDIC eventually tend toward those exhibited throughout by PIC. Our observation of a negligible net effect of compressibility on the rate or extent of reconnection with CDIC is consistent with the claims of Erlebacher *et al.* [73] and Lee *et al.* [74] that CDIC suppresses compressibility effects in fully developed turbulence. In contrast, PIC and CIC show a much lesser extent of vortex reconnection (at M_2), thus clearly indicating that compressibility effects are in fact important. Since CIC leads to strong wave-vortex interactions, an artifact of adjacent periodic domains, it is inappropriate for spectral simulations. From a physical standpoint, PIC is also preferable because experimental studies of compressible vortices [75] also suggest a polytropic relation between P and ρ in the vortex core. Consequently, we limit our discussion of reconnection dynamics in the following subsection to the PIC case. For further details regarding initial condition selection, see [76].

Compressibility effects in bridges. For PIC at M_2, the above-mentioned shocklet dissipates around t=1.5, and the rate of circulation transfer to π_d is subsequently reduced (see figure 29). Subsequently, compressibility effects are no longer significant in π_s, and the circulation transfer is predominantly due to viscous effects. Since the threads are much stronger compared to the incompressible case, we expected their influence in causing a nonconcentric bridge vorticity distribution to be more significant. However, we find that the vortex line twisting is clearly reduced, contrary to expectation, and the vorticity distribution in the bridges is more concentric than in the incompressible case.

The cause of a more concentric vorticity distribution in the bridges, despite stronger threads at M_2, can be understood in terms of physical effects represented by terms of the compressible enstrophy equation:

$$\frac{D(\omega\cdot\omega/2)}{Dt} = \underbrace{\omega\cdot(\omega\cdot\nabla)\mathbf{u}}_{\text{Term I}} - \underbrace{\omega\cdot\omega(\nabla\cdot\mathbf{u})}_{\text{Term II}} + \underbrace{\frac{1}{\gamma M_o^2}\omega\cdot(\nabla P\times\nabla V)}_{\text{Term III}} + \underbrace{\mu V\omega\cdot\nabla^2\omega}_{\text{Term IV}}$$

$$- \underbrace{\mu\omega\cdot(\nabla V\times(\nabla\times\omega))}_{\text{Term V}} + \underbrace{\frac{4}{3}\mu\omega\cdot(\nabla V\times\nabla(\nabla\cdot\mathbf{u}))}_{\text{Term VI}}. \tag{5-4}$$

Note that compressibility effects are represented by four new terms in this equation: dilatation production (II), baroclinic production (III), and compressible viscous effects (V,VI). At late times, only the first four terms in equation (5-4) are significant in the bridges, each of which is shown in figure 31 for both M_1 and M_2. At M_1, the vortex stretching term has a distribution similar to the incompressible case (shown in figure 27); it is negative over a portion of the cross section and has a large positive peak where the threads are closest to the bridges. Compared to vortex stretching, the dilatation and baroclinic production terms are negligible at M_1. At M_2, the vortex stretching is higher than at M_1, as expected from the stronger threads. In addition, the dilatation production is significant with a similar distribution, but with opposite sign, thereby causing a more concentric bridge cross-section.

As a consequence of these compressibility effects, the twisting of bridge vortex lines is noticeably reduced, as illustrated in figure 32. In each panel, portions of isovorticity surfaces are overlaid to emphasize their distinction from vortex surfaces, on which vortex lines lie; the vortex lines are started on two rakes along a $|\omega|$ contour in π_s and π_d. Although isovorticity surfaces mimic the vortex surface topology, they fail to reveal important details, such as formation of hairpin-like structures (region C in figure 32c) and the fact that threads do not connect to the vortex center away from the reconnection zone. The hairpin shape of vortex lines in region C of the M_1 flow (figure 32b) results from the flow induced by strong bridges. In contrast, the bridges and threads at M_2 are of comparable strength, thus preventing hairpin formation (figure 32c); similarly, hairpins are not present for M_1 at an earlier time (figure 32a) since strong bridges have not yet formed. It is important to note that these characteristics of the post-reconnection flow are lost if we use CDIC and are difficult to decipher for CIC due to multiple artificial wave-vortex interactions.

The key general implications for compressible reconnection are that bridging is initiated earlier due to shocklet formation, which in turn drastically slows subsequent reconnection. As a result, the reconnection timescale is longer and the associated energy cascade rate is slower in compressible flows. Additional details on the shocklet structure and its effect on circulation transfer are available in [77], where it is also shown that temperature dependences of viscosity and conductivity do not significantly influence the reconnection dynamics.

5.4 Definition of vortex reconnection

Up to this point, we have discussed vortex reconnection in considerable detail using a physical, rather qualitative notion of the phenomenon. Clearly, to understand its role in more complex turbulent flows, we need a mathematical framework to identify regions of reconnection activity. In the following, we briefly outline our approach to rigorously define vortex reconnection using a dynamical systems approach.

We begin by considering a 2D Hamiltonian system

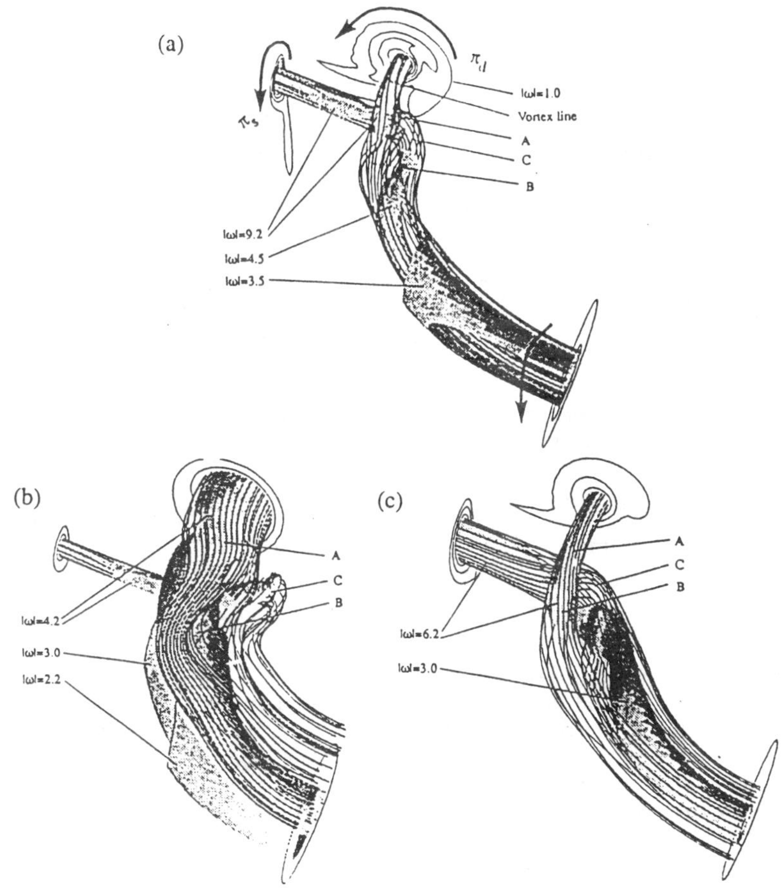

Figure 32. Vortex lines and portions of $|\omega|$ surfaces within the bridge and thread for M_1 at (a) t=4.5; (b) t=6; (c) M_2 at t=6.

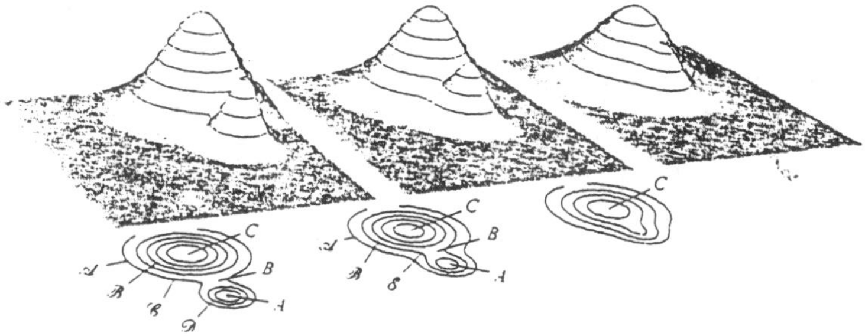

Figure 33. Carpet plots of the Hamiltonian H (top) and corresponding phase portraits (bottom) for three values of the bifurcation parameter μ.

$$\frac{dx}{d\sigma} = \frac{\partial H}{\partial y}, \quad \frac{dy}{d\sigma} = -\frac{\partial H}{\partial x} \tag{5-5}$$

where σ is the independent variable, $H(x,y;\mu)$ is the Hamiltonian and μ is a bifurcation parameter. Let H vanish at infinity and be a smooth function of x, y and μ. For illustrative purposes, consider the phase portrait at $\mu = \mu_0$ in figure 33(a), which from a topological standpoint, includes: critical points (A,B,C), separatrices, an island I_A enclosing A but not C, an island I_C enclosing C but not A, and an island I_{ABC} enclosing all three critical points. We now define the flux of the vector field $(dx/d\sigma, dy/d\sigma)$ through an oriented curve connecting any two points P and Q in plane (x,y) as $\Gamma_{PQ} = H(Q)-H(P)$. The island fluxes thus follow as: $\Gamma(I_A) = H(A)-H(B)$; $\Gamma(I_C) = H(C)-H(B)$; $\Gamma(I_{ABC}) = H(B)-H(\infty) = H(B)$.

Now, if H has a smooth dependence on μ, then the phase portrait in figure 33(a) is structurally stable, and the flux in the islands will change with μ (see figures 33b,c). Physically, flux changes can result from: annihilation, creation, or transfer of circulation from neighboring islands. *We define reconnection as the exchange of flux between neighboring islands in a structurally stable phase portrait* [78].

Thus, the μ dependence is particularly instructive if we can label and track individual orbits in a phase portrait. To apply this concept to reconnection, we define 3D physical space as the phase portrait, the vortex lines as the orbits, and time t as the bifurcation parameter μ. To date, we have successfully used this approach to describe reconnection and to calculate its rate in axisymmetric vortex rings with swirl, where axisymmetric vortex surfaces can be tracked [78]. It is indeed a major challenge to extend this description of reconnection to arbitrary 3D geometries.

6. Vortex core dynamics

As mentioned in §5.2, our study of vortex line twisting and strong axial flow *within* vortex cores – core dynamics – was first motivated by the behavior of bridges formed during vortex reconnection (figure 34). Our initial results (discussed in this section) indicate that core dynamics, a virtually unknown aspect of vortex dynamics, are an inherent feature of CS. Subsequent studies have produced dramatic revelations: core dynamics appear to play a key role in many enigmatic hydrodynamic phenomena, such as vorticity wavepacket motion, vortex breakdown, and shear flow transition (discussed in § 7).

Generally speaking, core dynamics arise from variations in vorticity magnitude along a vortex tube, which may arise by instability (as in the mixing layer) or local stretching by nearby CS (as for bridges). The subsequent evolution involves wave motion of vorticity along the vortex axis, brought about by helical twisting and untwisting of vortex lines within a nominally rectilinear vortex. In this sense, core dynamics are distinct from the more extensively studied vortex bending waves, which cause local displacement of the vortex axis itself.

In this section, we first analyze vortex core dynamics in a very idealized setting: an isolated, straight, axisymmetric vortex tube with a sinusoidal core size variation. In this way, we obtain clean evolutions for which the essential dynamics are readily tractable. A more detailed discussion of core dynamics has been presented in [29].

6.1 Swirl – meridional flow formulation

For axisymmetric flows, the Navier-Stokes equations in cylindrical coordinates (r,θ,z) can be rewritten in terms of the swirl $\xi = rv_\theta$ and the meridional flow $\eta = \omega_\theta/r$ as

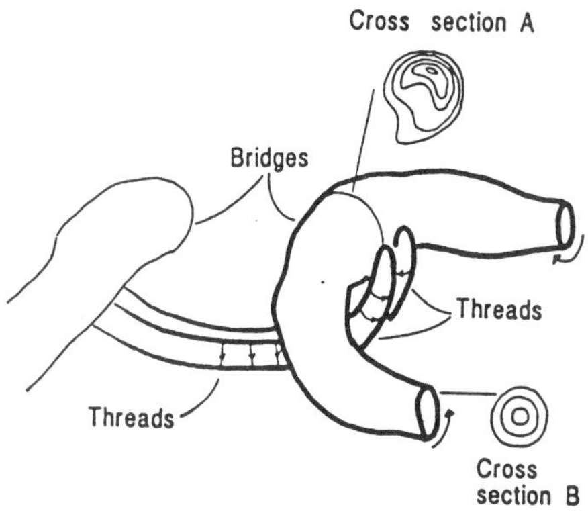

Figure 34. Schematic of vortex core nonuniformity in reconnection bridges.

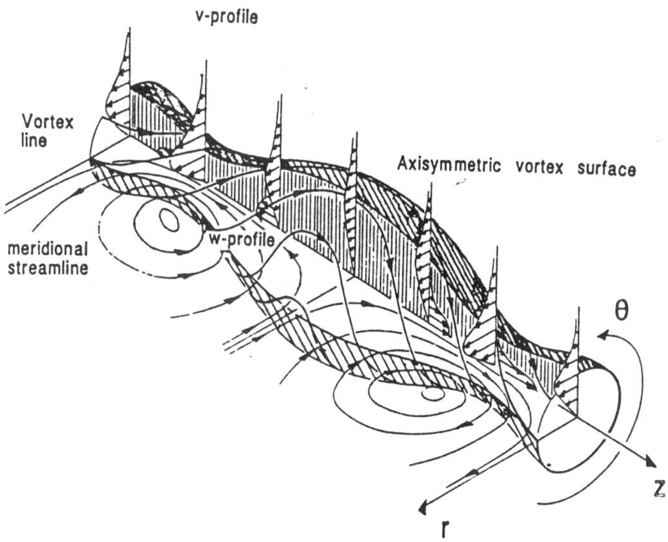

Figure 35. Definition sketch of important features of core dynamics: vortex surface nonuniformity, twisted vortex lines, and meridional flow.

$$\frac{D\xi}{Dt} = \frac{1}{Re}\left(\frac{\partial^2\xi}{\partial r^2} - \frac{1}{r}\frac{\partial\xi}{\partial r} + \frac{\partial^2\xi}{\partial z^2}\right) \tag{6-1}$$

$$\frac{D\eta}{Dt} = \frac{1}{r^4}\frac{\partial\xi^2}{\partial z} + \frac{1}{Re}\left(\frac{\partial^2\eta}{\partial r^2} + \frac{3}{r}\frac{\partial\eta}{\partial r} + \frac{\partial^2\eta}{\partial z^2}\right) \tag{6-2}$$

where $D/Dt \equiv \partial/\partial t + u_r\partial/\partial r + u_z\partial/\partial z$ represents the material derivative for the meridional flow. Since surfaces of constant ξ are axisymmetric vortex surfaces, on which vortex lines lie, this representation is very useful in interpreting vortex line twisting, meridional flow generation, and core oscillation.

6.2 Coupling between swirl and meridional flows

In equation (6-2), the vortex surface geometry (described by ξ) dynamically influences the coiling of vortex lines (described by η) through the coupling term $(1/r^4)\,\partial\xi^2/\partial z$. The physics represented by this coupling term are easily understood by its Taylor expansion for two points A and B on a vortex surface $\xi=\xi_o$:

$$\left.\frac{1}{r^4}\frac{\partial\xi^2}{\partial z}\right|_A = \omega_z(z_A, r_A)\lim_{B\to A}\frac{\dot\theta_B - \dot\theta_A}{z_B - z_A} \tag{6-3}$$

where $\dot\theta_B = \xi_o/r_B^2$, $\dot\theta_A = \xi_o/r_A^2$, and the subscripts denote the coordinates of ξ_o at A and B. Equation (6-3) indicates that the rate of η generation (*i.e.* the coupling term in equation 6-2) at a point is given by the differential rotation along the axisymmetric vortex surface passing through that point weighted by the axial vorticity magnitude, *i.e.* vortex line twisting.

Consequently, where an axisymmetric vortex surface converges in the z-direction (*i.e.* $\partial\xi^2/\partial z > 0$), there is generation of positive η due to local vortex line twisting. Likewise, negative η generation occurs where the vortex tubes diverge (*i.e.* $\partial\xi^2/\partial z < 0$) in the z-direction. We emphasize that the *instantaneous* value of η does not influence the *instantaneous* generation of η due to the coupling term, a fact that proves to be useful later in understanding the η-transport for core dynamics.

6.3 Initial conditions

To study core dynamics in the simplest possible configuration, we initialize a rectilinear vortex with a sinusoidal core size variation, with a self-similar axial vorticity profile at each z when scaled by the local core size. The initial radial vorticity component is determined such that the vorticity field is divergence free, while the azimuthal vorticity component is specified to be zero, so that no meridional flow exists initially. Specifically, the initial vorticity distribution is given by

$$\omega_r(r,\theta,z) = P\left(\frac{r}{r_o s(z)}\right)\frac{r}{s^3(z)}\frac{ds}{dz} \tag{6-4a}$$

$$\omega_\theta(r,\theta,z) = 0 \tag{6-4b}$$

$$\omega_z(r,\theta,z) = P\left(\frac{r}{r_o s(z)}\right)\frac{1}{s^2(z)} \tag{6-4c}$$

where P is a compact Gaussian function, r_o is the unperturbed vortex core radius, and $s(z) = [1-\mu\cos(2\pi z/\lambda)]$ is the shape of the initial axisymmetric vortex surfaces (λ is the length of the computational domain and μ is the perturbation amplitude). The evolution of this vorticity field is obtained by solving the complete Navier-Stokes equations with periodic boundary conditions and a fourth-order predictor-corrector scheme for time advancement.

The flow geometry is depicted in figure 35, which also provides an overview of the fairly complex internal dynamics produced by this simple core size perturbation. In particular, note the meridional flow generated by the twisting of initially straight vortex lines.

6.4 Vorticity wavepacket motion

The essence of core dynamics is best reflected in figure 36 by the evolution of vorticity magnitude for $Re=\Gamma/\nu=665$. Clearly, the initial $|\omega|$ peak splits into two equal peaks of considerably lower amplitude (compare figures 36a,b and recall the periodic boundary condition in z). The vorticity transport at these times must occur as traveling wavepackets, because the mass transport (given by ψ in figure 37) and vorticity transport are in opposite directions. This is possible because of the enstrophy production P_ω, which allows variations in vorticity magnitude to propagate along vortex lines; P_ω changes sign near each wave packet — enstrophy is produced at the front of and annihilated at the back of each packet, hence producing the packet's motion. The low-enstrophy bubble observed in figure 36(i) occurs at a location where P_ω is negative for a significant time. The disappearance of this bubble at later times is due to a change in the sign of P_ω and also "fill in" by viscous diffusion.

Significant insight into the evolution of P_ω can be obtained by studying the ψ distributions shown in figure 37. From equation (6-1), constant ξ surfaces (axisymmetric vortex surfaces) are convected with the meridional velocity in an inviscid flow. Hence the local cross-section of a vortex tube shrinks to increase the axial vorticity (angular momentum conservation) when the meridional flow converges on the axis, and vice versa. In figure 37(b), P_ω is thus negative near z=0, 2l at early times, because the meridional flow is directed away from the axis here. Furthermore, the meridional flow converges on the axis near z=l, making P_ω positive there due to vortex stretching. These estimates of P_ω from ψ contours are justified by the approximation

$$P_\omega \approx -4\,\omega_z^2 \frac{1}{r^2}\frac{\partial \psi}{\partial z}, \tag{6-5}$$

for inviscid flows, which is asymptotically correct near the axis (r=0).

In figure 37, we see two counter circulating cells — *primary cells* — per wavelength of the vortex. If these cells were to persist for all times, the vortex would continue to expand near z=0 and contract near z=l. However, the emergence of *secondary cells* (see figure 37e) with a reversed meridional flow arrests and eventually reverses the core size variation.

6.5 Vortex line twisting

The generation and reversal of meridional flow inherent to core dynamics is easily understood in terms of the vortex line evolution shown in figure 38. At t=0, all vortex lines lie in meridional planes and therefore have no torsion initially (figure 38a). Axial variation of the swirling motion ($\dot{\theta} \equiv \xi/r^2$) around a vortex surface causes vortex lines on this surface to become twisted, as indicated by the coupling term interpretation. Near z=l, the core vortex lines rotate more slowly around the vortex axis than near z=0, 2l, causing helical twisting of vortex lines. Meanwhile, the vortex core expands near z=0 (and 2l) and contracts near z=l due to the induced meridional flow, eventually causing the vortex line coiling to reverse its direction (figure 38g). This coiling and uncoiling of core vortex lines occurs cyclically, but dampens with time due to viscous effects, so that for this isolated vortex (*i.e.* without background shear), the axial flow and core dynamics ultimately decay.

The physics of the coupling between swirl and meridional flows is most easily understood in terms of vortex lines on an axisymmetric vortex tube. When the vortex lines are coiled on

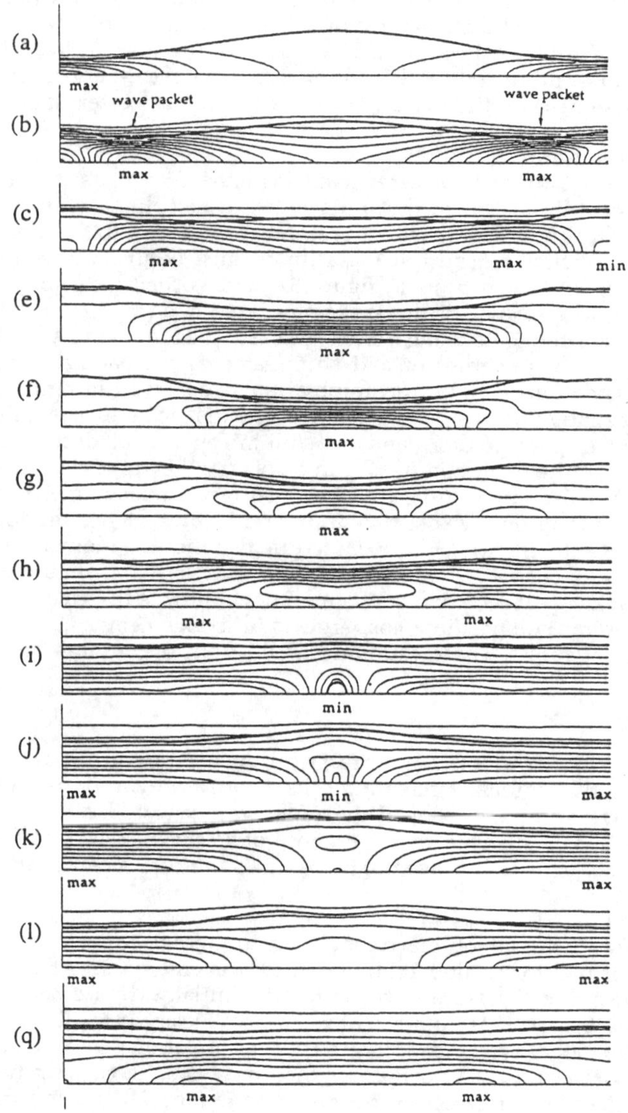

Figure 36. Evolution of $|\omega|$ in a meridional plane for core dynamics at (a) t=0; (b) t=0.5; (c) t=1.0; (e) t=2.0; (f) t=2.5; (g) t=3.0; (h) t=3.5; (i) t=4.0; (j) t=4.5; (k) t=5.0; (l) t=5.5; (q) t=8.0. The vortex core boundary as given by the kinematic vorticity number is shown as a bold line in each frame. The gaps in the labeling sequence indicate nonuniformity in the time difference between frames, with successive letters separated by 0.5 time units.

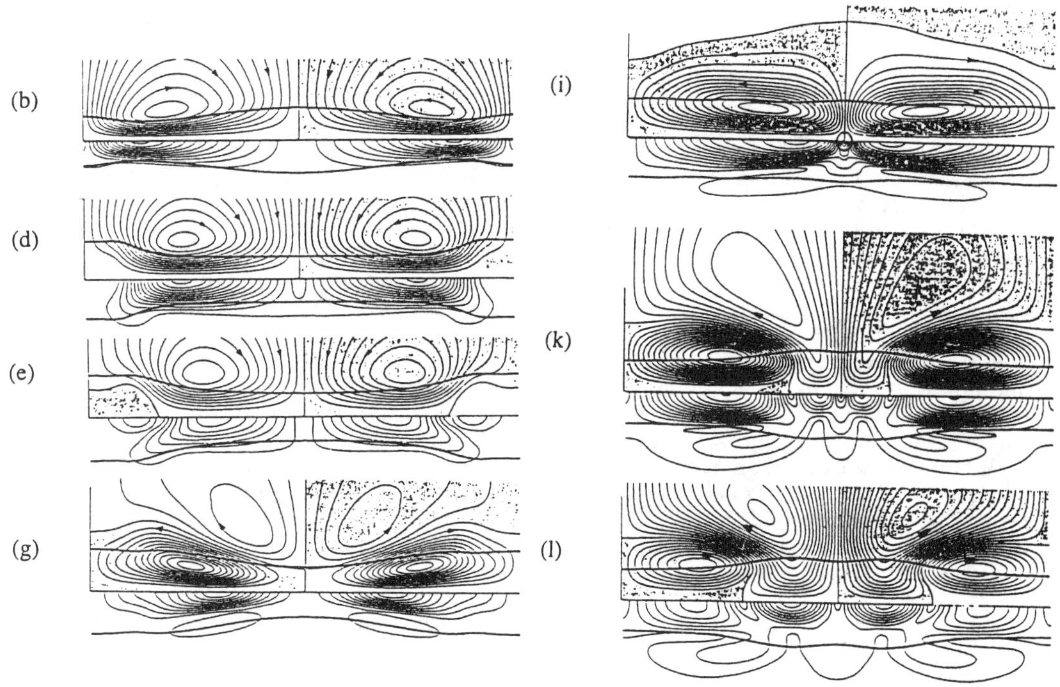

Figure 37. The evolution of the meridional flow (streamlines above the axis and meridional velocity magnitude below the axis) at selected times corresponding to those in figure 36. The vortex core given by the kinematic vorticity number is overlaid in each frame. ψ is normalized to be zero on the axis; thus the motion is counter clockwise in the shaded regions ($\psi<0$) and clockwise in the unshaded regions ($\psi>0$).

the tube as in figure 35, the vortex contains significant azimuthal vorticity and hence nonzero η. Note that the instantaneous winding of vortex lines on a vortex surface is not directly related to the shape of the surface. For example, it is possible to have the vortex lines wind in entirely different fashions on identical vortex surfaces. In this sense, the coiling of vortex lines on an axisymmetric vortex surface is independent of its instantaneous shape. However, as discussed earlier, the differential rotation is given by the shape of the vortex surface and acts to increase or decrease η through the coupling term.

To illustrate, consider the vortex in figure 39(a), at a phase when the differential rotation is decreasing the magnitude of η, *i.e.* causing uncoiling of the vortex lines. Clearly, there is a time delay before this differential rotation uncoils the vortex lines to change the sign of η (figure 39c). While the differential rotation is uncoiling the vortex lines, the meridional flow continues to distort the shape of the vortex tube, because η has not yet changed sign, and hence the orientation of the circulatory motion is preserved (figures 39a,b). Since the vortex core size is strongly nonuniform in figure 39(b), the differential rotation is large even though η is small. This results in renewed coiling of the vortex lines, with an opposite twist direction, so as to cause meridional flow reversal and subsequent reduction of core size nonuniformity (figure 39c). When the vortex core is virtually rectilinear, the vortex lines are still coiled (figure 39d), so that the induced meridional flow continues to shrink the core at z=l and expand it at z=0,2l, thereby continuing the cycle (figure 39e).

It is important to note that the η-transport differs significantly from the meridional flow's material transport. This is because there are two η-transport mechanisms: (i) advection of η by the meridional flow and (ii) material η-generation/annihilation by differential rotation (through the coupling term). At early times, these two effects compete in the primary cells: in the region between z=0 and z=l, the material transport of η by the meridional flow is towards z=0 (figures 37b,d), while the differential rotation annihilates η near z=0. The net η transport due to these two effects is toward z=0, against the meridional flow, as explained schematically in figure 40. In the secondary cells which form later near z=0 (figure 37e), the meridional advection and η generation both transport η in the same direction, namely away from z=0. The actual transport of η in these secondary cells is thus faster than the material transport.

To generalize, the η-distribution moves in the direction of the material transport when a cell is expanding, but faster than the meridional flow; in a contracting cell, η moves against the meridional flow, but slower than its transport in an expanding cell. In both cases, the η-transport is drastically different from simple material transport in the meridional flow. Instead, the η-transport consists of a combination of material transport and wave motion, the latter of which is a consequence of vortex line twisting/untwisting, or equivalently, an approximate 90° temporal phase shift between the cyclic evolutions of η-generation and η.

6.6 Core size oscillation

To provide a quantitative estimate of the core size evolution, we first define the vortex core as the spatial region where the *kinematic vorticity number* $N_k \equiv (\omega^2/2S_{ij}S_{ij})^{1/2} > 1$. The vortex core cross-sectional area A(z, t) can then be used to define the instantaneous local core size as

$$\sigma(z,t) = [A(z,t)/\pi]^{1/2}, \tag{6-6}$$

provided that the region of $N_k > 1$ is simply connected (if this is not true we specify $\sigma(z,t)$ as the radius of the outermost $N_k = 1$ contour).

The evolution of core size nonuniformity is shown in figure 41 by the standard deviation of $A(z,t_o)$ normalized by the instantaneous average area $\bar{A}(t_o) \equiv \pi\,\bar{\sigma}^2(t_o)$ for μ=0.5 and different Re. It is clear that the axial core size variations progressively disappear with increasing viscosity, suggesting that the damping is purely viscous. Moreover, the decay of the axial

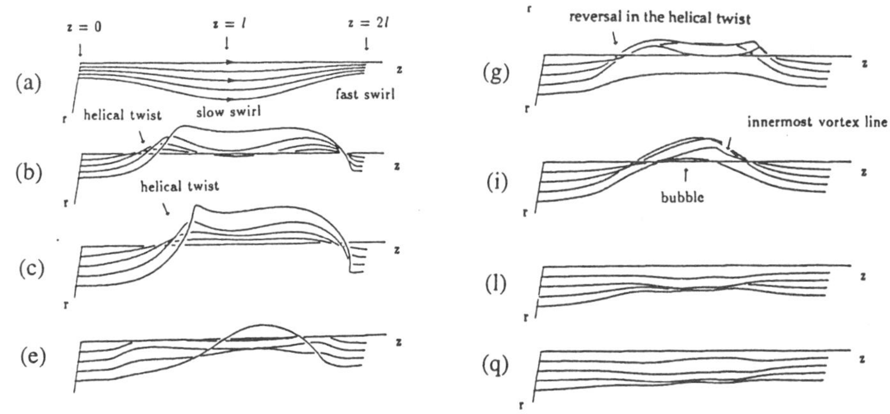

Figure 38. Helical twisting and untwisting of core vortex lines, at selected times corresponding to those in figure 36.

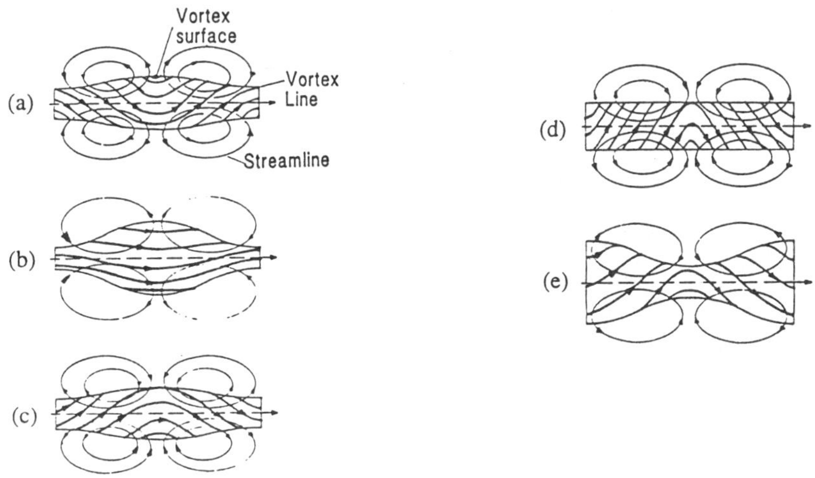

Figure 39. Schematic of core dynamics evolution, reflecting the coupling between swirl and meridional flow.

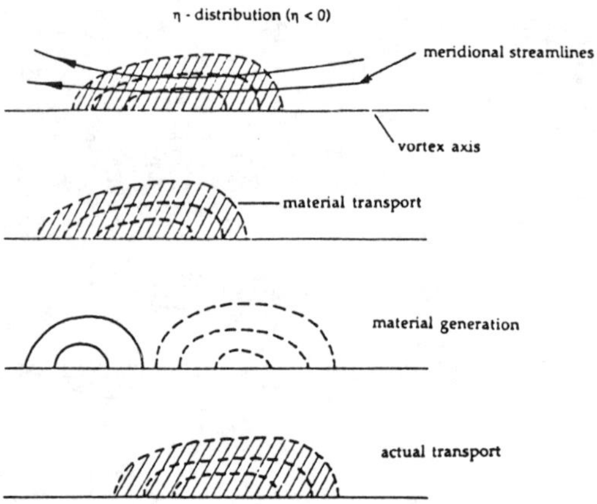

Figure 40. Schematic of η transport mechanisms, illustrating mechanism of upstream η transport.

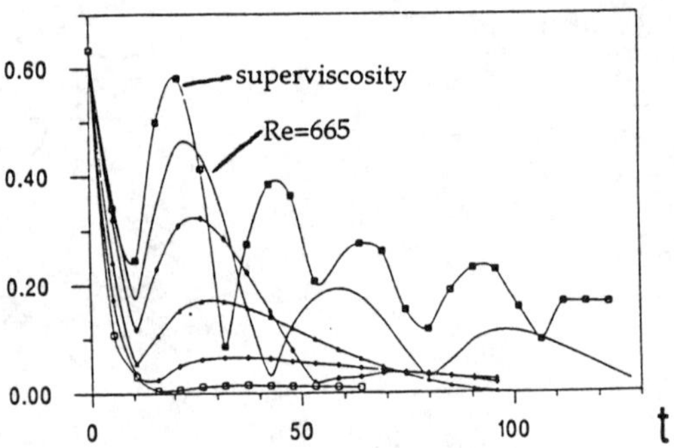

Figure 41. Evolution of the standard deviation of the core area normalized by the instantaneous average core area, i.e. $\left(\overline{[A(z,t) - \overline{A(z,t)}\,]^2} \right)^{0.5} / \overline{A(z,t)}$.

variations has an interesting dependence on Re. Namely, the decay is not monotonic, but rather like a damped oscillation, where both the amplitude and frequency depend on Re. The oscillation frequency increases with Re, but appears to have a finite high Re-limit; the 'super-viscosity' simulation is indicative of this limiting frequency. Therefore, core dynamics in this isolated, axisymmetric case behave like a neutral, oscillatory, inviscid instability, which is damped at finite Re by viscous effects.

To summarize, using an isolated axisymmetric vortex for simplicity, we have characterized in detail the fundamental mechanisms which sustain the large-amplitude oscillations of core size and meridional flow associated with core dynamics, in the context of vortex line twisting/untwisting. Our preliminary studies suggest that these mechanisms are essential to vortex breakdown (recall the low-enstrophy bubble formation), and, as shown in the following section, core dynamics are unstable for shear layer vortices and constitute a powerful agent for transition.

7. Core dynamics instability (CDI) and mixing layer transition

Up to this point, we have discussed core dynamics of an isolated axisymmetric vortex, which decay viscously in time. More importantly in a practical sense, we have found that in a mixing layer, similar (although nonaxisymmetric) core dynamics can occur within spanwise rolls and *amplify* as an instability – termed core dynamics instability (CDI) [28]. In this case, high-amplitude core dynamics eventually cause an explosive breakdown to turbulence.

This new CDI-based transition mechanism not only clearly illustrates the importance of internal CS dynamics, but also reconciles an important discrepancy between previous numerical results and experimental observations regarding mixing layer transition. Namely, the conventional, widely-accepted transition mechanism – the entrapment of ribs within pairing rolls [79,80] – has been shown by DNS to require artificially large initial rib perturbations (5 times more energetic than 2D modes responsible for rollup and pairing). If the 3D perturbation amplitude is moderate (comparable to that of the 2D modes initially), transition does not occur even after two pairings, and, in fact, small-scale vorticity becomes less prominent after the second pairing. In contrast, experimental studies clearly indicate that transition is delayed, but not forever suppressed, by weaker initial three-dimensionality. Hence, a different transition mechanism such as CDI must be operating when the 2D and 3D initial perturbation amplitudes are comparable, or when 2D modes are amplified by forcing. In the following, we will demonstrate that CDI does in fact initiate transition when the 2D and 3D modes are initially comparable.

7.1 Excitation of CDI

Analytically, CDI is a linear secondary (*i.e.* 3D) instability of the steady Stuart vortex solution, which models mixing layer vortices ('rolls') formed by Kelvin-Helmholtz instability [81]. Even for unsteady 2D basic flows studied via DNS, we have found that core dynamics still grow rapidly, even well into the nonlinear regime.

We have identified two distinct routes of CDI excitation which can occur during unsteady vortex rollup and roll pairing: (i) fundamental - CDI of fundamental wavelength rolls after rollup, and (ii) subharmonic - CDI of subharmonic wavelength coalesced rolls which commences after pairing. Fundamental CDI arises much like the core size perturbation considered in §6. The process of subharmonic CDI excitation is shown schematically in figure 42, with the 3D perturbation amplitude exaggerated for clarity. At rollup, the so-called 'helical pairing' class of secondary instability causes opposite spanwise undulations of neighboring rolls (figure 42a). Between rollup and pairing, these undulations amplify while, at the same

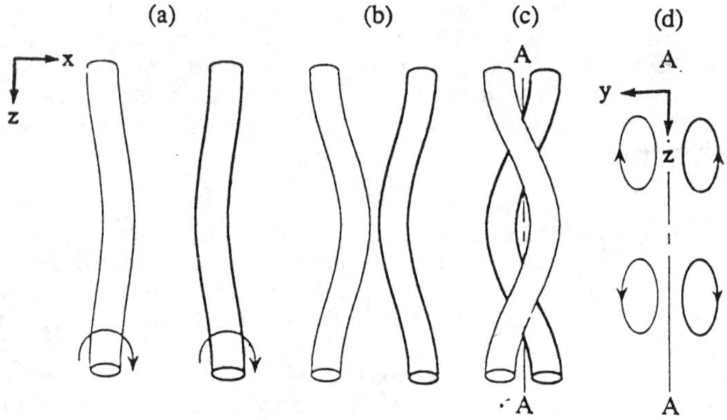

Figure 42. Schematic of subharmonic CDI excitation by pairing of bent rolls.

time, pairs of rolls approach due to (nonlinear) 2D pairing growth (figure 42b). Once the rolls begin to coalesce (figure 42c), their opposite undulations create the meridional flow characteristic of core dynamics, thus initiating CDI of the composite vortex pair. This subharmonic route of CDI excitation eventually produces an intense meridional flow within rolls that triggers transition, beginning from only moderate-amplitude 3D perturbations. In flows which undergo pairing, subharmonic CDI is more energetic than its fundamental mode counterpart and thus will be focused upon here.

7.2 Nonlinear CDI evolution

We now consider the nonlinear, post-pairing evolution of a subharmonic CDI mode, initialized with half the energy of the 2D modes which excite rollup and pairing. The flow geometry and the spanwise midplane z_π and endplane z_0 are defined in figure 43, which contains the evolution of i) constant $|\omega|$ vorticity surfaces and ii) two bundles of vortex lines passing through the two $|\omega|$ peaks in z_0. The locations of these distinct $|\omega|$ peaks are tracked in time and the corresponding vortex line bundles are distinguished by their line thickness (heavy and light for the downstream and upstream rolls respectively). Note that these vortex lines cannot be tracked in time as material lines in viscous flows.

The evolution through pairing is as illustrated schematically in figure 42. The $|\omega|$ field at t=18 illustrates the geometry of the pairing rolls, which have been bent by a helical pairing-type instability. The meridional flow induced by these bent portions of the rolls (figure 42d) excite finite-amplitude CDI after pairing. In the early stages of evolution (figures 43a-d), the rolls are simply rotating around each other as distinct vortices, with a planar 3D deformation (*i.e.* bending waves) of roll vortex lines. As evolution continues, a compact paired roll first appears near z_π (t=30), followed by rapid expansion and formation of a low-$|\omega|$ bubble within the paired roll core (t=33). At the same time, the roll vortex lines first develop a strong helical twist (t=30) and then sharp outward flaring near the bubble in z_π. This evolution illustrates that the CDI mode excited by this initially simple pairing configuration attains a high nonlinear amplitude, producing a strongly 3D internal roll structure.

The evolution of spanwise vorticity ω_z in the planes z_π and z_0 shown in figure 44 provides a clear perception of the 'paired' roll's core size oscillation. In contrast to 2D pairing dynamics, the initially coalesced roll ω_z in z_0 (t=21) weakens and 'unpairs' by t=27, due to the compression and outward ejection of roll fluid by the meridional flow cells. Between t=30 and t=36, the cells reverse sign, subsequently causing the rolls to recoalesce and intensify in z_0 to form a vortex sheet which undergoes secondary rollup by Kelvin-Helmholtz instability, reaching (by t=36) a peak ω_z of $4.4\Omega_o$, where Ω_o is the initial peak ω_z.

In z_π, the rolls have slipped past each other without pairing by t=21 (figure 44a). Subsequently, the rolls coalesce (t=24), forming essentially a vortex sheet (t=27), which is strengthened by the extensional strain rate induced by the meridional flow. After this sheet rolls up between t=27 and t=30 into a single vortex in z_π (by Kelvin-Helmholtz instability), the meridional flow reverses, causing the roll ω_z to weaken. The compressive flow in this plane is so strong that a shell of ω_z is formed around a bubble of nearly irrotational fluid (t>33). Expansion of this shell continues, forming an elongated and very thin sheath of ω_z surrounding a low-$|\omega|$ bubble (t=36).

The mechanism of sheath and bubble formation near z_π, illustrated schematically in figure 45, consists of a strong CDI-induced saddle flow centered at z_π which causes both compression and outward ejection of near-axis ω_z. It can be shown that ω_z subject to continual compression in z_π decays exponentially; the resulting low-enstrophy bubble appears first near the roll center (figure 45b) and then enlarges radially (figure 45c). A key feature of this mechanism shown in figure 45(c) is that at a sufficiently large radius, the meridional flow

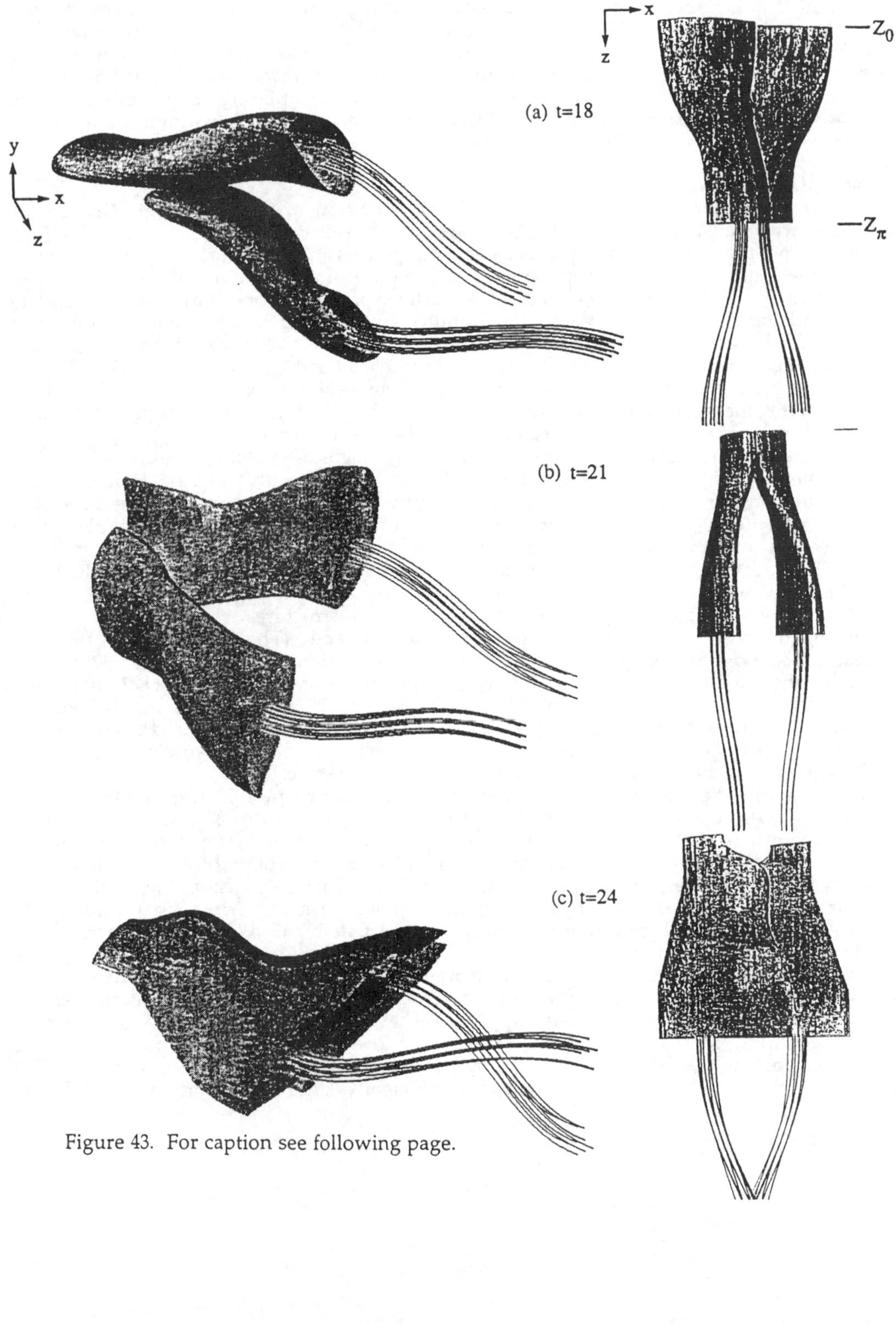

Figure 43. For caption see following page.

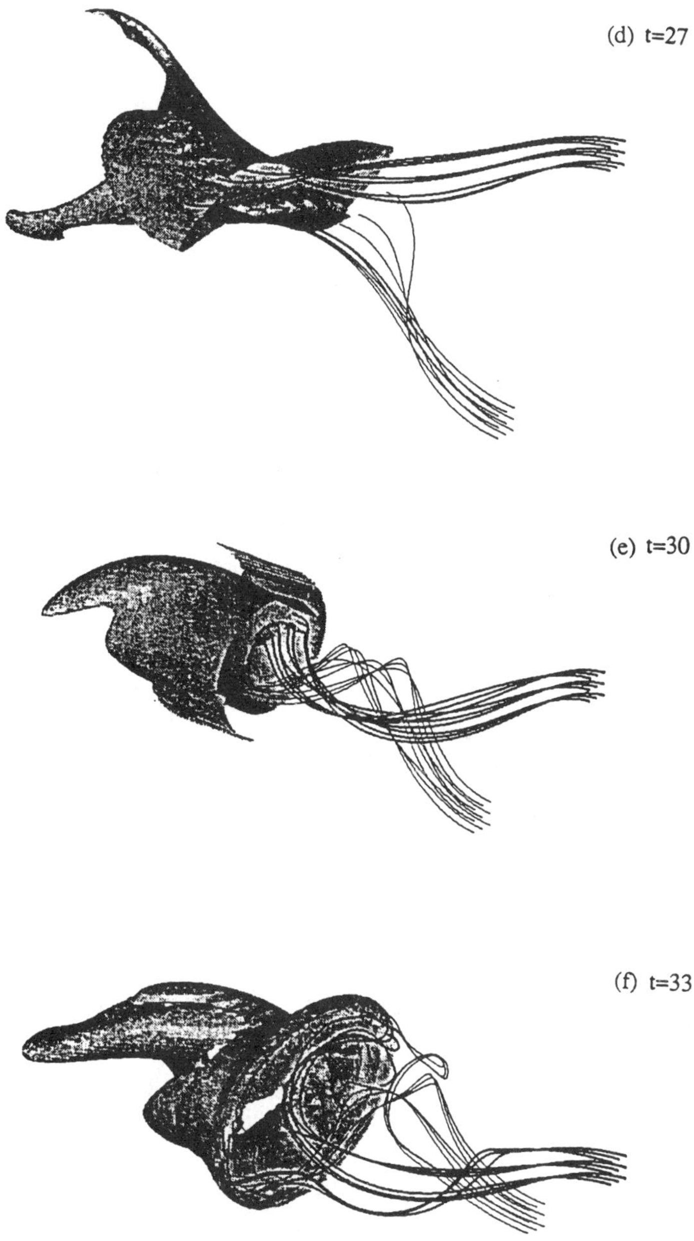

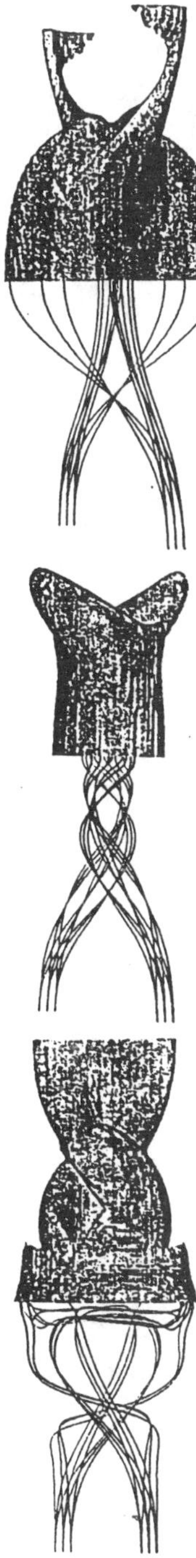

Figure 43. The evolution of $|\omega|$ surfaces and vortex lines within spanwise rolls during CDI-induced transition.

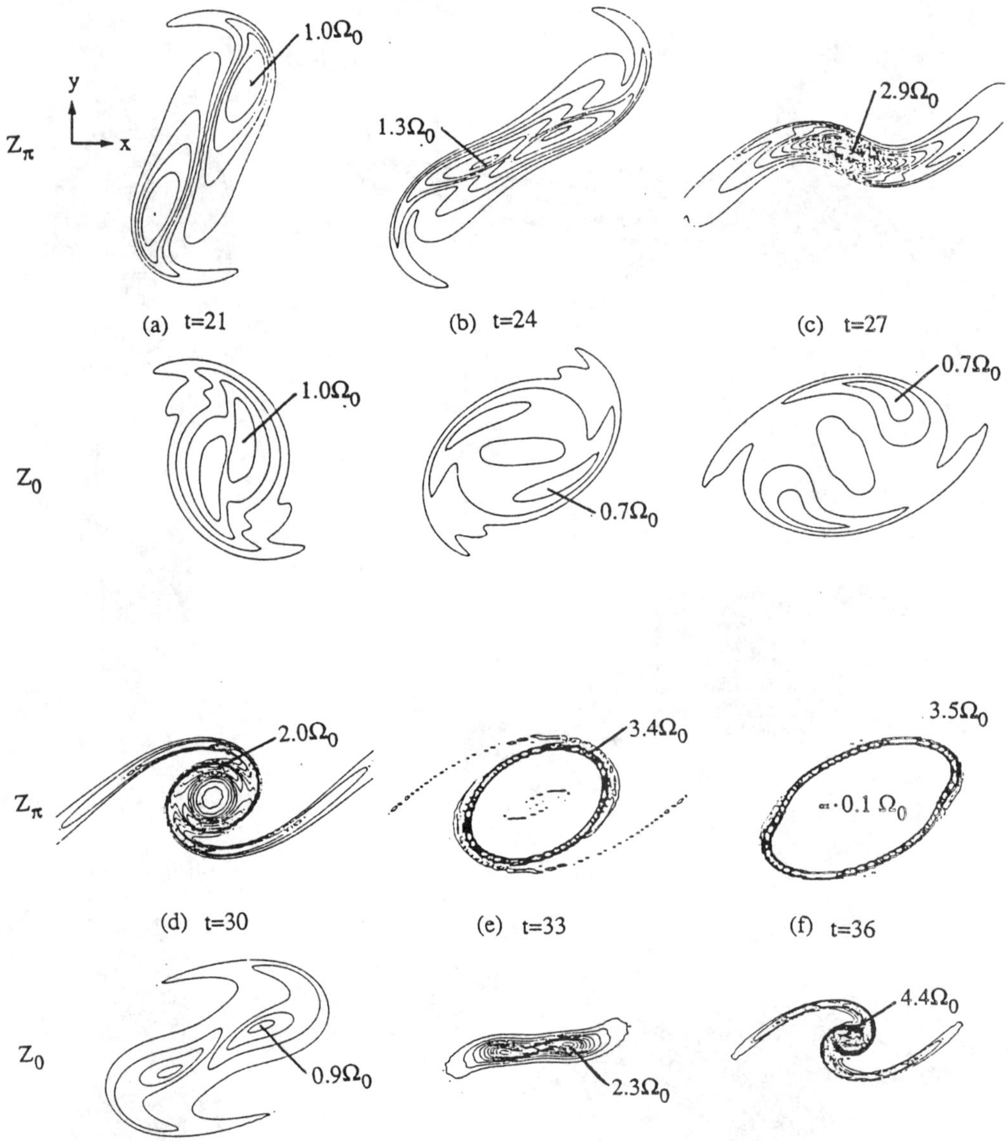

Figure 44. Contours of ω_z in z_π and z_0 during CDI evolution.

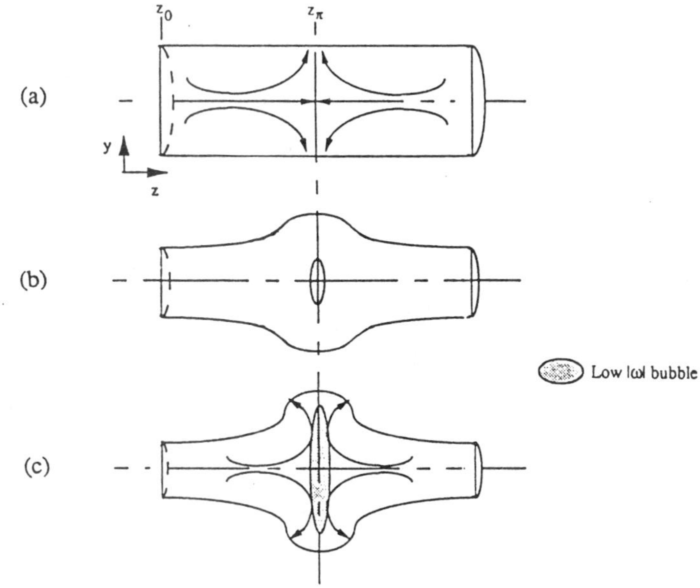

Figure 45. Schematic mechanism of sheath and bubble formation near z_π.

cells induce a strain rate opposite to that near the axis. Thus, once ω_z is radially ejected sufficiently far, it begins to amplify by stretching, thereby creating a fine-scale, high ω_z sheath as in figure 44(f).

The strong amplification of ω_z shown in figure 46(a) indicates that these core dynamics have reached a very high amplitude. Intense core dynamics are also reflected in figure 46(b) by the peak meridional flow velocity u_z at the vortex center in $z_{\pi/2}$. This spanwise flow within the roll oscillates according to the meridional flow strength and attains very large peak values, which in fact exceed the freestream U, as indicated in the figure. With regard to the meridional flow distribution (shown schematically in figure 45), it is now clear why such strong near-axis ω_z compression and outward ejection of roll core fluid (responsible for vortex sheath formation) occur near z_π between t~30-33.

7.3 Onset of transition

Although the local measures in figure 46 reflect the generation of small-scale vorticity their large values alone do not necessarily imply roll vorticity's internal intermittency - reflected by a granular vorticity distribution and characteristic of fully turbulent flows. For instance, the peak $-\omega_z$ in z_0 at t~36 (equal to $4.4\Omega_0$) results from a smaller-scale, yet well-organized roll near this plane (figure 44f). In this regard, the development of strong opposite-to-mean $+\omega_z$ is a useful indicator in that it illustrates a qualitative change in the vortex line geometry – from vortex lines which all run in the -z direction before transition to the appearance of folded vortex lines with S-shaped kinks which locally run opposite to the mean vorticity direction (*i.e.* +z). For this CDI flow, figure 46(a) demonstrates that the development of significant $+\omega_z$ begins around t=31 and eventually reaches a maximum of $-3\Omega_0$.

To determine where this opposite-signed ω_z develops, we show in figure 47(a) a bundle of vortex lines centered in the roll core along with the corresponding isosurface plot in figure 47(c) of $|\omega|$ (grayscale) overlaid by $+\omega_z$ (crosshatch) at t=37.5 (the time of maximum $+\omega_z$). Viewed together, figures 47(a,c) illustrate that $+\omega_z$ is localized around the vortex sheath near z_π and results from folded vortex lines which flare outward intensely here. It is interesting to note that between z_0 and $z_{\pi/2}$, the roll's vortex line and $|\omega|$ distributions are those of a laminar, albeit strongly 3D flow. Contour plots of ω_z near z_π (in the z plane containing the maximum $+\omega_z$) at t=33 and 37.5 in figures 48(a,b) respectively clearly illustrate the complex vorticity distribution within the roll resulting from this $+\omega_z$ development. Specifically, layers of $+\omega_z$ first appear within the roll (t=33), followed by development of intermingled regions of high-valued, fine-scale $-\omega_z$ and $+\omega_z$ within the roll (t=37.5). This suggests that strong core dynamics have initiated localized transition near the vortex sheath.

To ensure that this is in fact a turbulent transition, we doubled the initial 3D energy of this CDI mode. In this case, similar core dynamics occur, and the locally entangled vortex lines and fine-scale vorticity in figures 47(b,d) indicate an explosive transition to turbulence. The ω_z distribution near z_π in figure 48(c) shows strong internal intermittency and ensures that the fine scales in figure 47(d) are not numerical noise. Therefore, we will continue to analyze the more tractable transition dynamics of the lower amplitude case (discussed so far) with knowledge that a relatively small increase in the initial CDI disturbance amplitude will cause a very distinct transition.

The vortex line patterns responsible for the $+\omega_z$ regions in figure 48(b) are shown schematically in figures 49(c,d) respectively. Due to symmetry in z_π, $+\omega_z$ here is associated both with closed vortex line loops and also S-shaped folding of vortex lines. The key to both vortex line kinking and loop formation is the meridional flow pattern combined with its strong induced radial ejection near z_π at late times. As illustrated schematically in figure 49(a), the meridional flow cells induce u_z of opposite sign in the inner and outer core regions.

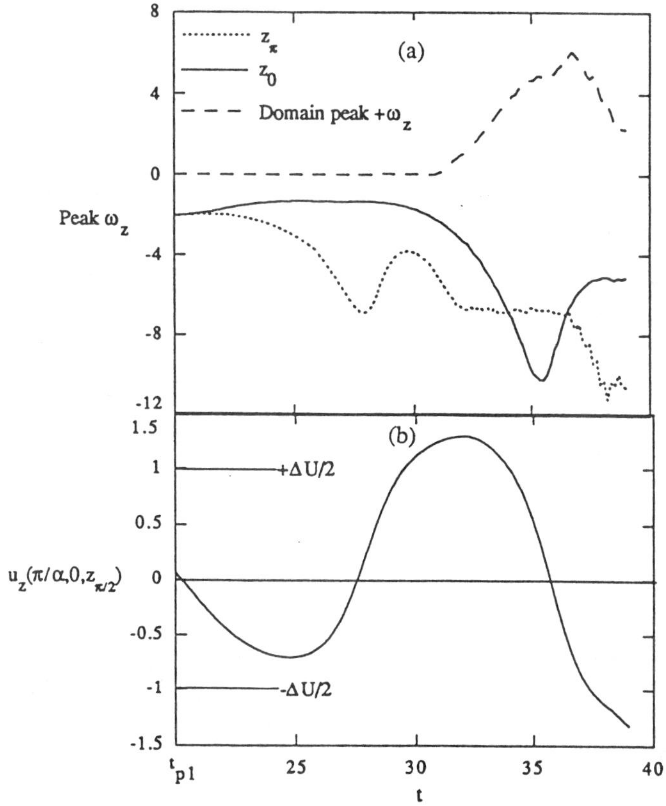

Figure 46. Evolution of (a) peak ω_z; (b) peak u_z, indicating the onset of transition.

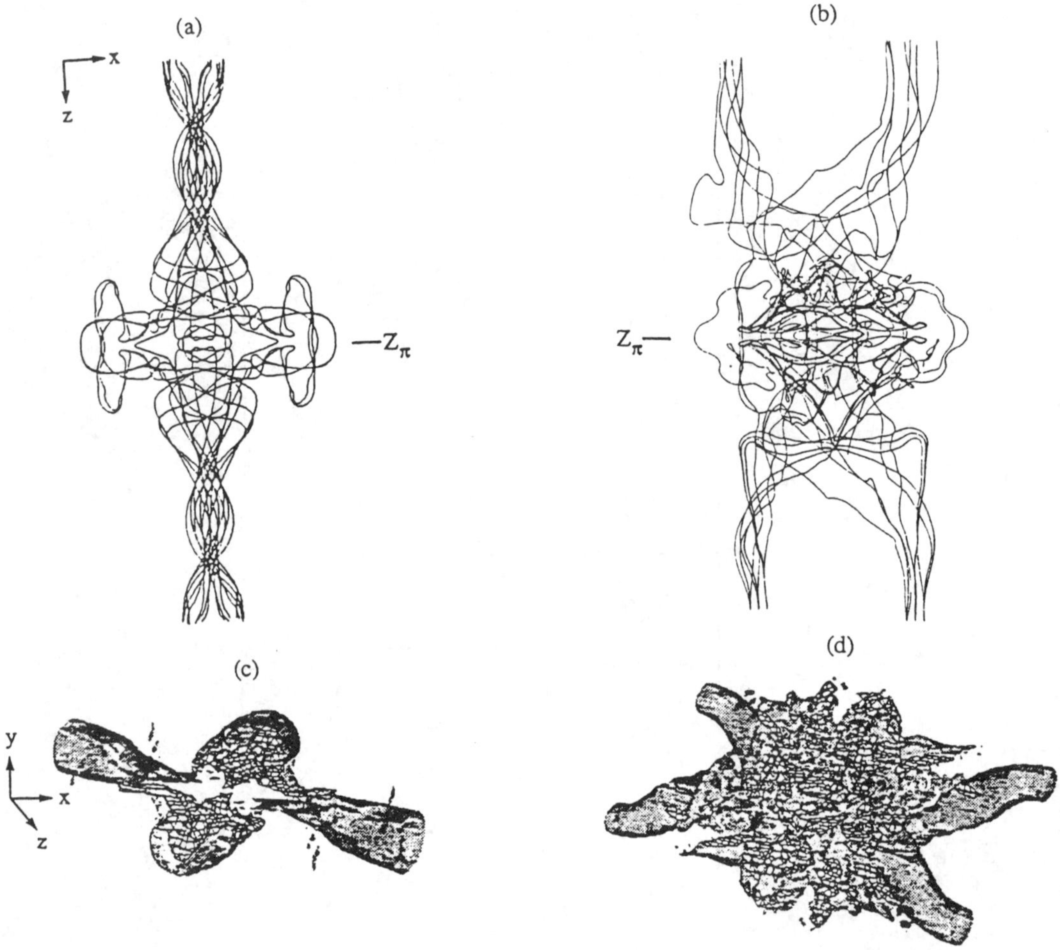

Figure 47. The structure of the post-transition flow, shown by (a,b) roll vortex lines; (c,d) $|\omega|$ and $\omega_z>0$ (cross hatch) surfaces, for (a,c) low initial 3D amplitude; (b,d) moderate initial 3D amplitude.

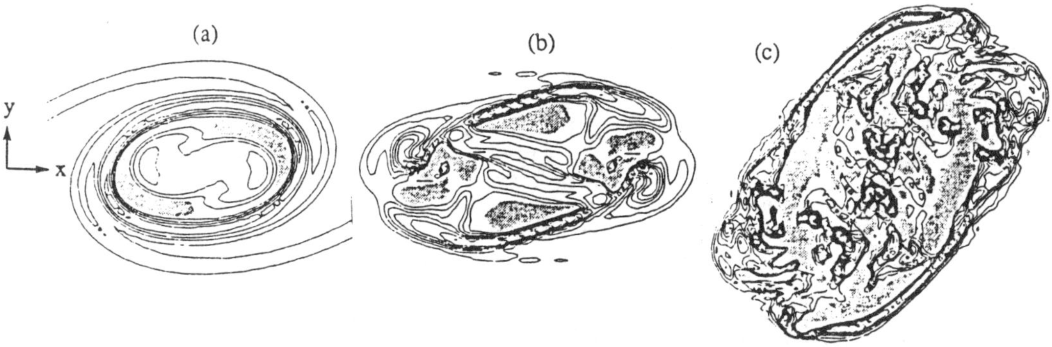

Figure 48. Contours of ω_z near z_π (a) before and (b) after transition for the low amplitude case. The post-transition flow for the moderate amplitude case is shown in (c).

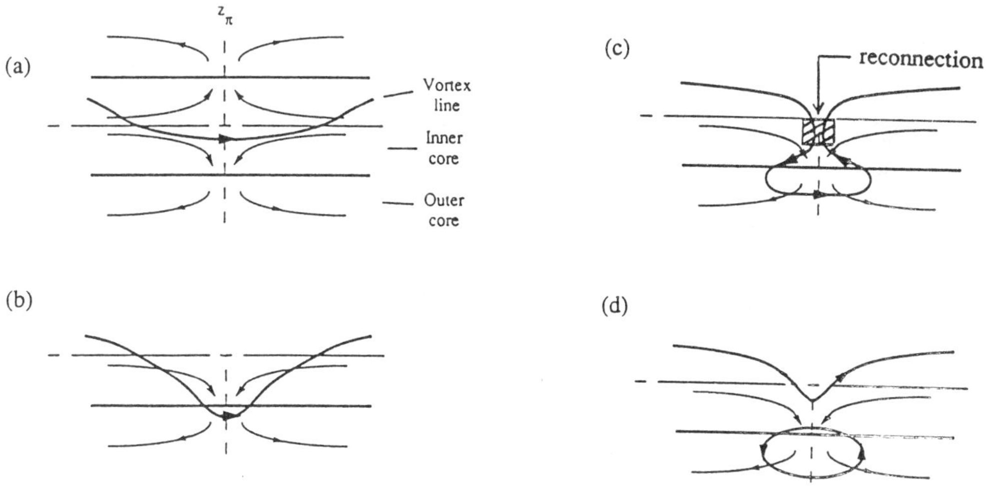

Figure 49. Schematic of folding and reconnection of vortex lines.

demarcated by a solid line. The vortex line sketched in figure 49(a) is a typical twisted inner core vortex line at t~30 (see figure 43e), when the meridional flow is nearly its strongest. The outward radial velocity near z_π causes a portion of this line to be ejected into the outer core region, as in figure 49(b). Note that the inviscid case is considered in figures 49(a-c) in order to track the evolution of a tagged vortex line for illustrative purposes. As additional ejection occurs, this line becomes part of the vortex sheath in z_π, and its legs are pressed together in the inner core region but stretched apart in the outer core to form an S-shaped kink with $+\omega_z$ (figure 49c). If the legs of a vortex filament surrounding this line are eventually pressed together in the inner core, viscous reconnection will form a closed loop, as diagrammed in figure 49(d). Note that closed vortex line loops passing through z_π can form only by this pinching-off since $\omega_x = \omega_y = 0$ in this plane because of symmetry, so that $+\omega_z$ can enter z_π only through the action of viscosity (*i.e.* reconnection). Thus, both characteristic vortex line patterns responsible for $+\omega_z$ arise as a consequence of the same mechanism, which causes vortex line kinking at an earlier stage (figure 49c) and reconnection to form several small-scale, closed vortex line loops at a later stage (figure 49d). It is interesting to recall that vortex loops were experimentally visualized to be pinched off and ejected outward in an axisymmetric mixing layer and suggested to be a mechanism of entrainment by Clark [82].

An interesting aspect of the spectral evolution in figure 50(a) is the fact that most of the small-scale generation occurs by t~33, even though little internal intermittency (*i.e.* spatial patchiness of vorticity *within* the roll) has been generated by this time (figure 48a). This indicates that spectral growth of high wavenumbers is not accompanied by internal intermittency. That is, sharp vorticity gradients reflected at high wavenumbers do not necessarily imply the presence of a granular 3D vorticity distribution, characteristic of turbulent flows. Therefore, in this case, cascade enhancement is not reliant upon a complex vorticity distribution (as in figures 48b,c), but instead generated by strong core dynamics. Once small-scale intense vorticity is produced, the vortex line geometry responsible is quickly disrupted, thus initiating internal intermittency but only slight additional small-scale energy (*cf.* t=33 and t=37.5 spectra in figure 50a).

To summarize, accelerated small-scale and internal intermittency generation indicate the onset of localized transition between the first and second roll pairings due to nonlinear evolution of a single CDI mode. As mentioned earlier, this scenario is significant in that it occurs from moderate 3D perturbation energy, 5 times smaller than that required for rib-induced transition. It is also important to note that this CDI evolution and transition are not reliant on the presence of special symmetries and have been shown to occur similarly even in the presence of moderate-strength ribs.

8. Helical wave decomposition (HWD)

As shown qualitatively in §6, core dynamics involves two vorticity wavepackets, which travel in opposite directions along the vortex axis. Nevertheless, traditional techniques are inadequate to interpret such wavepackets, such as to determine whether the packets reflect or pass through one another upon approach. For this purpose, we have developed a new technique – *helical wave decomposition* – which permits rigorous extraction of vorticity wavepackets. Additionally, HWD is quite effective in identifying local regions of helically twisted vortex lines, since it is based upon eigenfunctions of the curl operator. In fact, results discussed in §9 indicate a striking result: CS in turbulent flows have a natural tendency to be strongly polarized, *i.e.* to contain helically twisted vortex lines. For additional details regarding HWD, see [30].

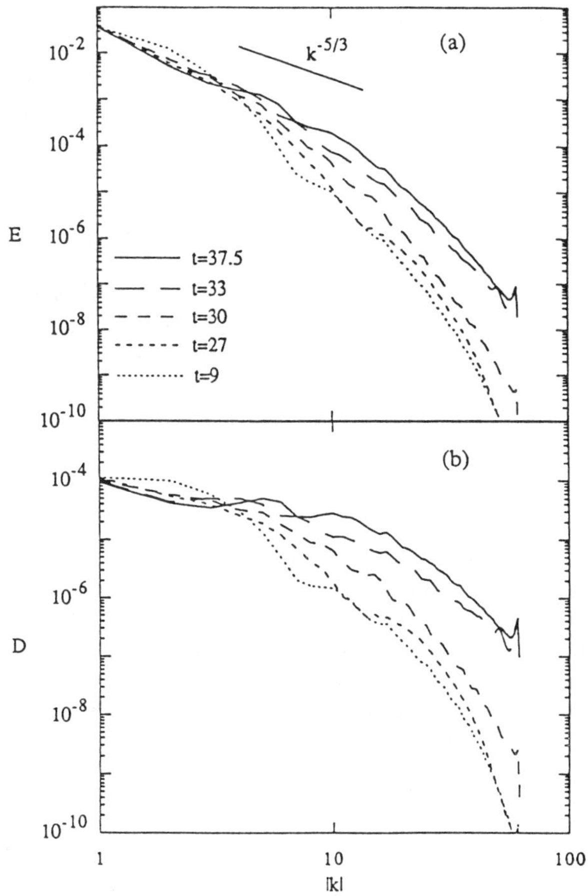

Figure 50. Evolution of (a) energy and (b) dissipation spectra during transition.

8.1 Analytical background

To construct the basis functions for HWD, we first construct an orthonormal basis in Fourier space as $\mathbf{a}(\mathbf{k}) = \frac{\mathbf{k}\mathrm{x}\mathbf{e}}{|\mathbf{k}\mathrm{x}\mathbf{e}|}$; $\mathbf{b}(\mathbf{k}) = \frac{\mathbf{k}\mathrm{x}\mathbf{a}}{|\mathbf{k}\mathrm{x}\mathbf{a}|}$; $\mathbf{c}(\mathbf{k}) = \frac{\mathbf{k}}{|\mathbf{k}|}$, where $\mathbf{e}$ is an arbitrary unit vector. Since a divergence-free vector function $\mathbf{F}(x,y,z)$ has Fourier coefficients $\hat{\mathbf{F}}(\mathbf{k})$ which satisfy $\mathbf{k}.\hat{\mathbf{F}}(\mathbf{k})=0$, $\hat{\mathbf{F}}(\mathbf{k})$ can be expressed as a linear combination of $\mathbf{a}(\mathbf{k})$ and $\mathbf{b}(\mathbf{k})$. In this way, Lesieur [83] first defined *complex helical waves,* which are a unique, Galilean invariant combination of $\mathbf{a}(\mathbf{k})$ and $\mathbf{b}(\mathbf{k})$, as

$$\mathbf{V}^{+}(\mathbf{k},\mathbf{x}) \equiv [\mathbf{b}(\mathbf{k}) - i\mathbf{a}(\mathbf{k})] \exp(i\mathbf{k}.\mathbf{x}), \tag{8-1a}$$

$$\mathbf{V}^{-}(\mathbf{k},\mathbf{x}) \equiv [\mathbf{b}(\mathbf{k}) + i\mathbf{a}(\mathbf{k})] \exp(i\mathbf{k}.\mathbf{x}), \tag{8-1b}$$

which are eigenfunctions of the curl operator corresponding to the eigenvalues $\pm|\mathbf{k}|\neq 0$, *i.e.* $\nabla\mathrm{x}\mathbf{V}^{+} = |\mathbf{k}|\mathbf{V}^{+}$ and $\nabla\mathrm{x}\mathbf{V}^{-} = -|\mathbf{k}|\mathbf{V}^{-}$. These eigenfunctions are orthogonal with respect to the inner product

$$\langle \mathbf{f},\mathbf{g}\rangle = \int \mathbf{f}\bullet\mathbf{g}^{*}\,d\mathbf{x}, \tag{8-2}$$

where '*' denotes complex conjugation and integration extends over the entire domain (or a periodic box). Using HWD, a divergence free velocity field can thus be decomposed as

$$\mathbf{u} = \int_{\mathbf{k}\neq 0} u^{+}(\mathbf{k},t)\mathbf{V}^{+}(\mathbf{k},\mathbf{x})\,d\mathbf{k} + \int_{\mathbf{k}\neq 0} u^{-}(\mathbf{k},t)\mathbf{V}^{-}(\mathbf{k},\mathbf{x})\,d\mathbf{k} + \nabla\phi = \mathbf{u}_R + \mathbf{u}_L + \nabla\phi. \tag{8-3}$$

In this equation, the first integral ($\mathbf{u}_R$) is the projection of $\mathbf{u}$ onto the vector space spanned by all eigenfunctions corresponding to positive eigenvalues ($+|\mathbf{k}|$) of the curl operator. We call this the *right-handed* component of $\mathbf{u}$ because the streamlines for $\mathbf{u}_R$ are right-handed helices. Likewise, the second integral ($\mathbf{u}_L$) is a linear combination of eigenfunctions corresponding to negative eigenvalues ($-|\mathbf{k}|$) and is designated the *left-handed* component of $\mathbf{u}$. In equation (8-3), $\nabla\phi$ represents potential flow, or equivalently the projection of $\mathbf{u}$ onto the null space of the curl operator. Similarly, for the vorticity field we have

$$\omega = \int_{\mathbf{k}\neq 0} |\mathbf{k}|\, u^{+}(\mathbf{k},t)\mathbf{V}^{+}(\mathbf{k},\mathbf{x})\,d\mathbf{k} - \int_{\mathbf{k}\neq 0} |\mathbf{k}|\, u^{-}(\mathbf{k},t)\mathbf{V}^{-}(\mathbf{k},\mathbf{x})\,d\mathbf{k} = \omega_R + \omega_L \tag{8-4}$$

where it is assumed that the fluid is not rotating at infinity. Computationally, HWD can be easily performed in Fourier space as

$$\hat{\mathbf{u}}_R(\mathbf{k},t) = \tfrac{1}{2}[\hat{\mathbf{u}}(\mathbf{k},t) + i\mathbf{c}(\mathbf{k})\times\hat{\mathbf{u}}(\mathbf{k},t)], \tag{8-5a}$$

$$\hat{\mathbf{u}}_L(\mathbf{k},t) = \tfrac{1}{2}[\hat{\mathbf{u}}(\mathbf{k},t) - i\mathbf{c}(\mathbf{k})\times\hat{\mathbf{u}}(\mathbf{k},t)], \tag{8-5b}$$

$$\hat{\omega}_R(\mathbf{k},t) = i\mathbf{k}\times\hat{\mathbf{u}}_R(\mathbf{k},t), \tag{8-6a}$$

$$\hat{\omega}_L(\mathbf{k},t) = i\mathbf{k}\times\hat{\mathbf{u}}_L(\mathbf{k},t). \tag{8-6b}$$

The physical space HWD components are then readily obtained by inverse Fourier transform.

8.2 Physical properties of HWD

It turns out that HWD is not only an effective analytical tool, but it also exhibits some important, physically relevant features. For instance, each individual eigenmode is itself a

Beltrami flow (*i.e.* parallel velocity and vorticity vectors), and any linear combination of eigenmodes corresponding to the same eigenvalue is also a Beltrami flow. In addition, a flow's total helicity can be expressed as

$$H=\int \omega\cdot\mathbf{u}\,d\tau = \int \omega_R\cdot\mathbf{u}_R\,d\tau + \int \omega_R\cdot\mathbf{u}_L\,d\tau + \int \omega_L\cdot\mathbf{u}_R\,d\tau + \int \omega_L\cdot\mathbf{u}_L\,d\tau + \int \omega_R\cdot\nabla\phi\,d\tau$$

$$= \int \omega_R\cdot\mathbf{u}_R\,d\tau + \int \omega_L\cdot\mathbf{u}_L\,d\tau = H_R + H_L, \tag{8-7}$$

where $d\tau$=dxdydz and the three integrals vanish because of the orthogonality of the curl operator eigenfunctions. Similarly, other global quantities such as total energy and enstrophy can be decomposed into left- and right-handed, Galilean invariant components.

8.3 HWD of a columnar vortex

To illustrate the physical nature of the two polarized components extracted by HWD, we first consider a rectilinear Lamb vortex $\omega(r) = e^{-r^2}\,\mathbf{e}_z$. In this simple case, HWD can be performed analytically. As illustrated in figure 51, the right-handed component has right-handed helical vortex lines, while the left-handed component consists of vortex lines with a left-handed helical twist. The corresponding polarized velocity fields are

$$\mathbf{u}_R = \frac{1}{2}e^{-r^2}\left(\frac{e^{r^2}-1}{2r}\mathbf{e}_\theta + \mathbf{e}_z\right) \tag{8-8a}$$

$$\mathbf{u}_L = \frac{1}{2}e^{-r^2}\left(\frac{e^{r^2}-1}{2r}\mathbf{e}_\theta - \mathbf{e}_z\right). \tag{8-8b}$$

A notable feature of these expressions is that the helical vortex lines propagate in opposite directions along the vortex column, a key predictive feature for flows in which one component dominates locally, *e.g.* core dynamics.

8.4 Interpretation of core dynamics using HWD

Unique insight into core dynamics is provided by the evolution of the polarized vorticity components in figure 52. Initially ω_R and ω_L have the same distribution, as for the rectilinear vortex above, since the core vortex lines are untwisted and no meridional flow is present initially. Note that $|\omega_R|$ is antisymmetric to $|\omega_L|$ in z with respect to z=l at all times, and in frame (i), the local minima of $|\omega_R|$ and $|\omega_L|$ coincide where the low-enstrophy bubble is observed. Helical vortex line twisting is reflected by the local dominance of one polarized component; where ω_R dominates, the twist is right-handed and vice versa. As seen in figure 52, the evolution of the polarized vorticity is quite simple, namely, the ω_R wavepacket moves to the left, while ω_L travels to the right. Thus, HWD extracts distinct wavepackets of polarized vorticity moving in opposite directions on the vortex — a striking result in light of the nonlinearity of the Navier-Stokes equations.

A particularly interesting feature of figure 52 is that the right-handed component moves to *left*. This is in contrast to the rightward motion of the right-handed component in isolation, shown in figure 53 using the right-handed component of the initial condition in figure 52. In order to study the nonlinear interactions between ω_R and ω_L responsible for the opposite translation direction with both components present, we introduce two projection operators P^+ and P^-. Here, P^+ is defined such that it extracts the right-handed component of a given vector field, while P^- produces the left-handed component. Using these projection operators, the vorticity transport equation can be split into two coupled evolution equations for ω_R and ω_L:

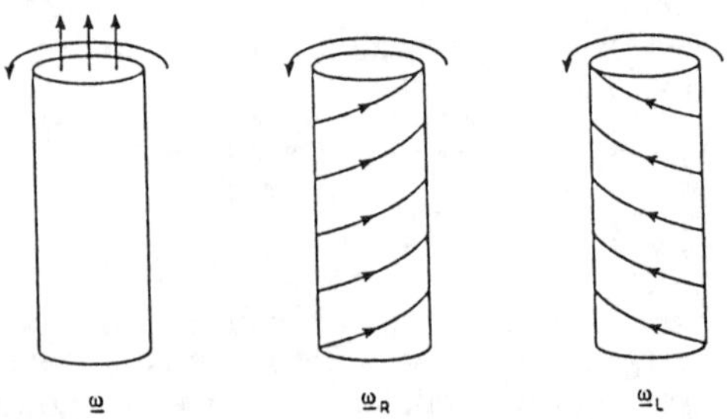

Figure 51. Schematic of the left- and right-polarized vorticity components of a rectilinear vortex.

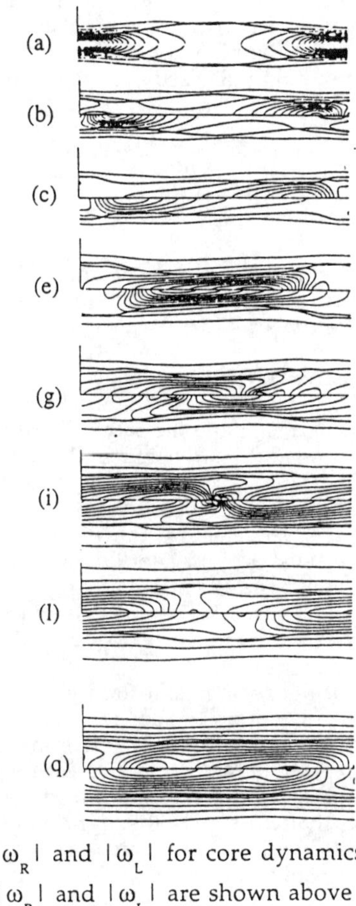

Figure 52. Evolution of $|\omega_R|$ and $|\omega_L|$ for core dynamics at times corresponding to those in figure 36. In each frame $|\omega_R|$ and $|\omega_L|$ are shown above and below the axis respectively. To indicate core size, the unity contour of the kinematic vorticity number is shown as a bold line in each frame.

$$\frac{\partial \omega_R}{\partial t} = - \nabla \times P^+ (\omega \times \mathbf{u}) + \frac{1}{Re} \Delta \omega_R \tag{8-9a}$$

$$\frac{\partial \omega_L}{\partial t} = - \nabla \times P^+ (\omega \times \mathbf{u}) + \frac{1}{Re} \Delta \omega_L . \tag{8-9b}$$

Here, we have used the fact that the Laplacian and the curl operator commute with the projection operators. In order to reveal the dynamics of polarized vorticity, we rewrite (8-9a) as

$$\begin{aligned}
\frac{\partial \omega_R}{\partial t} &= - \nabla \times (\omega_R \times \mathbf{u}_R) && \text{(term 1)}\\
&+ \nabla \times P^- (\omega_R \times \mathbf{u}_R) && \text{(term 2)}\\
&- P^+ [(\mathbf{u}_L \bullet \nabla)\omega_R] && \text{(term 3)}\\
&+ P^+ [(\omega_R \bullet \nabla)\mathbf{u}_L] && \text{(term 4)}\\
&- \nabla \times P^+ (\omega_L \times \mathbf{u}) && \text{(term 5)}\\
&+ \frac{1}{Re} \Delta \omega_R . && \text{(term 6)}
\end{aligned} \tag{8-10}$$

The analogous equation for ω_L is obtained by replacing 'R' by 'L' and P^+ by P^-. The first term in equation (8-10) represents the inviscid self-evolution of ω_R. The second term is the rate of generation of left-handed vorticity by the evolution of right-handed vorticity; examination of figure 53 suggests that this effect is small. Note that the sum of the first two terms equals $\nabla \times P^+ (\mathbf{u}_R \times \omega_R)$. The third term is the right-handed contribution to $\partial \omega_R / \partial t$ from advection of ω_R by the left-handed velocity field. The fourth term is the right-handed contribution from stretching of ω_R by the left-handed velocity field. The fifth term represents generation of right-handed vorticity by the evolution of ω_L; this includes generation of ω_R by the self-evolution of ω_L, the right-handed contribution from the stretching and advection of ω_L by the right-handed velocity field $\mathbf{u}_L$. Finally, the sixth term is simply the diffusion of ω_R.

To assess the relative importance of each effect represented in equation (8-10), we have calculated the scalar product of each of the first five terms with ω_R for the evolution in figure 52, where both left- and right-handed components are present initially. A few time frames are shown in figure 54, which are representative of the entire evolution. The order of importance of the terms is as follows (listed in decreasing magnitude): (i) transport by the velocity field of opposite polarity (term 3), (ii) self-evolution (term 1), (iii) transfer from ω_L (term 5), (iv) stretching of ω_R by $\mathbf{u}_L$ (term 4), and (v) the generation of left-handed vorticity by the self-evolution of ω_R (term 2), which is always negligible.

The spatial distributions of the third and fifth terms have a striking qualitative resemblance, as seen by comparing the corresponding frames of figure 54. Although the physical reason for this similarity is not yet clear, we can qualitatively describe the sum of the third and fifth terms as an advection in an enhanced left-handed velocity field. Hence, the evolution of ω_R is essentially a consequence of only two effects: self-evolution and advection in an enhanced left-handed velocity field. The self-evolution attempts to move the ω_R packet to the right by adding enstrophy at the front and subtracting it at the back of the packet (see term 1 distribution in figure 54). Note that the speed of the self-induced packet motion is significantly slower than the velocity field induced by the packet. On the other hand, the left-handed velocity advects the ω_R packet to the left at the speed of the $\mathbf{u}_L$ field. The fifth term enhances this leftward motion of the packet. Since the rightward self-induced motion of the ω_R packet is slower than the leftward advection by $\mathbf{u}_L$, the net result is leftward packet drift —

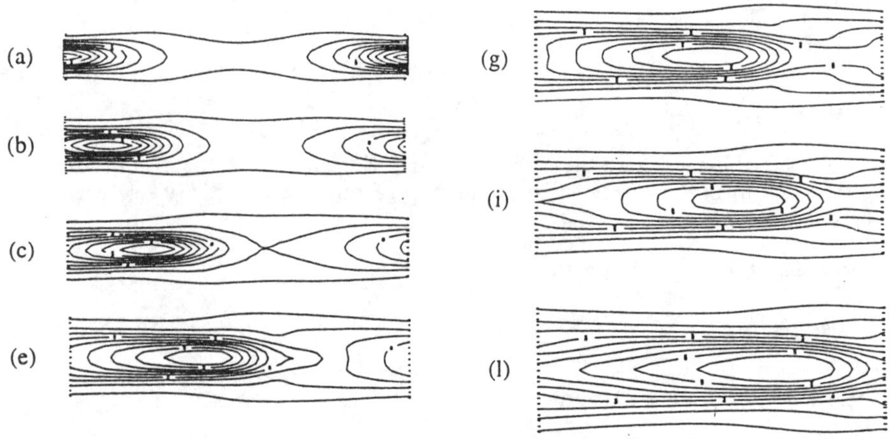

Figure 53. Isolated evolution of a $|\omega_R|$ wavepacket, at times corresponding to those in figure 52.

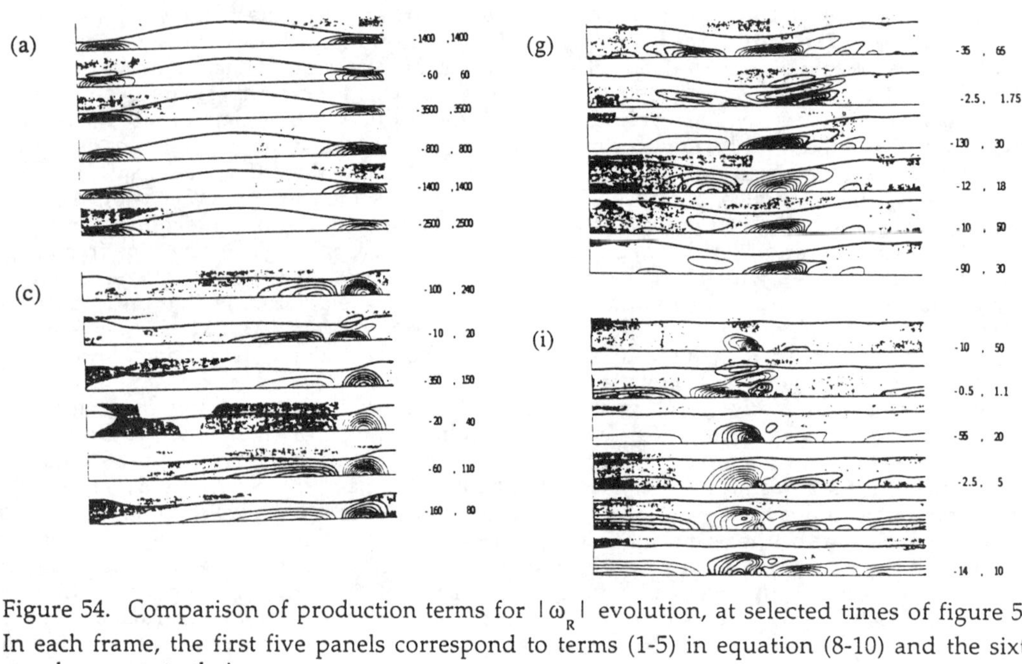

Figure 54. Comparison of production terms for $|\omega_R|$ evolution, at selected times of figure 52. In each frame, the first five panels correspond to terms (1-5) in equation (8-10) and the sixth panel represents their sum.

despite the rightward self-induced motion of the ω_R packet. Hence, advection by the velocity field of opposite polarity is the cause of the backward drift of the polarized packets in figure 52. The bubble that forms in front (with respect to the packet's propagation direction) of each polarized packet is also a result of advection by the velocity field of opposite polarity. Because the axial component of this velocity field is fastest on the axis, it pushes the packet backwards at a faster rate near the axis than away from it; consequently, a low-enstrophy bubble occurs in the composite field when the two polarized low-vorticity bubbles overlap.

This HWD approach provides a unique, rather simple interpretation of vortex core dynamics in terms of polarized vorticity wavepacket motion. Furthermore, results in the next section suggest that CS in turbulent flows are not only vortical, but typically also polarized; thus, HWD appears quite promising as a general interpretative tool, capable of providing fresh insight into turbulence physics and its modeling.

9. Interaction of CS with background turbulence

Up to this point, we have studied the evolutionary dynamics of CS in 'clean' environments free from background small-scale turbulence to clearly reveal the large-scale vortex dynamics. From a different viewpoint, the current physical-space knowledge of the interaction between CS and small-scale turbulence is quite limited, thus hampering attempts to model the influence of small scales. Invariably, theories of turbulent flows assume *local isotropy* of the small scales, a controversial assumption for flows dominated by CS. For instance, Hussain [84] speculates that in spite of wide scale separation, large- and fine-scale motions are intimately coupled, so that the presence of CS may prevent local isotropy and thus partially invalidate traditional statistical turbulence models.

With regard to current models of the CS/small-scale interaction, there is a fundamental conceptual difficulty. On one hand, CS generate a local shear, which, if sufficiently strong, can sustain nearby smaller-scale vortices via alignment and stretching of fine-scale vorticity (*i.e.* shear production). On the other hand, these small-scale vortices may significantly influence the evolution and internal dynamics of the CS itself. An example of this phenomenon is the axisymmetrization of a noncircular vortex in 2D turbulence [85], where small scales (filaments generated by the vortex) cause the large-scale vortex to become axisymmetric. Additionally, self-organization of small scales into increasingly larger scales – inverse cascade via inviscid merger – has been observed in 2D turbulence [86]. Thus, models of the small-scale influence involving only an enhanced turbulent diffusion (*e.g.* eddy viscosity, mixing length, and even sophisticated renormalization group models) obviously cannot capture these effects.

One might argue that such dynamics in 2D turbulence will be rare in fully 3D flows, where vorticity in all directions is present. However, we will show that a CS in a fully 3D flow can impose similar directional constraints on nearby smaller scales, which does in fact permit small-scale mergings. In this section, these and other aspects of CS/small-scale interaction are examined critically using carefully designed initial conditions and high resolution DNS. For additional results, see [87].

9.1 Initial conditions

To mimic CS-turbulence interaction, we initialize a single nonuniform core vortex (discussed in §6 above) as the CS, surrounded by a background flow initially consisting of random, fine-scale isotropic velocity fluctuations. In isolated evolution, the CS entails no self-induced transverse displacement of the vortex axis (as shown in §6); the CS dynamics consist entirely of core dynamics. This idealized setup allows us to focus directly on the large-

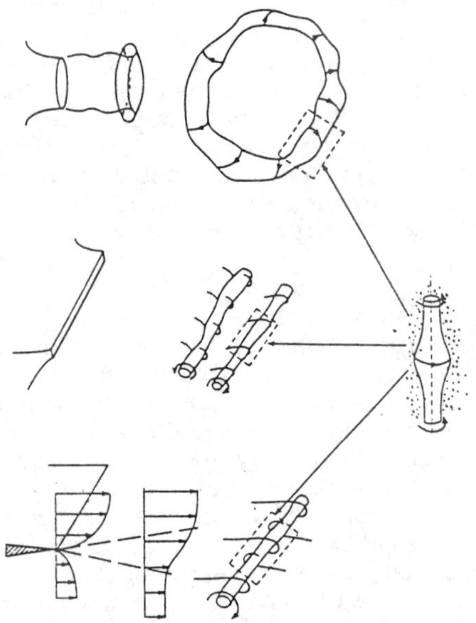

Figure 55. Sketch of a segment of CS in some prototypical shear flows: (a) circular jet; (b) plane jet; (c) mixing layer.

scale/small-scale coupling and is also an effective local model of CS in turbulent shear flows such as jets, wakes, and mixing layers (figure 55).

The initial turbulent background is isotropic, homogeneous, and consists of scales within a certain wavenumber interval, *i.e.* $|\mathbf{k}| \in [k_l, k_h]$, with random phases and a uniform probability distribution. The disturbance amplitudes are random with a smooth (C^∞) probability distribution and a normalization factor to provide the specified fluctuation level u'_{rms}. We select k_l such that there is clear gap in the energy spectrum between the large scales associated with the CS and the small scales of the background turbulence; the largest wavenumber k_h is determined by the maximum DNS resolution.

9.2 Small-scale organization

The evolution of the incoherent turbulence is illustrated in figure 56 by the vorticity ω_n, normal to a plane $\pi_{||}$ containing the axis of the CS (hence a meridional plane). Initially, the coherent vorticity does not contribute to ω_n (figure 56a) since there is no azimuthal vorticity within the CS at t=0. Thus, the size of regions in $\pi_{||}$ where ω_n is of one sign reveal visually the scales associated exclusively with the initial incoherent vorticity (see figure 56a). Subsequently, the CS acquires helical vortex lines, thereby contributing to ω_n with an odd symmetry about the axis and a larger scale than the incoherent background (figures 56a-q). It is important to note that the absence of intermediate-scale turbulence in the initial conditions ensures that the scale growth observed cannot be purely diffusive.

Although plots of ω_n effectively display the small-scale growth, the large-scale CS evolution is best illustrated by contours of $|\omega|$ in $\pi_{||}$ (figure 57). Initially, the incoherent vorticity is strong enough to mask the CS in figure 57(a). Nevertheless, away from the CS (where its shear is weak), the incoherent vorticity decreases drastically by enstrophy dissipation, so that the CS quickly becomes discernible (figure 57b). While the scales of the incoherent vorticity grow everywhere in the computational domain (figure 56), a large incoherent vorticity magnitude is sustained only immediately near the CS (figures 57g,i). The fact that the incoherent scale — unlike the vorticity magnitude — appears uniform throughout the flow in figure 56 is a consequence of the ratio of the CS size to the box size. In reality, incoherent vorticity far removed from the CS will not be influenced by the shear of the vortex and hence will not undergo such organization and scale increase.

In a relatively thin layer surrounding the coherent vortex, the shear induced by the CS causes small-scale vorticity to wind around it in a spiral fashion. This is clear from plots of $|\omega|$ in a cross-section perpendicular to the vortex axis (figure 58) and in the 3D isosurface plot in figure 59. The process of small-scale organization begins with this azimuthal alignment of the initially random vorticity in a layer surrounding the vortex. This alignment is obviously an inviscid effect since it is driven by the irrotational shear outside the coherent vortex core. Once the vorticity in the surrounding layer acquires a preferred direction (roughly parallel to the azimuthal direction), the small-scale dynamics are subjected to a geometrical constraint similar to that of 2D turbulence. Therefore, we will now investigate the possibility of pairing of small-scale structures as a possible inviscid mechanism for organization and scale growth. We must carefully distinguish between this inviscid mechanism and simple viscous decay, which obviously can also produce scale growth.

By meticulously following the evolution of individual small-scale structures in the surrounding layer, we have found many events that closely resemble pairings. A striking example is figure 60, which shows an expanded view of the amalgamation of two structures (identified by arrows in figure 57g) into a single vortical structure (indicated by an arrow in figure 57i). It is important to note that the cross-section of figure 60 is fixed in space, while the small structures are actually swirling around the CS; consequently, the pairing event appears

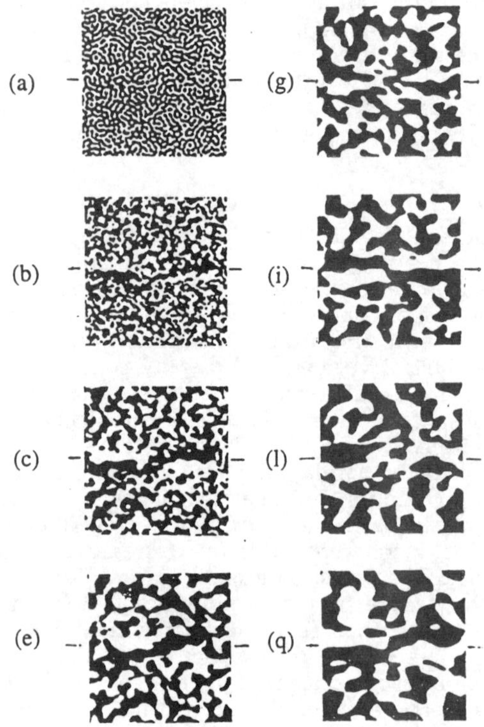

Figure 56. Evolution of positive and negative regions of ω_n in a plane containing the vortex axis, marked on the side of each frame. The frame labeling reflects the nonuniform time difference between frames; this labeling scheme is used throughout this section.

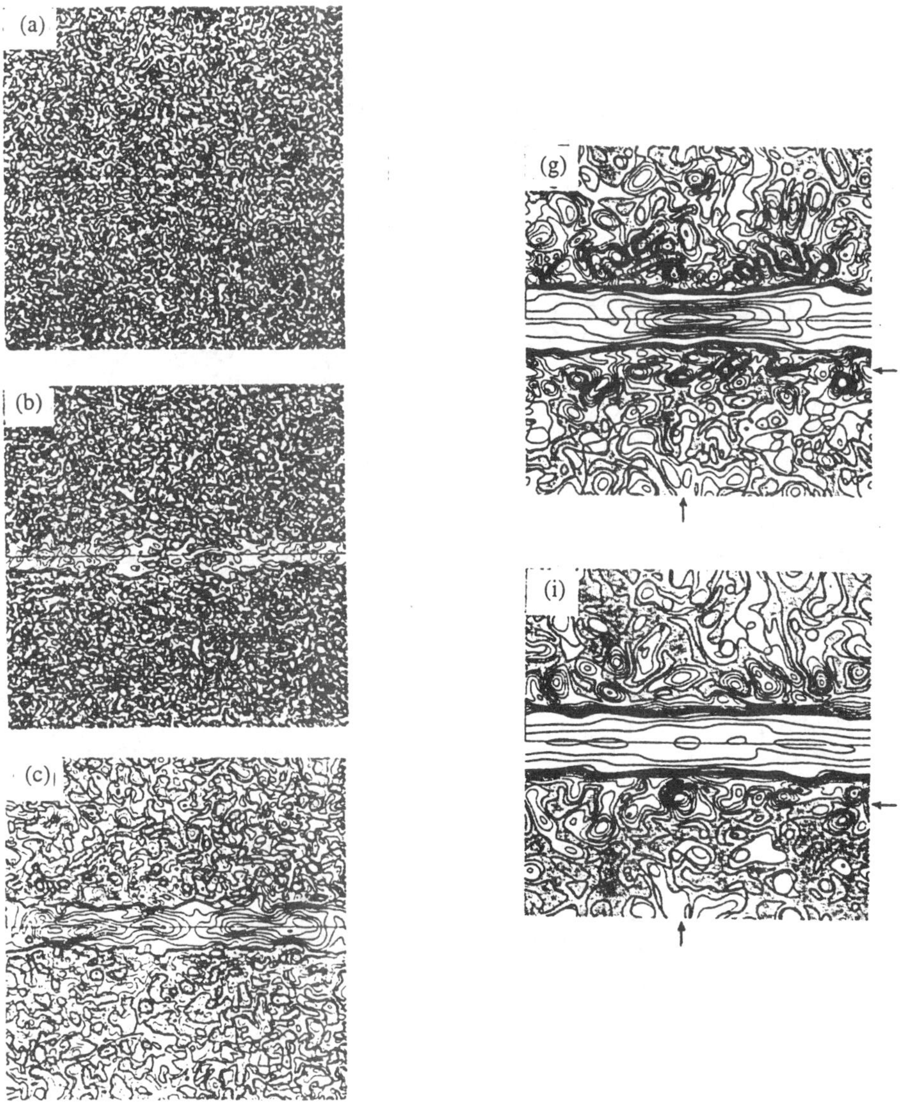

Figure 57. Evolution of $|\omega|$ in a plane containing the vortex axis at selected times of figure 56.

Figure 58. Cross-sectional contours of $|\omega|$ at stage (g). The lowest contour levels are shaded gray.

Figure 59. Isovorticity surface at stage (q), indicating secondary thread vortices surrounding the primary CS.

to be slower than it actually is. To ensure that this is indeed an inviscid pairing phenomenon which is only smoothed by viscous diffusion, we initialized a simulation with the flow field in figure 60(g), but at a much higher Re (with a "super-viscosity" viscous term), and found similar pairings and scale growth as in figure 60(i). This is clear evidence that the small-scale organization is an inviscid phenomenon driven by mergings of like-signed, smaller-scale vortices ('threads').

Under conditions of statistical equilibrium (not strictly true for the slowly decaying flow under consideration), this inverse cascade must be balanced by effects which generate new smaller scales in the surrounding layer. Many physical space scenarios of forward cascade are possible, such as (i) filamentation from the thread cores, (ii) peeling of vorticity from the CS by the threads, (iii) shedding of vorticity from the CS, and (iv) simple vortex stretching of threads by the coherent vortex.

The observed growth of incoherent scales was found to be typical over a wide range of k_I and u'_{rms}. However, as will be shown, the effect of the background turbulence on the evolution of the CS itself depends strongly on both the initial excitation level and the coherent Reynolds number ($Re_c \equiv \Gamma/\nu$, where Γ is the CS circulation).

9.3 Polarized internal intermittency

The spiral vortical structures surrounding the coherent vortex in figure 59 constitute an interesting new type of internal intermittency, consisting predominantly of axial vorticity within the CS core and strong azimuthal vorticity in a surrounding layer. Note that the spiral structures are threadlike rather than sheetlike, have high helicity density, and are highly polarized (in context of HWD, discussed in §8). HWD, applicable to both laminar and turbulent flows, reveals an important feature of the spiral threads; namely, individual structures tend to be either predominantly right-handed or left-handed (figure 61). When $|\omega_R|$ and $|\omega_L|$ surfaces are overlaid (figure 61), the polarized components clearly occupy distinct locations in the thread-containing layer. This indicates that spatial separation of the polarized components is an intrinsic feature of the small-scale organization process.

The degree of thread polarization, quantified as

$$R(x,y,z) \equiv \log_2 [\,|\omega_R(x,y,z)| \,/\, |\omega_L(x,y,z)|\,], \tag{9-1}$$

varies from -5 to 5 by the end of the simulation, indicating very strong local polarization. A careful comparison of the organized small scales and R shows that in the thread layer, the peaks of $|\omega|$ and R match closely. This correspondence is nontrivial in the sense that the left- or right-handed vorticity can be large even in a region where the total vorticity is small.

Interestingly, the right-handed threads' vorticity is aligned with the swirl, while the left-handed vorticity is antiparallel to the swirl direction, as indicated in figure 62(a). We can understand this phenomenon qualitatively by considering vector lines of ω_R and ω_L near the coherent vortex (note the distinction with vortex lines of ω). Inside the CS core, both ω_R and ω_L have a positive z-component because the CS has a positive circulation. Since the ω_R lines are locally right-handed helices, we can roughly expect an ω_R line which enters or exits the core to have a positive ω_θ component, as sketched in figure 62(b). By the same argument, ω_L lines should have negative ω_θ (figure 62c). Hence, we expect a tendency for

$$\omega_R \cdot \mathbf{e}_\theta > 0 \wedge \omega_L \cdot \mathbf{e}_\theta < 0 \tag{9-2}$$

in the thread layer; here $\mathbf{e}_\theta$ is a unit vector in the azimuthal direction. Figures 62(d,e) show that the above argument has merit, but with some exceptions. At some distance from the coherent vortex, there is no general trend, which is hardly surprising given the turbulent nature of the flow. More significantly, localized coiling of vortex lines around the thread axis can cause exceptions to the expected trend, as illustrated in figure 62(f).

Figure 60. Enlarged view of a pairing event, illustrated by contours of ω_n in a plane containing the vortex axis. Negative values are shaded, and this enlarged area contains the two threads indicated by arrows in figure 57(g,i).

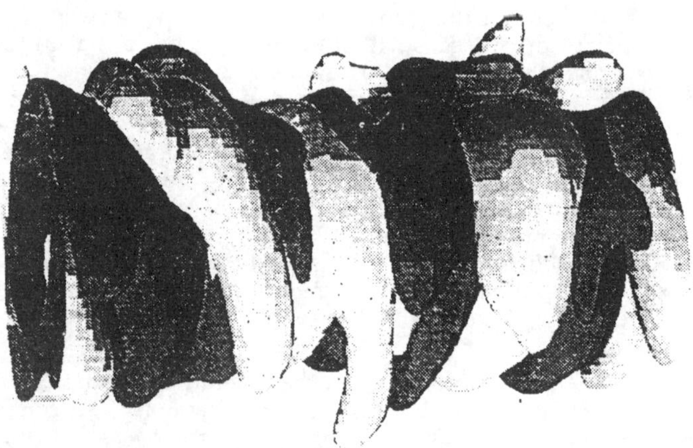

Figure 61. Isosurfaces of (a) right-handed (light) and (b) left-handed (dark) vorticity, illustrating the strong polarization of the CS and threads.

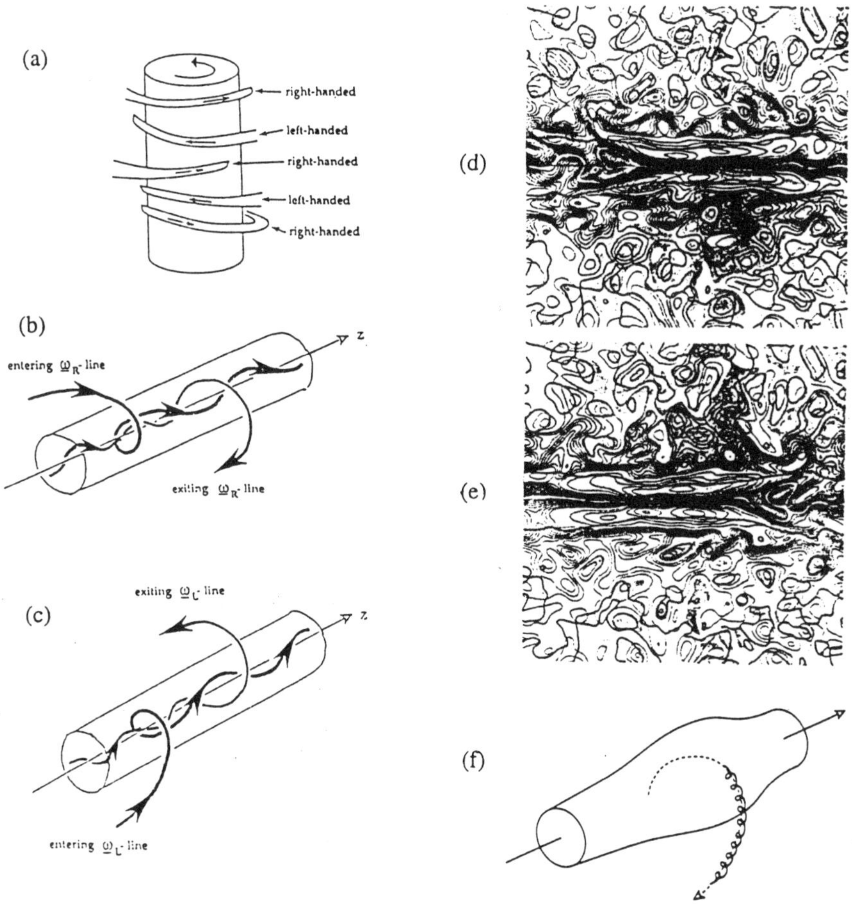

Figure 62. Sketches of (a) general orientation of the left- and right-handed small-scale vortical structures surrounding the coherent vortex; (b) right-handed vorticity vector lines; (c) left-handed vorticity component vector lines. (d,e) Contours of the normal component of the right- and left-handed vorticities respectively. (f) Schematic of localized vector line coiling.

9.4 Backscatter of small-scales on CS

The flow studied here is unforced (*i.e.* containing no background shear) and hence eventually relaminarizes on a time scale which depends on Re_c and also on the initial turbulent fluctuation level. We have observed three distinct regimes of evolution: (i) for low fluctuation levels, the CS recovers its axisymmetry completely, and although secondary structures do exist in the surrounding layer, their vorticity continuously decreases; (ii) for sufficiently strong fluctuations, the vortex is unable to survive, *i.e.* no signs of a coherent vortex appear even after long evolution; (iii) for intermediate fluctuation levels (as studied above), the CS survives and turbulence (*i.e.* spiral thread vortices) is sustained for a significant time.

In regime (iii), bending waves of the CS are excited, which influence the turbulence sustaining mechanism. As the threads align azimuthally around the vortex, their radial vorticity component diminishes and their vortex stretching thus decreases (*e.g.* a concentric vortex ring around a line vortex experiences no stretching). Upon excitation of bending waves, the CS position is continuously altered relative to the threads, thus permitting additional thread stretching.

To illustrate that the threads are responsible for these bending waves, we used the flow of figure 57(i) as an initial condition for two simulations at substantially higher Re; one case with a factor of 4 higher Re_c and another run with a super-viscosity term. This drastic reduction in viscous effects causes the small-scale threads to become more energetic since their energy losses from viscous dissipation are reduced. Consequently, the threads' vorticity magnitude increases and turbulence reappears. This has implications for the generation of bending waves, which can be excited by the induction from energetic small-scale structures. By comparing isovorticity surfaces at the end of these high Re simulations (figures 63a,b) with the initial condition (figure 63c), it is clear that the reenergized small scales have substantially bent the vortex axis. This phenomenon is the sole result of feedback (or backscatter) from small scales on the coherent vortex, an effect which obviously cannot be modeled by an enhanced eddy viscosity.

9.5 Conjecture regarding local anisotropy

Popular turbulence theories without exception invoke the assumption of local isotropy (*i.e.* no statistically preferred vorticity direction) for intermediate scales in the inertial subrange. Consequently, at high Reynolds number, anisotropy is permitted in this approach only for larger energy-containing flow scales (*i.e.* CS or integral scales). Note that the assumption of local isotropy does not imply that structures do not exist in the inertial subrange, nor does it imply anything about the shape of such structures. What it does imply, however, is that the orientation of small-scale structures is statistically random. In this sense, local isotropy is indeed a strong assumption, which, however, is very convenient for calculating the energy transfer between different scales. In essence, local isotropy implies a decoupling of the dynamics of large and fine scales. However, our results demonstrate that the traditional isotropy assumption is seriously flawed in this case, since: (i) the CS induces anisotropy of the surrounding small scales and (ii) the small scales strongly affect the CS evolution itself.

Based on our results, we now propose an alternative, inherently anisotropic description of small scales and a physical-space picture of cascade in turbulent flows dominated by CS. Obviously, the large scales are anisotropic due to the CS; this anisotropy can cascade to smaller scales via the following scenario. The coherent vortex first causes the formation of secondary vortical structures (threads) in the surrounding layer through alignment and subsequent amalgamation of small-scale vorticity. These secondary structures may then dominate the strain rate field in their immediate vicinity, provided that they are sufficiently strong. If the

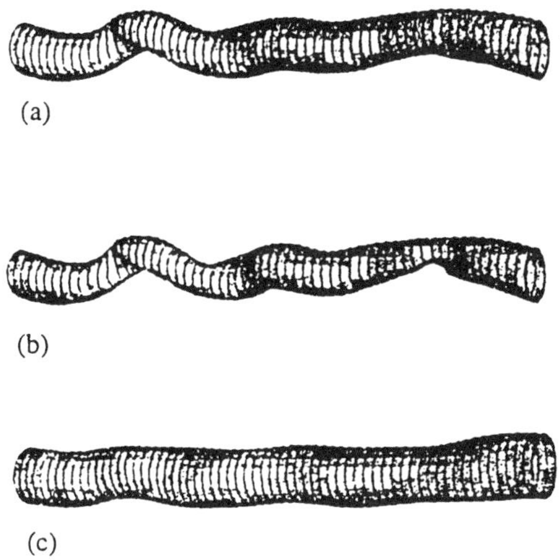

Figure 63. 65% $|\omega|$ surfaces for (a) factor of four higher Re and (b) superviscosity term, showing excitation of bending waves from the initial condition in (c).

Reynolds number is large enough to allow the existence of even smaller scales, we can then imagine tertiary vortices wound around the secondary threads. This line of reasoning suggests a hierarchy of increasingly smaller scale structures in a layer surrounding the CS, in which threads of a given scale are organized azimuthally around threads of the next larger scale. For modeling, this physical-space scale distribution could perhaps be represented in a tractable way using self-similar fractals.

Acknowledgements

Some of this research is part of our continuing collaboration with Prof. M.V. Melander. We are grateful to Dr. J. Jeong, Dr. D. Virk, and S. Narayanan for careful reviews of this manuscript. Preparation of this manuscript was supported by the National Science Foundation under grant CTS-9214818, the Air Force Office of Scientific Research under grant F49620-92-J-0200, and the Office of Naval Research under grant N00014-89-J-1361. Computer time for the simulations reported here was supplied by the NAS facility of NASA Ames Research Center.

References

1. Kline, S.J., Reynolds, W. D., Schraub, F.A. & Runstadler, P.W. 1967 The structure of turbulent boundary layers. *J. Fluid Mech.* **30**, 741.
2. Crow, S.C. & Champagne, F.H. 1971 Orderly structure in jet turbulence. *J. Fluid Mech.* **48**, 547.
3. Brown, G.L. & Roshko, A. 1974 On density effects and large structure in turbulent mixing layers. *J. Fluid Mech.* **64**, 775.
4. Laufer, J. 1983 Deterministic and stochastic aspects of turbulence. *J. Appl. Mech.* **50**, 1079.
5. Hussain, F. 1979 Investigations of coherent structures in free turbulent shear flows. *Symposium on Turbulence*, Univ. of Missouri-Rolla, p.207. Science Press.
6. Lumley, J. 1981 Coherent structures in turbulence. In *Transition and Turbulence* (ed. R. E. Meyer), p. 215. Academic.
7. Cantwell, B. 1981 Organized motion in turbulent flow. *Ann. Rev. Fluid Mech.* **13**, 457.
8. Fiedler, H. 1988 Coherent structures in turbulent flows. *Prog. Aerospace Sci.* **25**, p231.
9. Liu, J.T.C. 1989 Coherent structures in transitional and turbulent free shear flows. *Ann. Rev. Fluid Mech.* **21**, 285.
10. Robinson, S.K. 1991 Coherent motions in the turbulent boundary layer. *Ann. Rev. Fluid Mech.* **23**, 601.
11. Reynolds, W.C. & Hussain, F. 1972 The mechanics of an organized wave in turbulent shear flow. Part 3. Theoretical models and comparisons with experiment. *J. Fluid Mech.* **54**, 263.
12. Poje, A.C. & Lumley, J.L. 1995 A model for large-scale structures in turbulent shear flows. *J. Fluid Mech.* **285**, 349.
13. Goldshtik, M & Hussain, F. 1995 Structural approach to the modeling of a turbulent mixing layer. *Phys. Rev. E* (to appear).
14. Aubry, N., Holmes, P., Lumley, J.L. & Stone, E. 1988 The dynamics of coherent structures in the wall region of a turbulent boundary layer. *J. Fluid Mech.* **192**, 115.
15. Prabhu, A. & Narasimha, R. 1972 Turbulent non-equilibrium wakes. *J. Fluid Mech.* **54**, 19.
16. Chomaz, J.M., Huerre, P. & Redekopp, L.G. 1988 Bifurcations to local and global modes in spatially-developing flows. *Phys. Rev. Lett.* **60**, 25.

17. Frisch, U., Sulem, P.-L. & Nelkin, M. 1978 A simple dynamical model of intermittent fully developed turbulence. *J. Fluid Mech.* **87**, 719.
18. Broze, G. & Hussain, F. 1994 Nonlinear dynamics of forced transitional jets: periodic and chaotic attractors. *J. Fluid Mech.* **263**, 93.
19. Hussain, F. 1980 Coherent structures and studies of perturbed and unperturbed jets. *Lecture Notes in Physics* **136**, 252.
20. Hussain, F. & Melander, M.V. 1991 Understanding turbulence via vortex dynamics. In *Recent Developments in Turbulence* (eds. S. Sarker *et al.*), p. 157. Springer.
21. Hussain, F. 1981 Role of coherent structures in turbulent shear flows. *Proc. Indian Acad. Sci. (Engg. Sci.)* **4**, 129.
22. Bridges, J., Husain, H. & Hussain, F. 1989 Whither coherent structures? In *Whither Turbulence? Turbulence at the Crossroads* (ed. J. Lumley), p. 132. Springer.
23. Kida, S., Takaoka, M. & Hussain, F. 1991 Collision of two vortex rings. *J. Fluid Mech.* **230**, 583.
24. Meng , H. & Hussain, F. 1991 Holographic particle velocimetry: a 3D measurement technique for vortex interactions, coherent structures and turbulence. *Fluid Dyn. Res.* **8**, 33.
25. Meng, H. & Hussain, F. 1995 Instantaneous flow field in an unstable vortex ring measured by holographic particle velocimetry. *Phys. Fluids*, **7**, 9.
26. Adrian, R.J. 1991 Particle imaging techniques for experimental fluid mechanics. *Ann. Rev. Fluid Mech.* **23**, 261.
27. Dahm, W.J.A., Su, L.K. & Southerland, K.B. 1992 A scalar imaging velocimetry technique for fully resolved four-dimensional vector velocity field measurements in turbulent flows. *Phys. Fluids A* **7**, 2191.
28. Schoppa, W., Hussain, F. & Metcalfe, R.W. 1995 A new mechanism of small-scale transition in a plane mixing layer: core dynamics of spanwise vortices. *J. Fluid Mech.* (in press)
29. Melander, M.V. & Hussain, F. 1994 Core dynamics in a vortex column. *Fluid Dyn. Res.*, **13**, 1.
30. Melander, M.V. & Hussain, F. 1993 Polarized vorticity dynamics on a vortex column. *Phys. Fluids A* **5**, 1992.
31. Melander, M.V., Hussain, F. & Basu, A. 1991 Breakdown of a circular jet into turbulence. *Turb. Shear Flows 8*, p. 15.5.1-15.5.6. Munich.
32. Hussain, F. 1986 Coherent structures and turbulence. *J. Fluid Mech.* **173**, 303.
33. Hussain, F. 1983 Coherent structures — reality and myth. *Phys. Fluids.* **26**, 2816.
34. Cantwell, B. & Coles, D. 1983 An experimental study of entrainment and transport in the turbulent near wake of a circular cylinder. *J. Fluid Mech.* **136**, 321.
35. Hussain, F. & Zaman, K.B.M.Q. 1980 Vortex pairing in a circular jet under controlled excitation. Part 2. Coherent structure dynamics. *J. Fluid Mech.* **101**, 493.
36. Kleis, S.J., Hussain, F. & Sokolov, M. 1981 A 'turbulent spot' in an axisymmetric free shear layer. Part 3. Azimuthal structure and initiation mechanism. *J. Fluid Mech.* **111**, 87.
37. Tso, J. & Hussain, F. 1989 Organized motions in a fully developed turbulent jet. *J. Fluid Mech.* **203**, 425.
38. Hussain, F. & Hayakawa, M. 1987 Eduction of large-scale structures in a turbulent plane wake. *J. Fluid Mech.* **180**, 193.
39. Hayakawa, M. & Hussain, F. 1989 Three-dimensionality of organized structures in a plane turbulent wake. *J. Fluid Mech.* **206**, 375.

40. Metcalfe, R.W., Hussain, F., Menon, S. & Hayakawa, M. 1987 Coherent structures in a turbulent plane mixing layer: a comparison between direct numerical simulations and experiments. *Turb. Shear Flows* **5**, 110.
41. Hussain, F. & Zaman, K.B.M.Q. 1985 An experimental study of organized motions in the turbulent plane mixing layer. *J. Fluid Mech.* **159**, 85.
42. Hussain, F. 1983 Coherent stuctures and incoherent turbulence. In *Turbulence and Chaotic Phenomena in Fluids* (ed T. Tatsumi), p. 453. North-Holland.
43. Metcalfe, R.W., Orszag, S.A., Brachet, M.E., Menon, S. & Riley, J. 1987 Secondary instability of a temporally growing mixing layer. *J. Fluid Mech.* **184**, 207.
44. Breidenthal, R. 1981 Structure in turbulent mixing layers and wakes using a chemical reaction. *J. Fluid Mech.* **109**, 1.
45. Park, K.H., Metcalfe, R.W. & Hussain, F. 1994 Role of coherent structures in an isothermally reacting plane mixing layer. *Phys. Fluids* **6**, 885.
46. Bridges, J.E. & Hussain, F. 1987 Roles of initial condition and vortex pairing in jet noise. *J. Sound Vib.* **117**, 289.
47. Hasan, M.A.Z. & Hussain, F. 1982 The self-excited axisymmetric jet. *J. Fluid Mech.* **115**, 59.
48. Lugt, H.J. 1979 The dilemma of defining a vortex. In *Recent Developments in Theoretical and Experimental Fluid Mechanics*, p.309. Springer.
49. Bisset, D.K., Antonia, R.A. & Browne, L.W.B. 1990 Spatial organization of large structures in the turbulent far wake of a cylinder. *J. Fluid Mech.* **218**, 439.
50. Jimenez, J., Moin, P., Moser, R. & Keefe, L. 1988 Ejection mechanisms in the sublayer of a turbulent channel. *Phys. Fluids* **31**, 1311.
51. Chong, M.S., Perry, A.E. & Cantwell, B.J. 1990 A general classification of three-dimensional flow field. *Phys. Fluids A* **2**, 765.
52. Hunt, J.C.R., Wray, A.A. & Moin, P. 1988 Eddies, stream, and convergence zones in turbulent flows. *Center for Turbulence Research Rep.* CTR-S88, p.193.
53. Truesdell, C. 1953 *The Kinematics of Vorticity*. Indiana University.
54. Jeong, J. & Hussain, F. 1995 On the identification of a vortex. *J Fluid Mech.* **285**, 69.
55. Shtern, V. & Hussain, F. 1993 Hysteresis in a swirling jet as a model tornado. *Phys. Fluids A* **5**, 2183.
56. Kim, J., Moin, P. & Moser, R.D. 1987 Turbulence statistics in fully developed channel flow at low Reynolds number. *J. Fluid Mech.* **177**, 133.
57. Jeong, J. & Hussain, F. 1992 Coherent structure near the wall in a turbulent channel flow. *Proc. of Fifth Asian Congress of Fluid Mech.* , Taejon, Korea.
58. Bogard, D.G. & Tiederman, W.G. 1986 Burst detection with single-point velocity measurement. *J. Fluid Mech.* **162**, 389.
59. Johansson, A.V., Alfredsson, P.H. & Kim, J. 1991 Evolution and dynamics of shear-layer structures in near-wall turbulence. *J. Fluid Mech.* **224**, 579.
60. Jeong, J. 1994 A theoretical and numerical study of coherent structures. PhD dissertation, University of Houston.
61. Tennekes, T. & Lumley, J.L. 1972 *A First Course in Turbulence.* M.I.T. Press.
62. Kambe, T. & Minota, T. 1983 Acoustic wave radiated by head-on collision of two vortex rings. *Proc. R. Soc. Lond. A* **386**, 277.
63. Stanaway, S., Shariff, K. & Hussain, F. 1988 Head-on collision of vortex rings. *Center for Turbulence Research Rep.* CTR-S88, 287.
64. Crow, S.C. 1970 Stability theory of a pair of trailing vortices. *AIAA J.* **8**, 2172.
65. Moin, P., Leonard, A. & Kim, J. 1986 Evolution of a curved vortex filament into a vortex ring. *Phys. Fluids* **29**, 955.

66. Schatzle, P.R. 1987 An experimental study of fusion of vortex rings. PhD dissertation, California Institute of Technology.
67. Hussain, F. & Husain, H.S. 1989 Elliptic jets. Part I. Characteristics of unexcited and excited jets. *J. Fluid Mech.* **208**, 257.
68. Zabusky, N.J. & Melander, M.V. 1989 Three-dimensional vortex tube reconnection: morphology for orthogonally-offset tubes. *Physica D* **37**, 555.
69. Kida, S. & Takaoka, M. 1987 Bridging in vortex reconnection. *Phys. Fluids* **30**, 2911.
70. Melander, M.V. & Hussain, F. 1988 Cut-and-connect of two antiparallel vortex tubes. *Center for Turbulence Research Rep.* CTR-S88, 257.
71. Feiereisen, W., Reynolds, W.C. & Ferziger, J.H. 1981 Numerical simulation of a compressible homogeneous, turbulent shear flow. *Stanford U. Report* No. TF-13.
72. Passot, T. & Pouquet, A. 1987 Numerical simulation of compressible homogeneous flows in turbulent regime. *J. Fluid Mech.* **181**, 441.
73. Erlebacher, G., Hussaini, M.Y., Kreiss, H. & Sarkar, S. 1990 The analysis and simulation of compressible turbulence. *ICASE Rep.* No. 90-15.
74. Lee, S., Lele, S.K. & Moin, P. 1991 Eddy shocklets in decaying compressible turbulence. *Phys. Fluids A* **3**, 657.
75. Dosanjh, D.S. & Weeks, T. 1965 Interaction of a starting vortex as well as a vortex street with a traveling shock wave. *AIAA J.* **3**, 216.
76. Virk, D. & Hussain, F. 1993 Influence of initial conditions on compressible vorticity dynamics. *Theoret. Comput. Fluid Dyn.* **5**, 309.
77. Virk, D., Hussain, F. & Kerr, R. 1995 Compressible vortex reconnection. *J. Fluid Mech.* (to appear)
78. Melander, M.V. & Hussain, F. 1994 Topological vortex dynamics in axisymmetric viscous flows. *J. Fluid Mech.* **260**, 57.
79. Huang, L.-S. & Ho, C.M. 1990 Small-scale transition in a plane mixing layer. *J. Fluid Mech.* **210**, 475.
80. Moser, R.D. & Rogers, M.M. 1993 The three-dimensional evolution of a plane mixing layer: pairing and transition to turbulence. *J. Fluid Mech.* **247**, 275.
81. Pierrehumbert, R.T. & Widnall, S.E. 1982 The two- and three-dimensional instabilities of a spatially periodic shear lyer. *J. Fluid Mech.*, **114**, 59.
82. Clark, A.R. 1979 An experimental investigation of the coherent motions in an axisymmetric mixing layer. PhD dissertation, University of Houston.
83. Lesieur, M. 1990 *Turbulence in Fluids*, 2nd Edition, Kluwer Academic Publ.
84. Hussain, F. 1983 Coherent stuctures and incoherent turbulence. In *Turbulence and Chaotic Phenomena in Fluids* (ed T. Tatsumi), p. 453. North-Holland.
85. Melander, M.V., McWilliams, J.C. & Zabusky, N.J. 1987 Axisymmetrization and vorticity gradient intensification of an isolated two-dimensional vortex through filamentation. *J. Fluid Mech.* **178**, 137.
86. McWilliams, J.C. 1984 The emergence of isolated coherent vortices in turbulent flow. *J. Fluid Mech.* **146**, 21.
87. Melander, M.V. & Hussain, F. 1993 Coupling between a coherent structure and fine-scale turbulence. *Phys. Rev. E*, **48**, 2669.

STOCHASTIC ESTIMATION OF THE STRUCTURE OF TURBULENT FIELDS

R.J. Adrian
University of Illinois, Urbana, IL, USA

ABSTRACT

The stochastic estimation method educes structure by approximating an average field in terms of event data that are given. The estimated fields satisfy the continuity equation, and they possess the correct scales of length and/or time. The fundamental concepts of general stochastic estimation and the specific application of this technique to the estimation of conditional averages are discussed. Linear stochastic estimation of random fields and of their conditional averages is developed as the principal tool, and its accuracy is demonstrated. The linear stochastic estimate is expressible in terms of second order correlation functions between the given event data and the quantity being estimated. This establishes a simple link between conditional averages, the coherent structure that they represent and correlation functions. The related problems of selecting events and interpreting the estimates that result from a given set of events are explored by considering events of increasing complexity: single-point vectors, two-point vectors, local deformation tensors, multi-point vectors, space-time vectors, and space-wave-number events. General kinematic and statistical properties are derived, and stochastically estimated structures from various types of turbulent flows are described and related to the underlying coherent structures.

1. INTRODUCTION

1.1 Estimation of random processes

In general, stochastic estimation deals with the estimation of one random variable, say $\mathbf{y}$, in terms of a set of other variables $E_1,...,E_N$ about which some information is known. We shall denote this set as the *data vector* $\mathbf{E} = (E_1,...,E_N)$. The data vector $\mathbf{E}$ may or may not contain all of the variables that affect **y**, and it may even contain some that are irrelevant, but in any case it represents a list of the available information that is thought to be important. The problem is to find a function **F(E)** that approximates **y** in some sense. Estimating a quantity in terms of other parameters is, of course, ubiquitous in engineering and science, and for this reason stochastic estimation is a well known process, especially in communications and information theory-related disciplines [Papoulis 1984], [Deutsch 1965]. It also arises routinely in our daily lives, as for example when one makes the decision to walk across a street based on an estimate of the arrival time of the approaching automobiles in terms of visual image data and engine sounds, or when one estimates the future weather in terms of current conditions.

A common element in stochastic estimation methods is that they employ information on the relationships between the variables that is based largely on empirical experience as embodied in a statistical relationship between **y** and **E**. This should be contrasted to mathematical models in which empirical information is ultimately condensed into the form of the governing differential equations and the coefficients contained therein. Full statistical information is contained in the joint probability density function

$$f_{\mathbf{y},\mathbf{E}}(\psi,\varepsilon)d\psi d\varepsilon = \Pr ob\{\psi \le \mathbf{y} < \psi + d\psi \quad and \quad \varepsilon \le \mathbf{E} < \varepsilon + d\varepsilon\} \tag{1.1}$$

where ψ and ε are dummy variables in the *pdf.* The conditional probability density of **y** given **E** is defined by

$$f_{\mathbf{y}|\mathbf{E}}(\psi \mid \varepsilon) = f_{\mathbf{y},\mathbf{E}}(\psi,\varepsilon) / f_{\mathbf{E}}(\varepsilon) \tag{1.2}$$

Consider the plot of y versus E in Figure 1.1, where both are taken to be scalars for purposes of illustration. There are two widely used methods for estimating y in terms of E. Firstly, for a given value of E, the *maximum likelihood estimate* of y is defined as the most probable value, i.e that value y_{max} at which $f_{\mathbf{y}|\mathbf{E}}$ is a maximum. It is natural to think in terms of estimates which represent the most likely result given the available data.

The second method is *conditional averaging*. The conditional average of y given E is defined as the centroid of the conditional probability

$$\langle \mathbf{y} \mid \mathbf{E} \rangle = \int \psi f_{\mathbf{y}|\mathbf{E}}(\psi \mid \varepsilon) d\psi. \tag{1.3}$$

It represents the mean value of y found by averaging over the realizations in which the event E occurs. The degree to which $<y|E>$ is a good estimate of the state of y given E depends upon the dispersion of y around $<y|E>$. As with unconditional probability distributions, accuracy can be represented by the root mean square value of the conditional distribution:

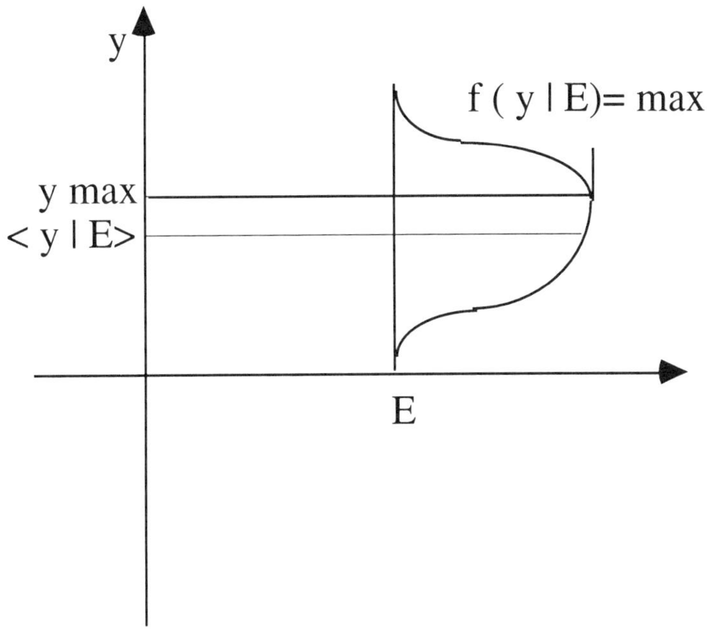

Figure 1.1. Maximum likelihood estimation and conditional averaging. The conditional average is the centroid of the conditional probability density function, whereas the maximum likelihood estimate of y is the location y_{max} at which the conditional probability density function reaches a maximum value.

$$\sigma^2_{y|E} = \int \left(\psi - \langle y \mid E \rangle\right)^2 f_{y|E}(\psi \mid E) d\psi . \tag{1.4}$$

If the conditional density is narrow, e.g. if

$$\sigma_{y|E} << \langle y \mid E \rangle \tag{1.5}$$

then both estimates are accurate. In general, either the maximum likelihood estimate or the conditional average is a useful form of estimation of y in terms of the available data E. They need not be the same, but in some situations they may be.

1.2 Stochastic estimation and turbulence structure.

The use of stochastic estimation to study turbulent flow is motivated by the desire to identify the "coherent structures" that are considered responsible for a significant part of the Reynolds stress and turbulent kinetic energy production, and by the need to specify a unique definition. Among the approaches generally used to search for and describe coherent structures in turbulent flows are:

- The study of individual realizations of the flow, in which one searches visually for patterns representing the "events".

- The study of spatial correlation functions:

$$R_{ij}(\mathbf{x},\mathbf{x}') = \langle u_i(\mathbf{x},t)u_j(\mathbf{x}',t)\rangle \tag{1.6a}$$

  or space-time correlation functions,

$$R_{ij}(\mathbf{x},\mathbf{x}',\tau) = \langle u_i(\mathbf{x},t)u_j(\mathbf{x}',t+\tau)\rangle \tag{1.6b}$$

  which are mathematically well-defined, but in which the features of coherent structures may be smoothed by the ensemble averaging operation.

- The study of conditional averages, which highlight the average velocity field associated with any chosen event. Conditional averages are precisely defined, and can be interpreted more easily than instantaneous realizations. The ensemble averaging process, of course, also smooths the data, but to a lesser extent than in correlations, since only a subset of the possible realizations is used in the conditional average.

Being well-defined mathematically, conditional averages can be a useful tool to understand and interpret correlations, thereby relating coherent structures to theory. They may, however, require large samples to obtain statistical convergence. Stochastic estimation is a way to estimate a conditional average given event vector **E** without actually performing conditional sampling.

The purpose of these lectures is to describe the application of stochastic estimation to the problem of educing structural information from empirical turbulence data obtained experimentally, or by numerical simulation. While maximum likelihood estimation is a useful method, it has proven difficult to use in turbulence. Consequently, the focus of these lectures will be on minimum mean square estimation, which includes conditional averages.

1.3 Linear and non-linear mean square estimation

In the context of turbulent structure identification the estimated variable is typically the velocity field $\mathbf{u}' = \mathbf{u}(\mathbf{x}',t')$ for some domain of positions and times, and the known variables are associated with the occurrence of certain events at one or more points in the field. Their totality is referred to as the *event data vector,* **E**. We shall denote the estimate of $\mathbf{u}'$ by $\hat{\mathbf{u}}'$. It will be seen that *the best mean square estimate of* $\mathbf{u}'$ *given* **E**, *is the conditional average* $<\mathbf{u}' \mid \mathbf{E}>$. That is, of all possible estimates $\hat{\mathbf{u}}' = \mathbf{F}(\mathbf{E})$, $< \mid \mathbf{u}' - \mathbf{F}(\mathbf{E}) \mid^2 >$ is a minimum when $\mathbf{F} = <\mathbf{u}' \mid \mathbf{E}>$. In general, $<\mathbf{u}' \mid \mathbf{E}>$ is a non-linear function of **E**, but under the condition that the elements of $\mathbf{u}'$ and **E** are joint normally distributed, it is well known that $<\mathbf{u}' \mid \mathbf{E}>$ is a linear function of **E** (Papoulis 1984). Often, this property is applied approximately to non-normal random variables by postulating an estimate of $\mathbf{u}'$ in the form of a linear combination of the event data. Then one speaks of *linear mean square estimation.* This is a well known subject, thoroughly discussed in the literature of stochastic theory (Papoulis 1984; Deutsch 1965). The mean square error of the linear estimate must be greater than or equal to that of the conditional average.

Instead of estimating $\mathbf{u}'$ as some function of **E**, it is more straightforward conceptually to estimate the conditional average of $\mathbf{u}'$, $<\mathbf{u}' \mid \mathbf{E}>$. The conditional average itself can always be interpreted as random variable since it is a function of **E,** and **E** is random. Then, it is natural to expand $<\mathbf{u}' \mid \mathbf{E}>$ in a Taylor series about **E=0**, and truncate at some level (Adrian 1977, Adrian 1979). When the series contains only first order terms, we refer to this as the *linear stochastic estimate (*LSE) to distinguish it from linear mean square estimation. In fact, it will be shown that the linear stochastic estimate of $<\mathbf{u}' \mid \mathbf{E}>$ and the linear mean square estimate of $\mathbf{u}'$ are numerically equal, so the principal difference is one of interpretation. For example, suppose that **E** that occurs at or near **x**. If $\mid \mathbf{x} - \mathbf{x}' \mid$ is large, $\mathbf{u}'$ is uncorrelated with **E**, and the mean square error of the linear estimate of $\mathbf{u}'$ must be large. On the other hand, under the same conditions, the error of the linear stochastic estimate of $<\mathbf{u}' \mid \mathbf{E}>$ may be small because $<\mathbf{u}' \mid \mathbf{E}>$ also vanishes as $\mid \mathbf{x} - \mathbf{x}' \mid$ becomes large.

The foregoing concepts can be illustrated simply by considering the estimation of a scalar random velocity $u' = u(x')$ given scalar event data E in the form of the value of $u(x)$ at x. We shall assume that $u(x')$ is a stationary, non-zero random process. Figure 1.2 shows a typical realization of $u(x')$ that passes through $u(x) = 1.5\sigma_1$., where σ_1 is the *rms* value. The conditional average $< u(x') \mid u(x) >$ is, of course, equal to u at $x = x'$, and therefore the error of the conditional average estimation vanishes at this point. As $|x - x'|$ increases, differences between $u(x')$ and $< u(x') \mid u(x) >$ appear, and the mean square error, $< [u(x') - < u(x') \mid u(x) >]^2 >$ also increases. As $|x - x'|$ becomes large, *u(x')* becomes statistically independent of *u(x)* and consequently $< u(x') \mid u(x) >$ approaches $< u(x') >$, which is zero by assumption. Hence, the mean square estimation error starts at zero at $|x - x'|$=0, cf. Fig. 1.3, and increases to the level of one rms value of $u(x)$.as $\mid x' - x \mid$ approaches infinity.

Clearly, one must exercise caution not to overinterpret results obtained by conditional averaging. Even though the conditional average is the *best* mean square estimate, it may commit serious errors in regions that are weakly related to the event data. Interpretation

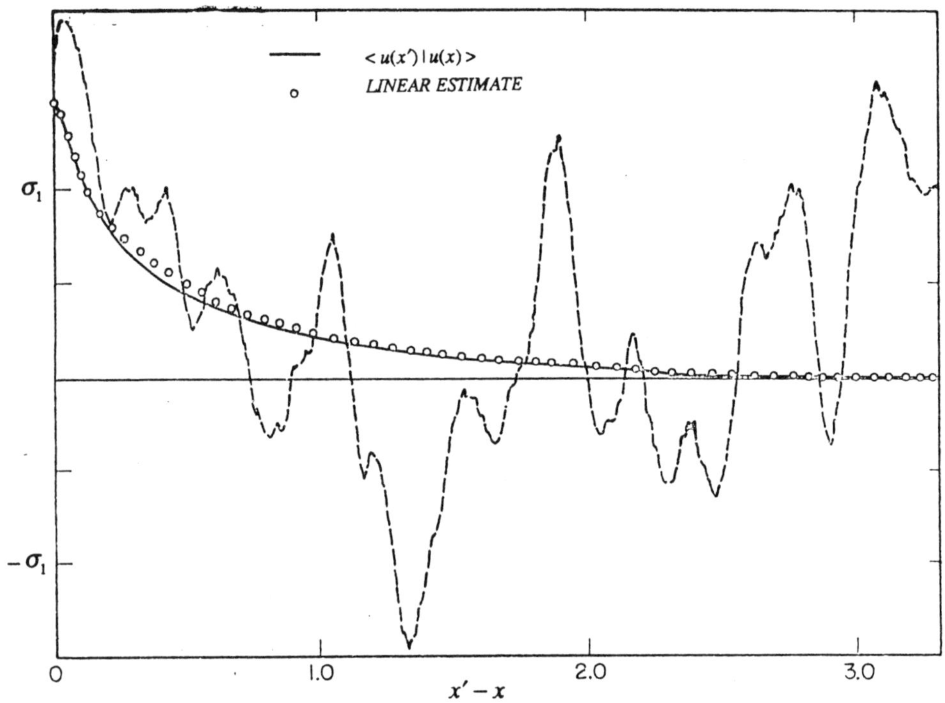

Figure 1.2. Conditional average of *u(x')* given *u(x)= 1.5 rms,* and linear stochastic estimate compared to a random realization.

should always be made with the understanding that the conditional average may not look very much like any particular random realization. It always represents an average of many realizations. If each realization is very similar to every other realization, then the conditional average will itself look like a single realization. If there is significant variation between realizations, then the conditional average can only represent a mean structure. Even in this case, there is still no reason why one should not use conditional averages to describe structure, much as we find means, moments, and correlation functions to be useful, albeit imperfect descriptors of turbulent structure. Since stochastic estimates are approximations to the conditional averages, all of the foregoing observations pertain to them, as well.

Conditional averages are not perfect estimates of the random process itself, the linear stochastic estimate can only be worse. For the problem at hand, the linear estimate of $u(x')$ given $E=u(x)$ is

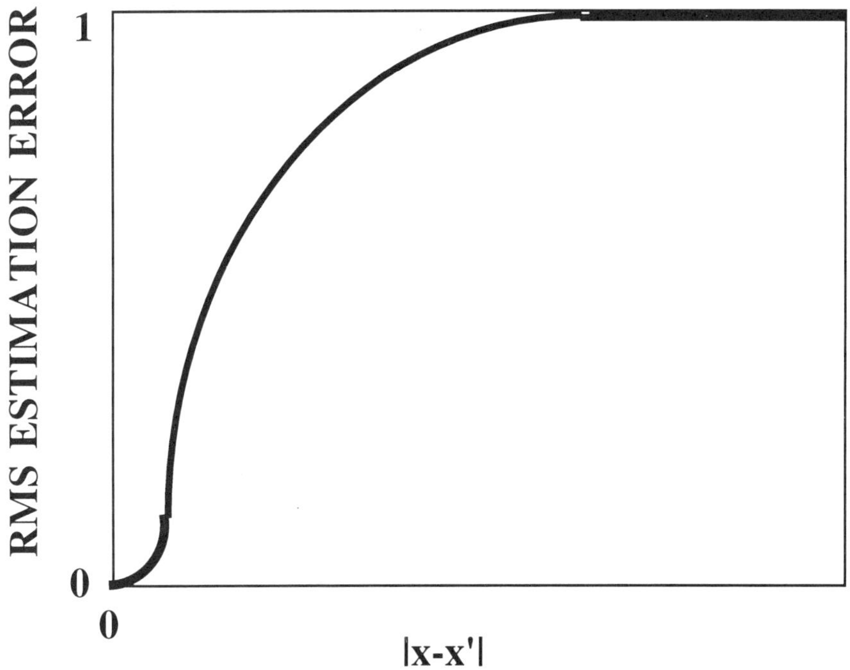

Figure 1.3. Mean square estimation error

The mean square error is

$$e = \left\langle (u(x') - Lu(x))^2 \right\rangle, \tag{1.8}$$

and it is minimized by setting

$$\frac{\partial e}{\partial L} = 2\left\langle (u(x') - Lu(x))(-u(x)) \right\rangle = 0. \tag{1.9}$$

Eq (1.9) is satisfied if

$$L = \frac{R_{uu}(x, x')}{\left\langle u^2(x) \right\rangle}. \tag{1.10}$$

The linear stochastic estimate is also shown in Fig. 1.2. Note that it is very similar to the conditional average.

If we view $u(x')$ as a random variable, then $< u(x') \mid u(x) >$ is also a random variable by virtue of being a deterministic function of a random variable. Hence, we can also apply the linear stochastic estimation procedure to it. Let

$$\hat{u}(x') = \text{linear estimate of } \langle u(x') | u(x) \rangle = L(x,x')u(x). \tag{1.11}$$

As before, the mean square error is

$$e = \left\langle [\langle u(x')|u(x)\rangle - \hat{u}(x')]^2 \right\rangle, \tag{1.12}$$

and it is minimized by setting $\partial e / \partial L = 0$, which leads to

$$L\left\langle u^2(x)\right\rangle = \langle\langle u(x') \mid u(x)\rangle u(x)\rangle \tag{1.13a}$$

$$= \langle\langle u(x')u(x) \mid (u(x)\rangle\rangle \tag{1.13b}$$

$$= \langle u(x)u(x')\rangle \tag{1.13c}$$

$$= R_{uu}(x,x') \tag{1.13d}$$

Hence,

$$L = \frac{R_{uu}(x,x')}{\langle u^2(x)\rangle} \tag{1.14}$$

In (1.13b), use has been made of the fact that *u(x)* is fixed by the conditional average so that it can be treated as a constant and taken inside the conditional brackets. Eq.(1.13c) is obtained by averaging over all *u(x)* to reduce the conditional average to an unconditional average.

From (1.10) and (1.14) we see that *the linear estimate of the conditional average* of $u(x')$ is identical to the linear estimate of $u(x')$. Later, it will be shown that this is always true. If so, why distinguish between the two estimates? The reason is clear from Figure 1.2. While it may be that neither estimate is a good approximation to $u(x')$ itself, we see that *the linear estimate is an excellent approximation to conditional average.* It will be shown later that this is also true for all turbulent processes that have been investigated thus far. It is, therefore, always best to think of linear estimates as estimates of the conditional averages. A conditional average and its linear estimate may provide the same level of detail as a random realization if the event data is very detailed. Interestingly, the linear estimate provides a link between conditional averages and correlation functions. This linkage is a powerful result that permits new interpretations of much of the body of common correlation and spectral analysis data.

The selection of event data has a profound influence on the conditional average and its linear estimate. How should **E** be specified? The correct approach is to design an event that corresponds to the physical question under investigation. Thus, if one selects **E** to be a large Reynolds stress event, $<\mathbf{u}' \mid \mathbf{E}(\mathbf{x})>$ gives the average vector field around the site of the large Reynolds stress. The most famous event of this type is the quadrant analysis conditional average in which **E** is the event that $(u_1(\mathbf{x}), u_2(\mathbf{x}))$ lies in one of the four quadrants in the $u_1 - u_2$ plane:

$$Q1:(u_1 > 0, u_2 > 0);\ Q2:(u_1 < 0, u_2 > 0);\ Q3:(u_1 < 0, u_2 < 0);\ Q4:(u_1 > 0, u_2 < 0). \quad (1.15)$$

Points lying in the Q2 and Q4 quadrants contribute to positive values of the Reynolds shear stress $-<u_1 u_2>$. A more specific condition is to specify the vector at **x** itself: **E**=**u**(**x**). This not only states which quadrant **u**(**x**) lies in, but also the magnitude of its contribution to $-<u_1 u_2>$.

To preview the sort of field that can be inferred from a linear stochastic estimate, consider the result in Fig. 1.4, obtained by estimating the vorticity field in the near wall region of low Reynolds number channel flow given a vector event **u**(**x**) that lies in the second quadrant of the *u*-*v* plane (Moin, Adrian and Kim 1987). The contours of constant magnitude of the fluctuating vorticity indicate a hairpin vortex with long legs trailing back along the wall. The legs are quasi-streamwise vortices, and their spacing is approximately one-hundred viscous wall units, corresponding to the well-known value for the spacing of low-speed streaks. It will suffice for now to note that this flow structure is very similar to the vortices observed in wall turbulence.

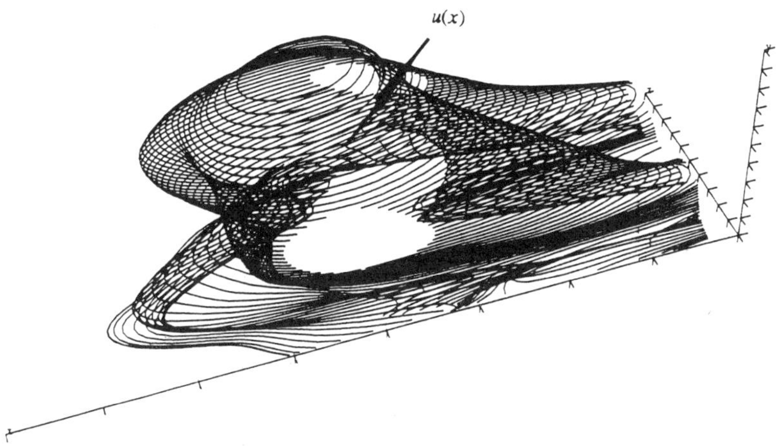

Figure 1.4 Stochastic estimate of the vorticity field of an eddy in low Reynolds number turbulent channel flow. The velocity at y^+=49 is specified to be a second quadrant event.

Events consisting of one or more vectors are natural in the probability density function equations of turbulence (c.f. Lundgren 1967, Monin 1967), as are events involving vorticity or deformation. *Table 1* lists various events that have been used in existing experimental studies. Note that events can become complicated very rapidly. For example, in Table 1, **d** is the deformation tensor, $d_{ij} = \partial u_i / \partial x_j$ which contains eight independent components in incompressible flow. In total, the event consisting of a three-dimensional vector **u** and all nine components of **d** spans an eleven dimensional space. One of the principal advantages of linear estimation of the conditional average is that it makes possible the use of relatively complicated events without sifting through very large spaces in search of realizations that satisfy the conditional event.

TABLE 1. Events used in analyses of turbulent fields by stochastic estimation

E	Estimated Property	Reference
$\mathbf{u}(\mathbf{x},t)$	$\mathbf{u}(\mathbf{x},t')$	Adrian (1979), Hassan (1980), Moin, Adrian and Kim (1987)
$\mathbf{u}(\mathbf{x},t)$	$\mathbf{u}(\mathbf{x}',t)$	Adrian (1978, 1979), Hassan (1980), Tung & Adrian (1980), Tung (1982), Chang (1985), Chin (1987)
$\mathbf{u}(\mathbf{x},t)$	$u_i(\mathbf{x}',t)u_j(\mathbf{x}',t)$	Tung (1982), Elam (1987)
$\mathbf{u}(\mathbf{x},t)$	$p(\mathbf{x}',t)$	Chang (1985)
$p_{wall}(\mathbf{x},t)$	$u(\mathbf{x}',t)$	Willmarth & Wooldridge (1963)
$p(\mathbf{x}',t)$	$\mathbf{u}(\mathbf{x}',t)$	Nithianandan (1980)
$\mathbf{u}(\mathbf{x}_1,t),\mathbf{u}(\mathbf{x}_2,t)$	$\mathbf{u}(\mathbf{x}',t),\omega(\mathbf{x}',t)$	Adrian, Moin & Moser (1987)
$\mathbf{u}(\mathbf{x},t),\mathbf{d}(\mathbf{x},t)$	$\mathbf{u}(\mathbf{x}',t),\omega(\mathbf{x}',t)$	Adrian & Moin (1988), Ditter (1987)
$\tau_{wall}(x,t)$	$\mathbf{u}(\mathbf{x}',t),\omega(\mathbf{x}',t)$	Guezennec (1989)
$\phi(\mathbf{x},t)$	$\phi(\mathbf{x}',t)$	Pecseli (1987, 1991)

2. MEAN SQUARE ESTIMATION OF A RANDOM VARIABLE

2.1 Optimal mean square estimation

In this section we will establish results that were stated without proof in the Introduction, and generalize them to cover vector variables. Let $\hat{\mathbf{y}}$ be the estimate of a vector random variable $\mathbf{y}$ in terms of available data $\mathbf{E}=(E_1,...E_N)$. In general

$$\hat{\mathbf{y}} = \mathbf{F}(\mathbf{E}) \tag{2.1}$$

where $\mathbf{F}$ is a non-linear function of the data. The best mean square estimator is found by selecting $\mathbf{F}$ such that the error found by averaging over all values of the data

$$e = \left\langle |\mathbf{y} - \mathbf{F}(\mathbf{E})|^2 \right\rangle \tag{2.2}$$

is a minimum. The ensemble average in (2.2) extends over $\mathbf{y}$ and $\mathbf{E}$. The necessary condition for e to be a minimum is that its first variation with respect to $\mathbf{F}$ must vanish for arbitrary variations $\delta\mathbf{F}$:

$$\delta e = \langle 2[y_i - F_i(\mathbf{E})]\delta F_i \rangle = 0 \tag{2.3a}$$

Eq. (2.3a) can be solved by writing the unconditional average as the average of $< [y_i - F_i(\mathbf{E})]\delta F_i \,|\, \mathbf{E} >$ and then noting that $\delta F_i(\mathbf{E})$ can be taken outside of the conditional average since $\mathbf{E}$ is fixed inside the conditional average:

$$\delta e = 2\langle\langle [y_i - F_i]\delta F_i \,|\, \mathbf{E}\rangle\rangle \tag{2.3b}$$

$$= 2\langle\langle [y_i - F_i] \,|\, \mathbf{E}\rangle \delta F_i(\mathbf{E})\rangle \tag{2.3c}$$

The inner bracket in (2.3) averages over $\mathbf{y}$ keeping $\mathbf{E}$ fixed, and the outer bracket averages over $\mathbf{E}$. Equation (2.3c) can be satisfied for any $\delta\mathbf{F}$ by setting $< y_i - F_i \,|\, \mathbf{E} >= 0$, or since $\langle \mathbf{F}(\mathbf{E}) \,|\, \mathbf{E}\rangle = \mathbf{F}(\mathbf{E})$,

$$\mathbf{F} = \langle \mathbf{y} \,|\, \mathbf{E}\rangle. \tag{2.4}$$

Thus, the best mean square estimate of $\mathbf{y}$ is its conditional average given all of the available data $\mathbf{E}$, and in general this average is a non-linear function of $\mathbf{E}$. It is clear that the mean square error will increase if not all of these data are used, or if the form of $\mathbf{F}$ is constrained.

The conditional average is related to the correlation functions by

$$\langle\langle y_i \,|\, \mathbf{E}\rangle E_j\rangle = \langle\langle y_i E_j \,|\, \mathbf{E}\rangle\rangle = \langle y_i E_j \rangle \tag{2.5}$$

Thus, *knowledge of* $<\mathbf{y}\,|\,\mathbf{E}>$ *and the probability density function of f(***E***) is sufficient to determine* $<y_iE_j>$. The converse is not true: *Knowledge of the correlations* $<y_iE_j>$ *and f(***E***) is not generally sufficient to determine* $<\mathbf{y}\,|\,\mathbf{E}>$.

2.2 Sub-optimal non-linear mean square estimation

Suppose the function **F**(**E**) is constrained to have some particular form

$$\mathbf{F}(\mathbf{E}) = \mathbf{F}_o(\mathbf{E},\mathbf{p}) \tag{2.6}$$

where the vector **p** contains the parameters that appear in the form $\mathbf{F}_0$. For example, $\mathbf{F}_0$ could be a polynomial and the vector **p** could be a list of the coeffiecients of the polynomial. $\mathbf{F}_0$ may be non-linear, but since it is constrained to a form that may not coincide with $<\mathbf{y}\,|\,\mathbf{E}>$, the mean square estimation error must be greater than that of the optimal estimate given by the conditional average. The variational problem to minimize

$$e = \left\langle |\,\mathbf{y} - \mathbf{F}_0\,|^2 \right\rangle \tag{2.7}$$

is

$$\frac{\partial e}{\partial p_j} = 0, \qquad j = 1,...J. \tag{2.8}$$

where J is the number of adjustable parameters in $\mathbf{F}_0$. The reader will recognize the problem of finding a non-linear, sub-optimal estimate in this way as the Rayleigh-Ritz method from variational calculus.

2.3 Linear mean square estimation

Suppose that the form of $\mathbf{F}_0$ is restricted further to be linear in the event data:

$$\hat{y}_i = F_{oi}(\mathbf{E}) = A_i + L_{ij}E_j, \qquad i = 1,...,I \tag{2.9}$$

(The summation convention applies to repeated Latin subscripts.) One expects this restriction to further increase the estimation error. The necessary conditions for minimum mean square error are

$$\frac{\partial e}{\partial A_k} = 0, \qquad k = 1,..,I \tag{2.10a}$$

$$\frac{\partial e}{\partial L_{il}} = 0, \qquad i = 1,\ldots,I, \qquad l = 1,\ldots,N \tag{2.10b}$$

Substituting (2.9) into (2.10) and applying these conditions gives the following set of equations:

$$A_i + \langle E_j \rangle L_{ij} = \langle y_i \rangle, \quad i = 1,..,I \;\; and \quad l = 1,..,N \tag{2.11a}$$

$$A_i\langle E_l \rangle + \langle E_j E_l \rangle L_{ij} = \langle y_i E_l \rangle, \quad i = 1,..,I, \quad j,l = 1,..,N \tag{2.11b}$$

These equations are in terms of variables **y** and **E** that may have non-zero mean values. They become decoupled if **E** and **y** are decomposed into mean and fluctuating parts. Then, equation (2.11b) reduces to the following system of linear algebraic equations for L_{ij}

$$\langle (E_j - \langle E_j \rangle)(E_l - \langle E_l \rangle) \rangle L_{ij} = \langle (y_i - \langle y_i \rangle)(E_l - \langle E_l \rangle) \rangle \tag{2.12}$$

whence equation (2.11a) can be solved for A_i.

From (2.12) it is clear that the mean values of the event data have no influence on the coefficients L_{ij}. If we agree to use only zero mean event data, the equations for L_{ij} simplifly to

$$\langle E_j E_l \rangle L_{ij} = \langle y_i E_l \rangle, \qquad i = 1,\ldots I \quad and \; j,l = 1,\ldots N. \tag{2.13}$$

These equations are known as the *Yule-Walker* equations (Papoulis 1984). They give the estimation coefficients in terms of the correlation functions $\langle E_j E_l \rangle$ between each of the data, and the correlation function $\langle y_i E_l \rangle$ between the data and the estimated variable. Hereafter we shall assume the $\langle E_l \rangle = 0$.

The *Orthogonality Principle* states that the condition for minimizing the mean square error of the linear estimate is that the errors must be *statistically orthogonal* to the data (i.e. they must be statistically uncorrelated). With $\langle E_l \rangle = 0$ a simple rearrangement of (2.13) shows that the error y_i-$L_{ij}E_j$ is indeed orthogonal to E_l:

$$\langle (y_i - L_{ij}E_j)E_l \rangle = 0, \qquad i = 1,\ldots I, \qquad j,l = 1,..,N. \tag{2.14}$$

The mean square error of the linear estimate is found from

$$e = \langle (y_i - L_{ij}E_j)(y_i - L_{ik}E_k)\rangle \tag{2.15a}$$

$$= \langle (y_i - L_{ij}E_j)y_i\rangle = \langle y_i y_i\rangle - \langle \hat{y}_i\, y_i\rangle . \tag{2.15b}$$

Suppose that $\mathbf{y}$ is correlated with the data, but the data are uncorrelated with each other. Then

$$\langle E_j E_l\rangle = \langle E_j^2\rangle \delta_{jl}$$

and the data correlation matrix $< E_j E_l >$ becomes diagonal, leading immediately to the simple solution

$$L_{il} = \langle y_i E_l\rangle / \langle E_l^2\rangle \tag{2.16}$$

If $\mathbf{y}$ is statistically independent of every datum, $< y_i E_j >= 0$, then (2.13) becomes a homogeneous set of equations for L_{ij}, and if

$$\det\langle E_j E_l\rangle \neq 0, \tag{2.17}$$

the only solution for L_{ij} is the trivial solution. Then $\hat{\mathbf{y}} =< \mathbf{y} >$, which shows that a non-trivial linear estimate cannot be obtained unless the given data are correlated with **y.**

It is helpful to consider a simple example in which the variable to be estimated is $y = u' = u(x')$ and the data consist of the value of u at x, as in Fig. 1.2. Then

$$\hat{u}' = L(x, x')u(x) \tag{2.18}$$

and the orthogonality principle requires

$$\langle (u' - Lu(x))\, u(x)\rangle = 0 \tag{2.19}$$

or

$$L = \frac{R_{uu}(x, x')}{R_{uu}(x, x)} \tag{2.20}$$

where

$$R_{uu}(x, x') = \langle u(x)u(x')\rangle . \tag{2.21}$$

Hence

$$\hat{u}(x') = \frac{R_{uu}(x,x')}{R_{uu}(x,x)}\, u(x). \tag{2.22}$$

The normalized mean error is

$$\frac{e}{\langle u'^2 \rangle} = 1 - \frac{\langle u' \hat{u}' \rangle}{\langle u'^2 \rangle} = 1 - \frac{R_{uu}{}^2(x,x')}{R_{uu}(x',x')R_{uu}(x,x)}. \tag{2.23}$$

Suppose u(x) is statistically homogeneous in x with Taylor microscale λ_τ defined by

$$\frac{R_{uu}(x-x')}{R_{uu}(x,x)} \cong 1 - \frac{|x-x'|^2}{\lambda_\tau{}^2}. \tag{2.24}$$

Then the error approaches zero as

$$\frac{e}{\langle u^2 \rangle} \sim \frac{2|x-x'|^2}{\lambda_\tau{}^2} \tag{2.25}$$

Thus, the linear estimate of a random variable *u(x)* is accurate only for separations between the estimated variable and the data which are smaller than the Taylor microscale.

3. MEAN SQUARE ESTIMATION OF CONDITIONAL AVERAGES

3.1 Governing Equations

The concept of using mean square estimation to approximate a *conditional average* was introduced in Section 1. In this section, this idea will be explored in more detail, and developed for general event vectors.

Suppose that $\mathbf{y}' = \mathbf{y}(\mathbf{x}',t')$is any quantity of interest. The conditional average of $\mathbf{y}'$ given $\mathbf{E}$ must be a function of the variables $(\mathbf{E},\mathbf{x}',t')$ plus, implicitly, any space-time locations that are involved in specifying the event data. $<\mathbf{y}'|\mathbf{E}>$ is a deterministic function of these variables, and since it is an average, it tends to be a smoother function of $(\mathbf{x}',t')$than the underlying random field. One also expects that $<\mathbf{y}'|\mathbf{E}>$ would depend smoothly upon the data, e.g., continuous changes in **E** produce continuous changes in the conditional average. (For arguments concerning when this may not always be the case, c.f. Novikov 1993.) Assuming that $<\mathbf{y}'|\mathbf{E}>$ is a continuous function of **E**, one can expand it in a Taylor series, the natural points of expansion being the most probable value of **E**, or, more conveniently, the mean value of **E**. If we assume that the event vector represents fluctuations with respect to the mean, then Taylor series expansion about **E=0** corresponds to analysis of *weak* fluctuations:

$$\langle y_i' \mid \mathbf{E}\rangle = A_{ij}E_j + B_{ijk}E_jE_k + C_{ijkl}E_jE_kE_l + 0(E^4) \tag{3.1}$$

The coefficients *A, B, C,...* are undetermined functions of $(\mathbf{x},'t')$ and the locations of the event data. Other forms of expansion could be used, but the foregoing is natural because it leads to results that involve standard central moments of **E**.

Minimizing the mean square error leads to the following set of equations for the coefficients

$$\langle[\langle y_i' \mid \mathbf{E}\rangle - A_{ij}E_j - B_{ijk}E_jE_k - ...]E_l\rangle = 0 \tag{3.2a}$$

$$\langle[\langle y_i' \mid \mathbf{E}\rangle - A_{ij}E_j - B_{ijk}E_jE_k - ...]E_lE_m\rangle = 0, \tag{3.2b}$$

and so on. These equations may or may not be coupled, depending on the statistical correlations $< E_jE_l >, < E_jE_kE_l >$, etc. Also, because of the additional terms in the power series, the coefficient A_{ij} of the linear term may not be equal to the linear estimation coefficient L_{ij}.

If the estimate is restricted to a *linear form*, the governing equation for the linear estimation coefficients are

$$\langle E_jE_l\rangle A_{ij} = \langle y_i'E_l\rangle, \quad i = 1,...I \; \; and \; \; j,l = 1,...,N \tag{3.3}$$

These are identical to the Yule-Walker equations for estimation of **y** in terms of **E**, so we see that $A_{ij}=L_{ij}$. if the estimate is linear.

Although there is no numerical difference between these estimates, their interpretations are quite different, as discussed previously. When estimating **y** in terms of **E**, one might hope that the estimate would be a good approximation to $\mathbf{y}(\mathbf{x}',t)$. But, of course, as $\mathbf{y}(\mathbf{x}',t)$ is moved away from the data locations, so that correlation between **E** and **y** is lost, the accuracy of approximation decreases. Thus, when estimating $< \mathbf{y}' \mid \mathbf{E} >$ directly, one realizes that $< \mathbf{y}' \mid \mathbf{E} >$ itself is an estimate of $\mathbf{y}'$ that has limited accuracy. Deviations of $< \mathbf{y}' \mid \mathbf{E} >$ from $\mathbf{y}'$ are accepted as part of the information loss that occurs when one characterizes random fields by statistical quantities.

The more interesting aspect of stochastic estimation is the fact that conditional data are found from the unconditional moments. Put another way, one can find conditionally averaged behavior, without performing conditional sampling. The practical implications of this property will be discussed later.

3.2 Joint normal random processes.

The accuracy of stochastic estimation of conditional averages is, of course, an important issue. In this regard, the class of joint normally distributed random variables provides an important reference case, because *if the random variables* $\mathbf{y}'$ *and* $\mathbf{E} = (E_1,...E_N)$ *are joint normally distributed, then the conditional average* $<\mathbf{y}'|\mathbf{E}>$ *is given exactly by the linear estimate, as specified by the leading term in (3.1) and (3.3).* Since joint normal probability distributions are determined completely by the first and second moments of the random variables, it should not be surprising that the conditional average can be found through equations like (3.3) which involve only $<E_l E_m>$ and $<y_i' E_l>$.

The proof of the foregoing result is instructive. Suppose without loss of generality that $<y_i'>=<\mathbf{E}>=0$. If the variables E_j are joint normally distributed, then the sum $L_{ij}E_j$ is also a normal random variable (Papoulis 1984). Likewise, if y_i' is joint normally distributed with respect to E_j, then the difference, $y_i' - L_{ij}E_j$ is also a normal random variable. For linear estimation, the coefficients L_{ij} are chosen to satisfy

$$\langle [y_i' - L_{ij}E_j]E_k \rangle = 0, \quad i = 1,...I \ \ and \ \ j,k = 1,...N \tag{3.4}$$

which states that $y_i' - B_{ij}E_j$ is uncorrelated with each E_k. But, joint normal random variables that are uncorrelated are also statistically independent (Papoulis 1984). Hence, if $y_i' - L_{ij}E_j$ is statistically independent of $\mathbf{E}$, then

$$\left\langle y_i' - L_{ij}E_j \,|\, \mathbf{E} \right\rangle = \left\langle y_i' - L_{ij}E_j \right\rangle \tag{3.5a}$$

$$= \left\langle y_i' \right\rangle - L_{ij}\left\langle E_j \right\rangle \tag{3.5b}$$

$$= 0 \tag{3.5c}$$

because $=<y_i'>=<E_j>=0$ by assumption. But, also

$$\left\langle y_i' - L_{ij}E_j \,|\, \mathbf{E} \right\rangle = \left\langle y_i' \,|\, \mathbf{E} \right\rangle - L_{ij}E_j \tag{3.6}$$

so, upon combining (3.5c) and (3.6) we have exactly

$$\left\langle y_i' \,|\, \mathbf{E} \right\rangle = L_{ij}E_j . \tag{3.7}$$

It should be noted that the key property of joint normal processes that is needed to make the foregoing proof is the fact that joint normality implies that uncorrelated random variables are statistically independent.

3.3 Non-normal random processes

Any other process for which uncorrelatedness implies independence would also have a conditional average given exactly by the linear stochastic estimate. Joint normality implies zero error for LSE, but this does not mean that LSE must be inaccurate if the processes are not joint normal. To illustrate, a situation in which the conditional average of a non-normal random process can be exactly linear, consider Figure 3.1. Fig. 3.1(a) shows the elliptical contours that are characteristic of a joint normal *pdf* relationship between y' and E. The conditional average, being the centroid of the *pdf* at fixed E, is a linear function of E (see also Figure 1.1). Next, consider the decidedly non-Gaussian contours of the non-normal *pdf* in Figure 3.1(b). It has been drawn so that its centroidal masses are also located on a linear function of E, with the consequence that the linear form of the conditional average would also be correct for this distribution.

The accuracy of LSE has been tested in various turbulent flows by measuring the conditional average and the correlation functions needed to estimate them. Figure 3.2 compares the linear estimate to the conditional average of the time-delayed streamwise velocity $< u_1(x,t+\tau) \mid u_1(x,t) >$ in grid turbulence for several different values the velocity at zero time delay. These curves correspond to Figure 1.2. Although the agreement is good, one might argue that grid turbulence is an easy test because it is weak.

Figure 3.3 compares the conditional average of the time-delayed streamwise velocity on the centerline of an axisymmetric jet to its LSE. In this flow, the turbulence intensity is much higher than in grid turbulence, and the turbulence statistics are decidedly non-normal. The curves are normalized by the data $E = u_1(x,t)$ to make the linearity of the relationship apparent. Agreement is generally good, although the data at $x_1/D = 3$ shows some non-linear dependence upon the event data.

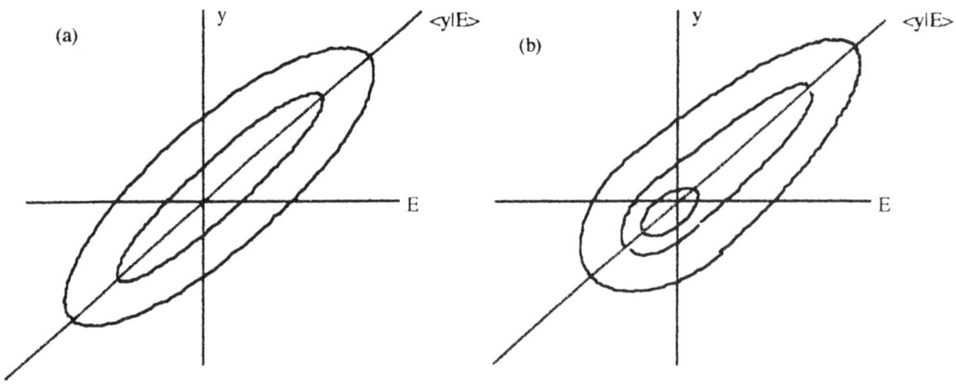

Figure 3.1 (a) Conditional average of joint normal variables (y,E) is linear; (b) a linear conditional average is also possible for non-normal random variables.

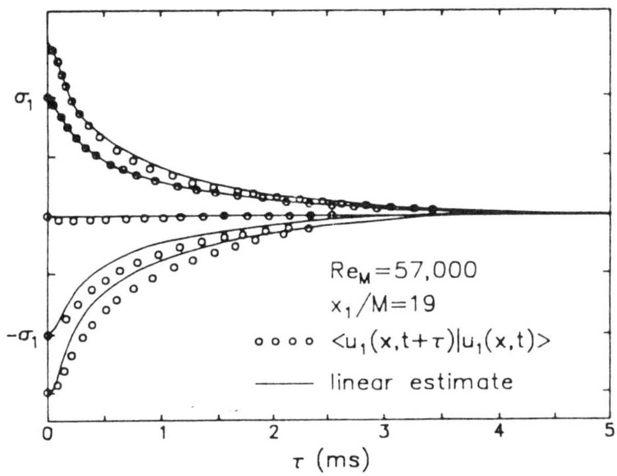

Figure 3.2 The conditionally averaged, time-delayed conditional velocity and linear estimate in grid turbulence for $u_1(\mathbf{x},t)/\sigma_1 = -1.5, -1.0, 0, 1.0,$ *and* 1.5. *[Adrian et al. 1989].*

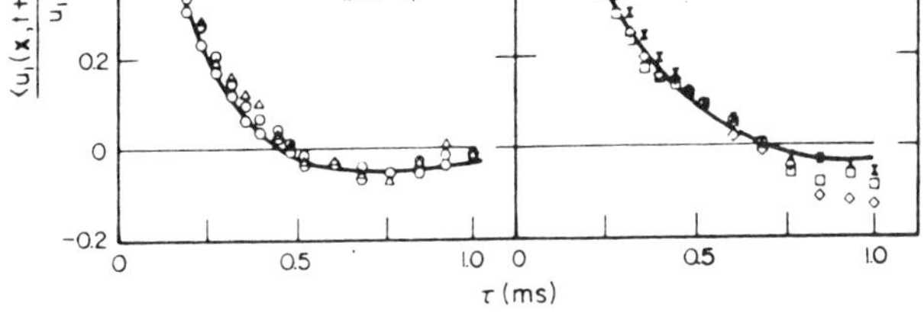

Figure 3.3 A comparison of conditionally averaged streamwise velocity to its linear estimate on the centerline of an axisymmetric shear layer. *[from Adrian et al. 1989].*

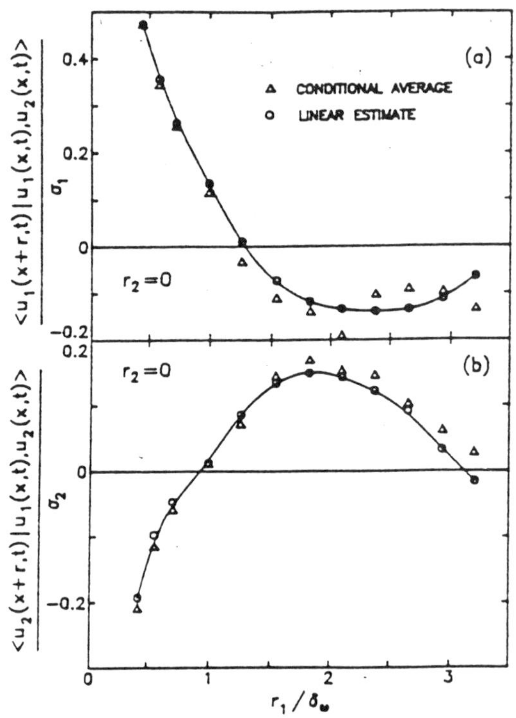

Figure 3.4. Conditional average and linear estimates on the centerline of a plane turbulent shear layer. (a) streamwise velocity; b) cross-stream velocity *[from Adrian et al. 1989]*.

Figures 3.4 and 3.5 each contain estimates of velocities that are separated from the event data in space, instead of time. Figure 3.4 contains measurements from a two-stream mixing layer in which the conditional event consists of two events: the streamwise velocity and cross-stream velocity on the shear layer centerline. The turbulence intensity is also large.

Figure 3.5 shows results for the streamwise velocity in pipe flow. The measurements in this experiment suffered from insufficient time averaging, resulting in some sampling error noise in the correlation functions and in the conditional averages. This noise is manifested in the small amplitude irregularities of the curves in Fig. 3.5. Interestingly, the sampling noise appears to be similar in both quantities.

To assess the significance of non-linear terms in the estimates, Figure 3.5 compares the linear estimate to an estimate containing quadratic terms. There is relatively little difference in this particular flow, a pipe flow, but as pointed out by Guezennec (1989), the non-linear terms are necessary if one hopes to see any difference in the conditional patterns due to a change in the sign of the event data. Thus for example, to within the limits of linear estimation there is no difference between the conditional average given a Q2 event vector and and identical but opposite Q4 event vector, aside from a change in the sign of the conditional field. Addition of quadratic or higher order terms alters this situation, but the results of

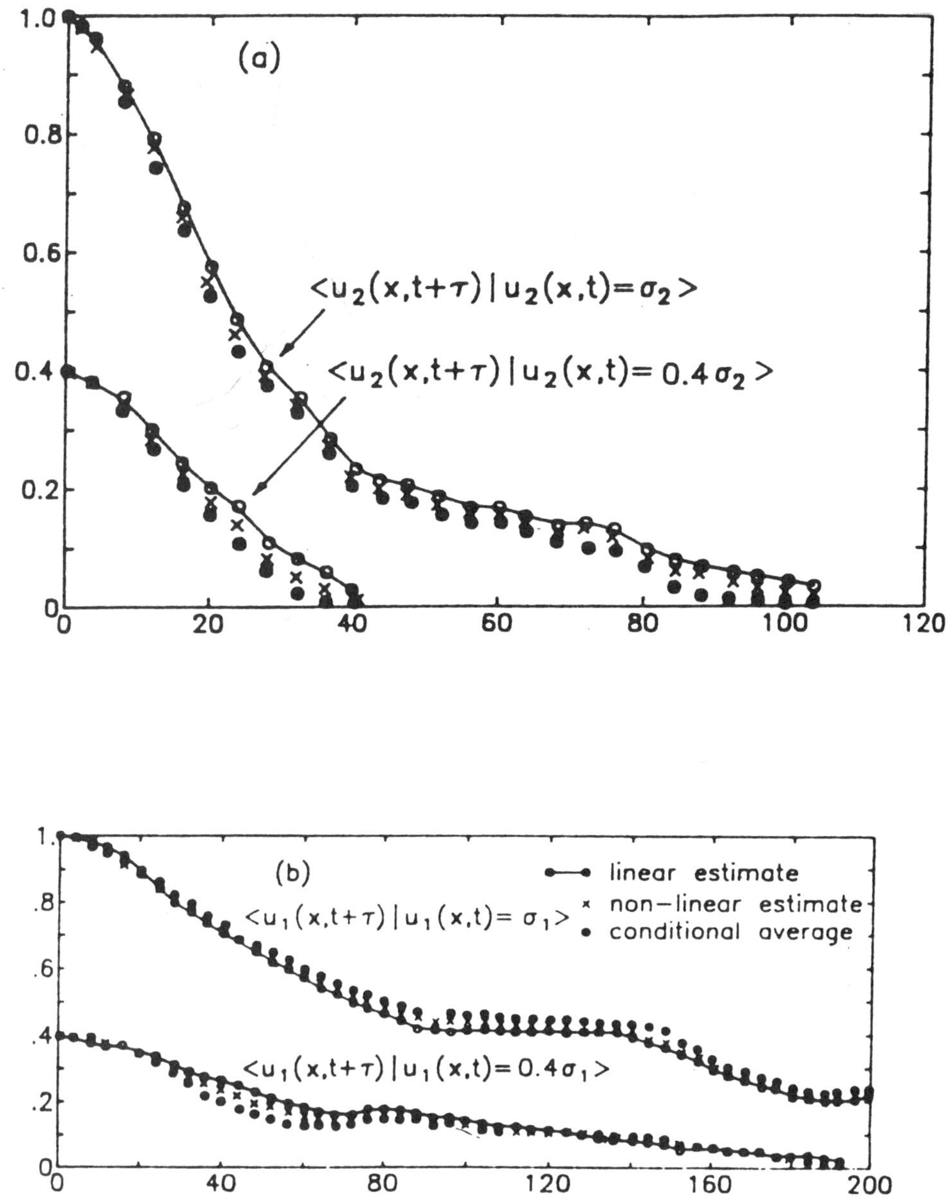

Figure 3.5 Conditional averages and linear estimates in the outer region of turbulent pipe flow at 50,000 Reynolds number. *[From Hassan 1980, c.f. also Adrian et al. 1989].*

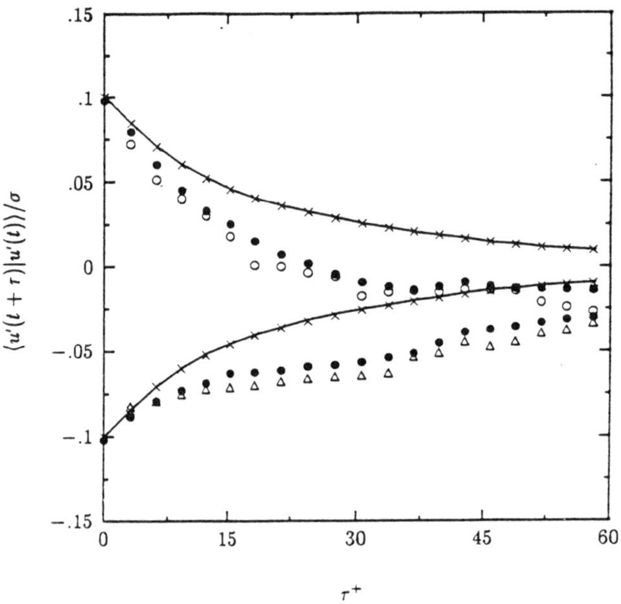

Figure 3.6. Improved approximation of conditional averages given low amplitude events. $o, < u(t+\tau) \mid u(t) = 0.1\sigma_1 >$; $\Delta, < u(t+\tau) \mid u(t) = -0.1\sigma_1 >$; $\times, LSE$; –, quadratic MSE;

• Laurent series MSE. *[From Brereton 1992].*

Guezennec (1989) do not produce noticeable change. It may be that the principal difference associated with a change in the sign of the event vector occurs when the conditional average of the fluctuating velocity is added to the mean flow.

In general, the experimental evidence supports the idea that linear estimation of conditional averages is a reasonably accurate procedure, even when the random processes involved are non-normal. Some steps have been made to improve on LSE. For example, Brereton (1992) has observed that the expansion may be postulated to contain a term like $S_{ij}E_j^{-1}$ so that the expansion can emphasize the importance of low amplitude events. Evidence that such a Laurent-like series offers improved approximations was obtained in a boundary layer, Figure 3.6.

3.4 Summary

Conditional averages of turbulent quantities can be approximated rather well by the linear and non-linear mean square estimation. In general, linear estimation is adequate, but non-linear estimation offers more improvement as well as some conceptual advantages. The estimation results are expressed in terms of correlation functions which must be known from some other mean such as experiment. The advantages of stochastic estimation relative to conditional averaging are:

- Complicated events can be used with the estimation procedure.
- Stochastic estimates are more stable than conditional averages because they use all of the data in forming the correlation functions.
- Once the stochastic estimates are formed, the numerical values of the conditional events may be changed easily to fully explore the event space.

4. CONDITIONAL EDDIES

4.1 Definition

It is widely recognized that the use of a conditional average $<\mathbf{u}' \mid \mathbf{E}>$ to characterize coherent motions runs the risk of obtaining results that are more representative of the conditional event than of the coherent structure. This is a legitimate concern, and for this reason it is, in the author's opinion, a mistake to interpret a conditional average as always being truly representative of instantaneous structure. Rather, one should only interpret it as an average field associated with the occurrence of the event - no more and no less. In certain instances the conditional average may correspond very closely to the instantaneous structures. This occurs, for example if the field contains many structures that all look similar. But, if the field consists of many different types of structures that each satisfy the conditional event, then the conditional average will be an average over all of the different types. Hence, in general, it is best to treat conditional averages as statistical quantities that characterize certain average features of a flow field, just as the mean, rms, pdf, correlation and power spectrum each characterize different aspects. Being conditional does not guarantee, by itself, that the average will give a very realistic picture of the coherent structure, and indeed, it has been shown in Section 1 that substantial differences do occur.

The fidelity with which a conditional average can represent turbulent structure depends strongly upon the information contained in the event. Given sufficiently complex data, it is possible to perfectly reconstruct the entire turbulent velocity field. But, this would not be a coherent structure. Given too little data, it is possible to get a field that bears little resemblance to a coherent realization. Between these limits, there may or may not be an event or set of events that faithfully reproduces all aspects of a coherent realization. Thus, in general, the conditional average is only an imperfect representative of coherent structure, and it is best to avoid implications to the contrary by carefully distinguishing between coherent structures and conditional fields $<\mathbf{u}(\mathbf{x}',t') \mid \mathbf{E}>$. To do so, we shall refer to the latter as *conditional eddies* (Adrian 1979). The intention behind introducing this term is to define the conditional eddies as mathematical entities, and thereby separate the issue of existence from the issue of how well they correspond to a coherent structure. For example, while one might doubt the existence of coherent structure in a highly chaotic, high Reynolds number flow, there can be no argument concerning the existence of conditional eddies; they must exist because it must be possible to define a conditional average of $\mathbf{u}(\mathbf{x}',t')$.

The conditional eddy field depends upon the type of event $\mathbf{E}$, and the proper definition of $\mathbf{E}$ is clearly critical to the formation of a useful conditional eddy concept. In fact, it is fair to say that many of the research questions concerning conditional eddies revolve around the definition of $\mathbf{E}$. Fortunately, the equations of fluid mechanics suggest natural forms.

Consider, for example, the one-point probability density function of the turbulent velocity at $\mathbf{x}$:

$$f_1(\mathbf{v},\mathbf{x},t) = \mathrm{Prob}\{\mathbf{v} \le \mathbf{u}(\mathbf{x},t) < \mathbf{v}+\mathbf{dv}\}d\mathbf{v}\,. \tag{4.1}$$

The Lundgren-Monin equation (Lundgren 1967, Monin 1967) for this probability density can be wriitten in the form

$$\frac{\partial f_1}{\partial t} = \frac{\partial}{\partial v_i}\left\{\left\langle \frac{\partial u_i}{\partial t}|\mathbf{v}\right\rangle f_1\,(\mathbf{v},\mathbf{x},t)\right\} \tag{4.2}$$

where the conditional time derivative is given by

$$\left\langle \frac{\partial u_i}{\partial t}\middle|\mathbf{v}\right\rangle = \lim_{\mathbf{x}'\to\mathbf{x}}\left\{-v_j\frac{\partial}{\partial x_j'}\langle u_i' \mid \mathbf{v}\,\rangle - \frac{1}{\rho}\frac{\partial}{\partial x_i'}\langle p' \mid \mathbf{v}\rangle + \nu\frac{\partial^2}{\partial x_l'^2}\langle u_i'|\mathbf{v}\rangle\right\} \tag{4.3}$$

It is well known that equation (4.2) contains pressure and viscous terms that are unclosed because they depend upon the two-point *pdf* of the turbulent velocities. As shown by equation (4.3) this dependence occurs fundamentally because the conditional stress terms on the right hand side of (4.3) each require a conditionally averaged field to be evaluated, either $< p' \mid \mathbf{v} >$ or $< \mathbf{u}' \mid \mathbf{v} >$. These terms correspond to the one-point conditional eddy. Thus, the closure problem for the one-point *pdf* equation is essentially one of modeling the conditional eddy given the velocity at one point. The problem is complicated somewhat by the fact that the conditional pressure is given by the solution of the Poisson equation

$$\nabla'^2\langle p'|\mathbf{v}\,\rangle = -\rho\frac{\partial^2}{\partial x_i'\partial x_j'}\langle u_i'u_j'|\mathbf{v}\rangle \tag{4.4}$$

which involves the conditional Reynolds stress as well as the conditional eddy.

More generally the N-point *pdf* equation depends upon the N+1-point *pdf*. It has also been shown (Adrian 1977) that the unclosed terms, aside from the pressure term, always have the form of a conditional average given the turbulent velocities at N points, e.g.,

$$< \mathbf{u}(\mathbf{x}',t) \mid \mathbf{u}(\mathbf{x},t) = \mathbf{v}_1,...,\mathbf{u}(\mathbf{x}_N,t) = \mathbf{v}_N >. \tag{4.5}$$

Thus, at every level in the hierarchy of *pdf* equations, a conditional eddy appears which is increasingly more detailed as the level of the hierarchy increases. The locations $\{ \mathbf{x}_1,\mathbf{x}_2,...,\mathbf{x}_N \}$ are variables in the *pdf* equation, and in particular, one point may be made to coincide with another so that an N-point conditional eddy can be reduced to an N-1 point conditional eddy. A *one-point* conditional eddy is of the form $< \mathbf{u}(\mathbf{x}',t) \mid \mathbf{u}(\mathbf{x},t) >$ where the data event $\mathbf{u}(\mathbf{x}_1,t)$ is located at $\mathbf{x}_1 = \mathbf{x}$.

A second type of conditional eddy is defined by using the local vorticity, $\omega(\mathbf{x})$ (Novikov 1967) or more generally the local deformation tensor

$$d_{ij} = \frac{\partial u}{\partial xj} \tag{4.6}$$

as events which specify the local rotation and deformation of the flow (Adrian and Moin 1987). The local velocity $\mathbf{u}(\mathbf{x},t)$ plus the local deformation tensor are sufficient to specify the state of motion at $\mathbf{x}$ in term of translation, rotation, and strain rate.

4.2 Properties of conditional eddies

Consider a general conditional eddy $< \mathbf{u}' \mid \mathbf{E} >$. The following reduction properties are statistical in nature

$$\langle\langle \mathbf{u}' \mid \mathbf{E} \rangle\rangle = \langle \mathbf{u}' \rangle \tag{4.7a}$$

$$\langle\langle \mathbf{u}' \mid \mathbf{E} \rangle \mid \mathbf{E}_s \rangle = \langle \mathbf{u}' \mid \mathbf{E}_s \rangle \tag{4.7b}$$

where $\mathbf{E}_s$ is a subset of $\mathbf{E}$. They describe the reduction of a conditional eddy to another conditional eddy with simpler event. If the elements of $\mathbf{E}$ are located at $\mathbf{x}$, then

$$\lim_{\mid \mathbf{x}' - \mathbf{x} \mid \to \infty} \langle \mathbf{u}' \mid \mathbf{E} \rangle = \langle \mathbf{u}' \rangle \tag{4.8}$$

because $\mathbf{u}'$ becomes statistically independent of $\mathbf{E}(\mathbf{x})$ as $\mid \mathbf{x}' - \mathbf{x} \mid$ becomes large. More generally a relation like (4.8) holds if $\mathbf{E}$ consists of multipoint events that are clustered around $\mathbf{x}$ within a sphere of finite radius.

Suppose $\mathbf{E}$ contains an event like $\mathbf{u}(\mathbf{x})=\mathbf{v}$. Then

$$\lim_{\mathbf{x}' \to \mathbf{x}} \langle \mathbf{u}'(\mathbf{x}) \mid \mathbf{E} \rangle = \mathbf{u}(\mathbf{x}) = \mathbf{v} \tag{4.9}$$

If **E** contains multipoint events $\mathbf{E} = \{\mathbf{u}(\mathbf{x}_\alpha) = \mathbf{v}_\alpha,\ \alpha = 1,\ldots,N\}$ then a relation similar to (4.9) holds for every point. Thus, the conditional eddy is like an interpolating function that is pinned to the data points.

Since **E** is independent of $\mathbf{x}'$, the operations of conditional averaging and differentiation with respect to $\mathbf{x}'$ commute. Then

$$\left\langle d_{ij}{}' \mid \mathbf{E} \right\rangle = \frac{\partial}{\partial x_j'} \langle u_i' \mid \mathbf{E} \rangle \tag{4.10}$$

i.e., the conditional average of the deformation field is the deformation of the conditional eddy velocity field. Clearly, a similar statement pertains to the vorticity, and if the flow is incompressible, the conditional eddy field is solenoidal:

$$\frac{\partial}{\partial x_i'} \langle u_i' \mid \mathbf{E} \rangle = 0. \tag{4.11}$$

Given a conditional eddy, the unconditional average of the product of **u'** with any function **Q(E)** can be found by averaging:

$$\langle \mathbf{Q}(\mathbf{E}) \langle \mathbf{u}' \mid \mathbf{E} \rangle \rangle = \langle \langle \mathbf{Q}\mathbf{u}' \mid \mathbf{E} \rangle \rangle \tag{4.12a}$$

$$= \langle \mathbf{Q}(\mathbf{E}) \mathbf{u}' \rangle \tag{4.12b}$$

The averaging requires, in principle, the *pdf* $f_\mathbf{E}$. The function Q can contain any combination of any of the quantities in **E.**

The conditional eddy specifies a vector velocity field given an event. If the event is uniquely associated with the conditional eddy, then the probability of a conditional eddy is related one-to-one to the probability of **E**. Thus, $< \mathbf{u}' \mid \mathbf{E} >$ gives the eddy shape and $f_\mathbf{E}$ gives the likelihood of occurrence. Unfortunately, the commonly used events such as quadrant analysis, VITA, and single point vectors do not guarantee a one-to-one relationship with a coherent structure. For example, a single coherent structure could have the same velocity vector at several points, or it could satisfy VITA or quadrant analysis criteria at several different points.

4.3 Properties of Linear Stochastic Estimates

The linear stochastic estimate of $< \mathbf{u}' \mid \mathbf{E} >$ shares many properties in common with the conditional eddy itself. Taking

$$\hat{u}_i' = A_i + L_{ij} E_j \tag{4.13}$$

where <**E**> is assumed to be zero, we see that (4.7a) follows immediately if $A_i = \langle u_i' \rangle$. (Recall that $L_{ij} = A_{ij}$). The reduction property (4.7b) is not, however, satisfied. The separation property (4.8), the coincidence property (4.9), and continuity (4.11) are each satisfied. To see that continuity is satisfied, take the divergence of $\hat{u}_i'$

$$\frac{\partial \hat{u}_i'}{\partial x_i'} = \frac{\partial L_{ij}}{\partial x_i'} E_j \tag{4.14}$$

From (3.3), the divergence of L_{ij} satisfies

$$\langle E_j E_l \rangle \frac{\partial L_{ij}}{\partial x_i'} = \left\langle \frac{\partial u_i'}{\partial x_i} E_l \right\rangle = 0 \tag{4.15}$$

But if (3.3) has solutions, then $\det < E_j E_l > \neq 0$, in which case the only solution of (4.15) is $\partial A_{ij} / \partial x_i' = 0$. Hence, $\partial \hat{u}_i' / \partial x_i' = 0$, as assured.

The continuity of the stochastic estimate of the conditional average is a very useful property. If means that the stochastic estimate is at the least a viable initial condition for the Navier-Stokes equation.

The essential difference between the conditional average and its LSE is the failure of the LSE to satisfy the reduction property (4.7b) in general. The special case $\mathbf{E} = \mathbf{u}(\mathbf{x})$ is an exception. The exact relationship for this case is

$$\langle u_i \langle u_j' \mid \mathbf{u} \rangle \rangle = \langle u_i u_j' \rangle . \tag{4.16}$$

The corresponding relationship obtained using the linear estimate of the conditional eddy is

$$\langle u_i L_{jl} u_l \rangle = L_{jl} \langle u_i u_l \rangle \tag{4.17a}$$

$$= \langle u_j' u_i \rangle \tag{4.17b}$$

where (4.17b) follows from (3.3). Hence the correct two-point spatial correlation function is retrieved identically from the LSE, as should be expected since the LSE takes the two-point spatial correlation function as input. For any other **Q**(**E**) the linear estimation does not guarantee reduction of <**Q**(**E**)<**u'**|**E**>> to <**Q**(**E**)**u'**>. This is important in regard to the use of LSE for closure approximations.

5. CONDITIONAL EDDIES AND COHERENT STRUCTURES

5.1 Isotropic Turbulence

Isotropic turbulence is defined by invariance of its statistics with respect to translations of the coordinate origin, reflections about planes through the origin and arbitrary rotations about the origin. It is understood to be the least structured of all turbulent flows. In this secton we shall summarize results concerning the structure of stochastically estimated eddies in isotropic turbulence. We will see that although the field is isotropic, the conditional eddies are anisotropic structures, owing to the imposition of anisotropic events.

The first studies of the structure of isotropic tubulence (Adrian 1979, Tung and Adrian 1980) concentrated on using the single-point velocity vector $\mathbf{E} = \mathbf{u}(\mathbf{x},t)$ to define the flow condition at $\mathbf{x}$. It was shown in general that the vector field of the conditional eddy $<\mathbf{u}'|\mathbf{u}>$ must be axisymmetric about $\mathbf{u}(\mathbf{x},t)$ for any isotropic turbulent flow. Linear stochastic estimates of $<\mathbf{u}'|\mathbf{u}>$ relate it approximately to the two-point spatial correlation, and they showed that the axisymmetric field has the form a vortex ring that is centered on $\mathbf{u}(\mathbf{x},t)$ and oriented with its plane perpendicular to it. Non-linear stochastic estimates revealed that corrections to the vortex ring due to higher order terms are significant only if the magnitude of the local velocity fluctuation $|\mathbf{u}(\mathbf{x},t)|$ is improbably large (Tung and Adrian 1980). For probable levels of the fluctuation the conditional eddy is a slightly perturbed vortex ring, suggesting a certain robustness of the ring structure in turbulence.

Ditter (1987) considered the conditional eddy given the local kinematics, defined by $<\mathbf{u}'|\mathbf{u},\mathbf{d}>$, the conditional average given both the velocity $\mathbf{u}(\mathbf{x},t)$ and the deformation tensor

$$d_{ij}(\mathbf{x},t) = \partial u_i / \partial x_j \tag{5.1}$$

Interest in this more complicated conditional event stemmed from the fact that $\mathbf{u}(\mathbf{x},t)$ and $\mathbf{d}(\mathbf{x},t)$ jointly specify the complete kinematic state of the fluid motion at $(\mathbf{x},t)$ in terms of the translation, rotation and rate-of-strain at that point. Consequently, more detailed and more accurate estimates of the surrounding structure were expected by virtue of having more copletely described the field at $(\mathbf{x},t)$. The following results draw heavily from the thesis of Ditter (1987) and Ditter and Adrian (1988).

In homogeneous turbulence the form of the linear stochastic estimate of $< u_i'(\mathbf{x}') \mid \mathbf{u}(\mathbf{x}), \mathbf{d}(\mathbf{x}) >$ is

$$\hat{u}_i' = \hat{u}'(\mathbf{x}') \quad = A_{ij} u_j(\mathbf{x}) + B_{ijk} d_{jk}(\mathbf{x}), i, j, k = 1,2,3 \tag{5.2}$$

where $\mathbf{x}'$ is any point around $\mathbf{x}$, and the estimation coefficients $\mathbf{A}$ and $\mathbf{B}$ are functions of

$$\mathbf{r} = \mathbf{x}' - \mathbf{x} \tag{5.3}$$

for homogeneous turbulence. (Hereafter, the time dependence will be suppressed).

The linear stochastic estimate in (5.2) is obtained by expanding the conditional average $< \mathbf{u}'|\mathbf{u},\mathbf{d} >$ in a first order Taylor series about $\mathbf{u}(\mathbf{x}) = 0$ and $\mathbf{d}(\mathbf{x})=\mathbf{0}$ using the fact that the mean value <**u**> vanishes in isotropic turbulence to show that the constant term in the expansion is zero.

The coefficients in the expansion are determined, as usual, by minimizing the mean square errors, computed for each component of u_i' by averaging over all possible random values of **u** and **d**.

$$e_i = \left\langle [u_i' - \langle u_i' \mid \mathbf{u},\mathbf{d}\rangle]^2 \right\rangle = \min imum, \quad i = 1,2,3. \tag{5.4}$$

Setting $\partial e_i / \partial A_{i\ell} = 0$ and $\partial e_i / \partial B_{i\ell m} = 0 \; for \; i,\ell,m = 1,2,3$ leads to the following equations for **A** and **B**.

$$A_{ij}(\mathbf{r})\, R_{j\ell}(\mathbf{0}) + B_{ijk}(\mathbf{r})\, R_{\ell j,k}(\mathbf{0}) = R_{\ell i}(\mathbf{r}) \tag{5.5}$$

$$-A_{ij}(\mathbf{r})\, R_{lj,m}(\mathbf{0}) + B_{ijk}(\mathbf{r})\, R_{\ell j,km}(\mathbf{0}) = R_{\ell i,m}(\mathbf{r}) \tag{5.6}$$

for i,j,k, ℓ,m = 1,2,3. Here $R_{\ell j}(\mathbf{r})$ is the two-point spatial correlation tensor.

$$R_{\ell j}(\mathbf{r}) = <u_\ell(\mathbf{x})\, u_j(\mathbf{x}+r)>, \tag{5.7}$$

and

$$-R_{\ell j,k}(\mathbf{0}) = [-\frac{\partial R_{\ell j}}{\partial r_k}]_{\,0} = < u_j d_{\ell k} > \tag{5.8}$$

$$-R_{\ell i,m}(\mathbf{r}) = [\frac{\partial R_{\ell i}}{\partial r_{m\; r}}]_{\mathbf{r}} = < u_i'\, d_{\ell m} > \tag{5.9}$$

$$-R_{\ell j,km}(\mathbf{0}) = [\frac{\partial^2 R_{\ell j}}{\partial r_k \partial r_m}]_{\mathbf{0}} = < d_{\ell m} d_{jk} > \tag{5.10}$$

give the correlations between the velocity and its derivatives in terms of $R_{\ell i}(\mathbf{r})$. Equations (5.5) and (5.6) constitute three independent sets (i=1,2,3) of twelve equations.

In isotropic turbulence $< u_\ell u_j > = u^2 \delta_{\ell j}$ where $u^2 \;=\; < u_i u_i > / 3$ is the mean square fluctuation, and $R_{\ell j,k}(\mathbf{0}) = 0$ from reflectional invariance. Hence, (5.5) and (5.6) decouple, and the following solution for the first estimation coefficient is obtained from (5.5) immediately:

$$A_{i\ell} = R_{\ell i}(\mathbf{r}) / u^2. \tag{5.11}$$

This is exactly the same as the solution found for the linear estimate of $< u_i' | \mathbf{u} >$.

The solution for $B_{ijk}(\mathbf{r})$ is complicated by the fact that

$$d_{11} + d_{22} + d_{33} = 0 \tag{5.12}$$

in incompressible flow. Consequently, one of the nine components of the deformation data is statistically dependent on the others, and the system of equations for B_{ijk} does not yield unique solutions. The non-uniqueness in (5.12) is made evident by noting that incompressibility requires that $R_{\ell i,\ell}(\mathbf{r}) = 0$. Hence, one of the equations is linearly dependent. In Adrian and Moin (1988) the non-uniqueness of **B** was eliminated by deleting d33 from the list of data given at **x** and performing numerical solutions of (5.5) and (5.6) for all i,j,k, ℓ,m = 1,2,3 *except* ℓ=m=3, j=k=3. This approach is best for situations requiring numerical solutions, but in the present case, where analytical solution is possible due to the high degree of symmetry in the statistics of the flow, it is easier to carry d33 and obtain non-unique solutions for B_{ijk}.This non-uniqueness is not troublesome because it vanishes when B_{ijk} is combined with djk to form the estimate of velocity.

Using $R_{\ell j,m}(\mathbf{0}) = 0$, (5.6) reduces to

$$B_{ijk} \; R_{\ell j,km}(\mathbf{0}) = R_{\ell i,m}(\mathbf{r}). \tag{5.13}$$

For isotropic, incompressible turbulence the correlation on the left-hand side is given by

$$R_{\ell j,km}(0) = \begin{cases} -u^2/\lambda^2 \;, & j = k = \ell = m \\ -u^2/\lambda^2 \;, & \ell = m \neq j = k \; or \; j = k \neq \ell = m \\ -2u^2/\lambda^2 \;, & k = \ell \neq j = m \\ 0 \;. & \text{otherwise,} \end{cases} \tag{5.14}$$

where λ is the longitudinal Taylor microscale. The solution to (5.13) becomes

$$B_{ill}(\mathbf{r}) = \text{arbitrary}, \tag{5.15a}$$

$$B_{i22}(\mathbf{r}) = -\frac{\lambda^2}{u^2}\{\frac{4}{3}R_{2i,2}(\mathbf{r}) + \frac{2}{3}R_{3i,3}(\mathbf{r})\} \tag{5.15b}$$

$$B_{i33}(\mathbf{r}) = -\frac{\lambda^2}{u^2}\{\frac{2}{3}R_{2i,2}(\mathbf{r}) + \frac{4}{3}R_{3i,3}(\mathbf{r})\} + B_{ill}(\mathbf{r}), \qquad (5.15c)$$

and

$$B_{i\ell m}(\mathbf{r}) = -\frac{2\lambda^2}{15u^2}\{R_{mi,\ell}(\mathbf{r}) + 4R_{\ell i,m}(\mathbf{r})\} \qquad (5.15d)$$

for $\ell \neq$m. Details can be found in Ditter (1987).

The complete stochastic estimate is

$$\hat{u}_i(\mathbf{x}+\mathbf{r}) = \frac{R_{ji}(\mathbf{r})}{u^2}u_j(x) - \frac{2\lambda^2}{15u^2}\{R_{ki,j}(\mathbf{r}) + 4R_{ji,k}(\mathbf{r})\}d_{jk}(x). \qquad (5.16)$$

In this equation the unknown function $B_{ill}(\mathbf{r})$, drops out of the product $B_{ijk}(\mathbf{r})\,d_{jk}(\mathbf{x})$ due to the incompressibility condition d_{ii} = 0. Each term in (5.16) can be evaluated in terms of the longitudinal correlation coefficient

$$f(r) = \langle u_1(\mathbf{x})u_1(\mathbf{x}+r\hat{\mathbf{e}}_1)\rangle / u^2 \qquad (5.17)$$

by means of the isotropic relation (Batchelor 1986)

$$R_{ji}(\mathbf{r}) = u^2\{-\frac{1}{2}f'(r)\frac{r_i r_j}{r} + [f + \frac{1}{2}r\,f']\,\delta_{ij}\}. \qquad (5.18)$$

The grid turbulence measurements of Van Atta and Chen (1968) were used to evaluate *f(r)* at 48 mesh spacings downstream of a bi-plane grid. The mesh spacing was M= 25.4 mm and the conditions at x_1/M = 48 were u=0.24 ms^{-1}, L_{11} (longitudinal integral length scale) = 13 mm, λ = 3.9 mm, and uL_{11}/ν = 200. For the results concerning large scale structure the longitudinal correlation coefficient *f(r)* is represented adequately by the curve fit

$$r(r^*) = (1 - 0.435r^* + 0.188r^{*2}$$

$$-0.026r^{*3})\,e^{-0.626r^*} \qquad (5.19)$$

where r* = r/L_{11}. This representation is not valid for r less than approximately one Taylor microscale; but at large Reynolds number λ/L_{11} <<1, so the representation is accurate everywhere except in a small region about r = 0. (For Van Atta and Chen's experiment $\lambda / L_{11} = 0.3$ *at* $x/M = 48$. The longitudinal correlation coefficient f(r) and the associated transverse correlation coefficient g(r) are shown in Fig. 5.1. Note that the data of Van Atta and Chen are distinctive in that f(r) becomes negative at large separations, with the

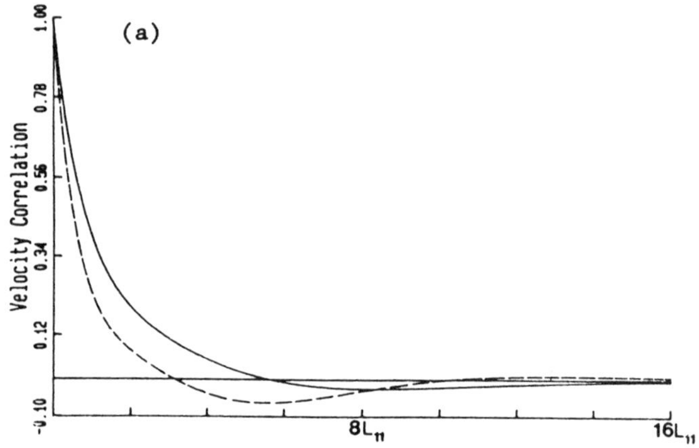

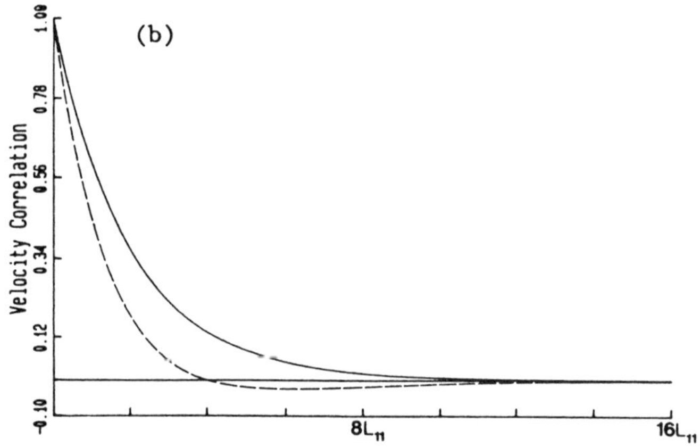

Figure 5.1. Velocity correlation data of Van Atta and Chen (1968): f(r)———; g(r) - - - --.

consequence that g(r) has two zeroes. The implications of this behavior in regard to the structure of the conditional eddy $<\mathbf{u}'|\mathbf{u}>$ will be discussed shortly.

The results will be presented in terms of the following dimensionless variables:

$$u^* = \mathbf{u}/u, \; \mathbf{d}^+ = \lambda \mathbf{d}/u \tag{5.20}$$

$$R^*_{ij} = R_{ij}/u^2, \quad R^*_{ij,k} = L_{11}R_{ij,k}/u^2, \tag{5.21}$$

where
$$\lambda = u/\sqrt{<(\partial u_1/\partial x_1)^2>}.$$

In terms of these variables,

$$\hat{u}_i^*(\mathbf{x}^*+\mathbf{r}^*) = R_{ji}(r^*)\,u_j^*(\mathbf{x}^*) - \frac{2\lambda}{15L_{11}}\{R^*_{ki,j}(\mathbf{r}^*) + 4\,R^*_{ji,k}(\mathbf{r}^*)\}\,d^+_{jk}(\mathbf{x}^*). \quad (5.22)$$

The non-dimensionalization of **d** by u/λ is chosen to make the elements of $\mathbf{d}^+$ of order unity when their dimensional values are of the order of the root mean square spatial derivative. The nondimensional derivatives of the correlation function $R^*_{ki,j}$ are of order one when r is of the order of the integral length scale.

5.2 Pure translation event

In pure translation, u(**x**) = (|**u**(**x**)|,0,0). and **d**(**x**) = **0**. The linear stochastic estimate is a pair of vortex rings centered on **x** and axisymmetric about **u**(**x**), Fig. 5.2. The outer ring is a consequence of the negative tail in van Atta and Chen's measurements of f(r). The core of the inner ring is centered on the first zero of the transverse correlation function, g(r), and the core of the outer ring is centered on the second zero of g(r). It moves off to infinity if the negative region of f(r) is reduced to zero, e.g., if f(r) decreases monotonically.

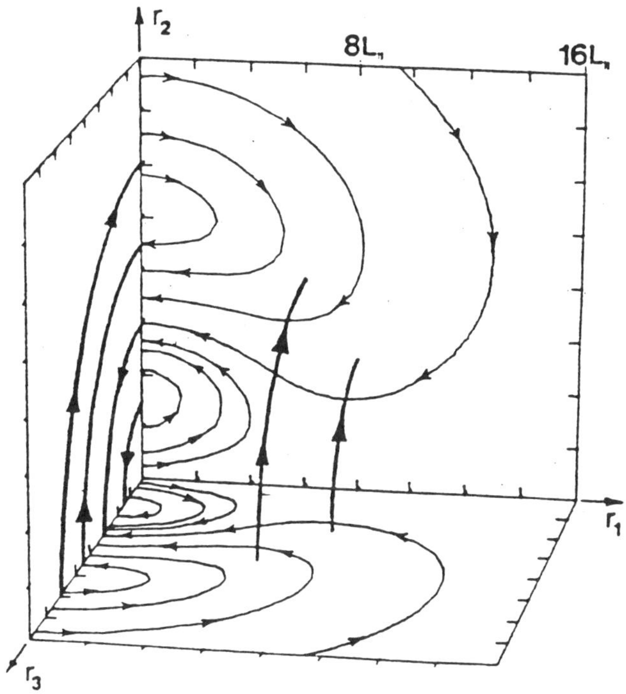

Figure 5.2. Conditional eddy given pure translation. Streamlines (light ink) and vortex lines (heavy ink).*[from Ditter and Adrian 1988]*

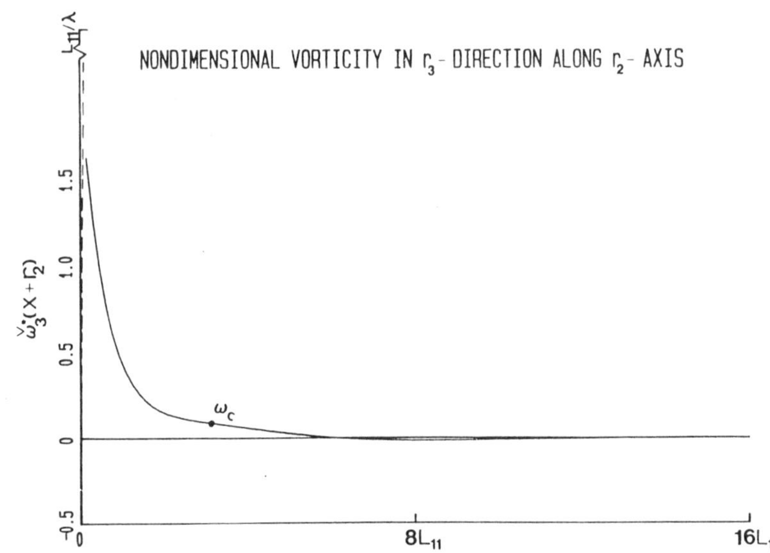

Figure 5.3. Distribution of $\hat{\omega}_3^*(\mathbf{x}+\mathbf{r})$ along the r_2 axis for a ring vortex conditioned on $\mathbf{u}(\mathbf{x})$ = (u, 0,0). The dashed line is calculated from $f \doteq 1 - r^2/2\lambda^2$. The vorticity at the center of the inner vortex ring is ω_c .*[from Ditter and Adrian 1988]*

The magnitude of the velocity field within the inner ring is of the order of $\mathbf{u}(\mathbf{x})$, and if $\mathbf{u}(\mathbf{x})$ were strong a random realization similar to this field would be expected to be strong enough to persist as an organized motion, despite the presence of other random fields surrounding it. In contrast, the outer ring is much weaker. Because it is unlikely that such a weak structure could maintain itself for long against the disturbances of the surrounding random environment, it is perhaps more reasonable to interpret the inner ring as the mean field of an organized structure, and the outer ring as the mean of all of the other organized structures that surround it. The fact that the mean surrounding field is non-zero implies that there is an interaction between the inner vortex ring and surrounding organized structures which may, themselves, be ring-like.

The vorticity distribution along a radial line in the $r_2 - r_3$ plane of the ring is shown in Fig. 5.3. The solid line is calculated from (5.18), (5.19), and (5.22); it represents the high Reynolds number behavior of the large, inviscid scales of motion, $r_2 \sim L_{11} >> \lambda$. In this region, which covers most of the vortex ring, the vorticity scales with $u(\mathbf{x})/L_{11}$. For small values of r, the Taylor series expansion

$$f(r) = 1 - r^2/2\lambda^2 + \ldots \tag{5.23}$$

combined with (5.18) and (5.22) predicts that the vorticity varies like $5|\mathbf{u}(\mathbf{x})|/\lambda$. indicated by the dashed line in Fig. 5.3. The vorticity of the purely translational conditional eddy vanishes at r = 0, and increases rapidly to a maximum value before decaying according to the solid line in Fig. 5.3. Subsidiary calculations using a more complete model of f(r) show that the maximum vorticity is of order $u(\mathbf{x})/\lambda$, and that it occurs at a radius of order λ. For small

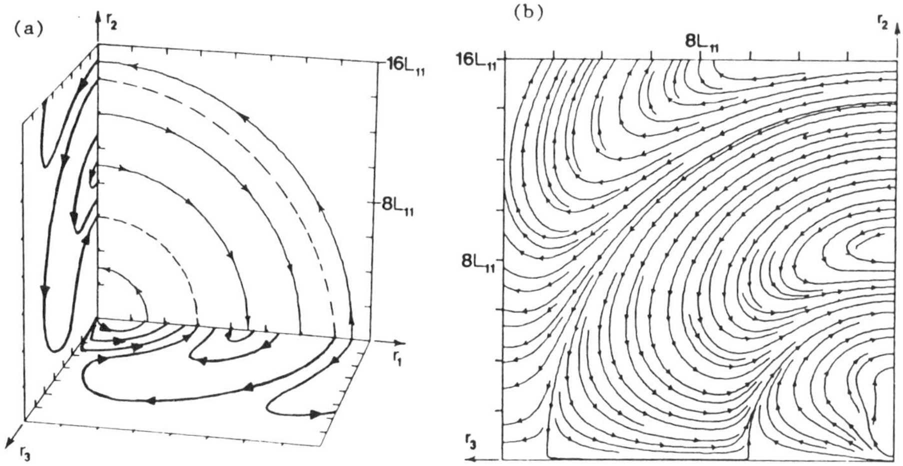

Figure 5.4. Conditional eddy given pure rotation about the r_3 axis: a) streamlines in $r_1 - r_2$plane and vortex lines in $r_1 - r_3$ and $r_2 - r_3$ planes demonstrating symmetry about r_3 −axis; b) vortex lines in $r_2 - r_3$plane. *[from Ditter and Adrian 1988].*

radii the flow is affected by viscous stresses, and the vorticity is primarily due to an axisymmetric shearing motion, as opposed to a circulating vortical motion. The ratio of the maximum vorticity to the vorticity in the core of the inner vortex ring, indicated by ω_c in Fig. 5.3, is proportional to L_{11} / λ. Thus, at low Reynolds the vorticity of the core is comparable to the maximum vorticity, whereas at high Reynolds number it is much less than the maximum, and most of the vorticity is concentrated in a region of radius λ. However, the *circulation* of the core of the ring is of order $[\mathrm{u}(\mathbf{x})/ L_{11}]L_{11} = \mathrm{u}(\mathbf{x})$. Thus, it is comparable to the circulation of the small scale region, $[\mathrm{u}(\mathbf{x})/ \lambda]\,\lambda \sim \mathrm{u}(\mathbf{x})$.

5.3 Pure deformation events

Four conditions in which the state of the flow at x is one of pure deformation are considered here: $\mathrm{u}(\mathbf{x}) = \mathbf{0}$ and $\mathbf{d}(\mathbf{x}) \neq \mathbf{0}$. They are pure rotation, Fig. 5.4; axisymmetric strain, Fig. 5.5; plane strain, Fig. 5.6; and plane shear, Fig. 5.7. In each oblique view the light lines indicate streamlines and the heavy lines indicate vortex lines.

The state of pure rotation about the r3-axis in Fig. 5.4 is given by $\mathbf{u}(\mathbf{x}) = \mathbf{0}$, $d_{12} = -d_{21} = -\,\omega_3(\mathbf{x})$, and $d_{ij}(\mathbf{x}) = 0$, otherwise. The field is axisymmetric about r_3, and it consists of three counter-rotating shells. If vortex rings are the dominant organized structure in isotropic turbulence, one might expect that specifying the vorticity at **x** would place **x** on the core of the ring with its *tangentially oriented* vorticity aligned with the given vorticity. However, merely specifying $\omega(\mathbf{x})$ is insufficient to identify the orientation of a ring; the

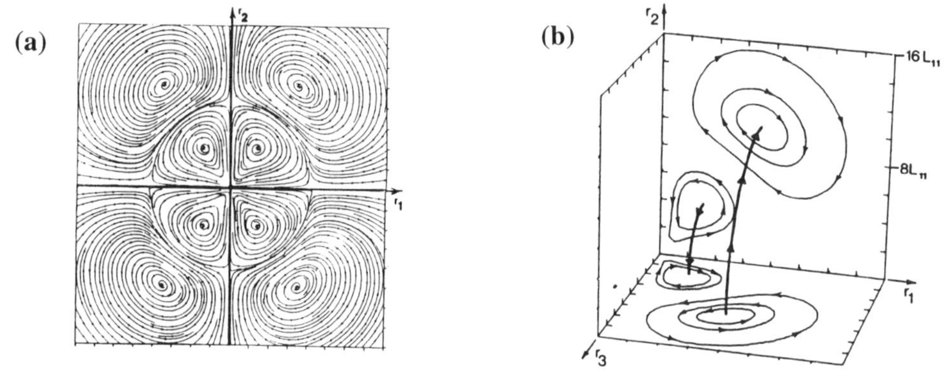

Figure 5.5. Conditional eddy given axisymmetric strain.(a) streamlines in $r_1 - r_2$ plane;(b)streamlines (light ink) in the $r_1 - r_2$ and $r_1 - r_3$ planes with vortex lines (heavy ink) demonstrating the axisymmetry of the flow about the r_1-direction.*[from Ditter &Adrian 1988]*

vorticity could correspond to any ring whose central velocity u(**x**) is perpendicular to $\omega(\mathbf{x})$. Thus, the conditional eddy is Fig. 5.4 is rotationally symmetric about ω, and corresponds, qualitatively, to an average of translational vortex rings rotated about $\omega(\mathbf{x})$.

Contours of constant $|\omega|$ for pure rotation indicate a long, rod-like zone of high vorticity aligned with the r_3 axis. The length of this rod is of order several integral length scales and its diameter is of order of the Taylor microscale. The strength of the vorticity in the rod is of the order of the given vorticity, while that outside of the rod, i.e., most of the region shown in Fig. 5.4, is of order $|\omega|\lambda/L_{11}$, i.e., much smaller. The intense rod-like structure may be related to the vortical "worms" that have been observed in direct numerical simulation of isotropic turbulence.

The state of pure axisymmetric strain at **x** with principal axis of positive extension aligned with the r_1-direction is given by $d_{11}(\mathbf{x}) = -2d_{22}(\mathbf{x}) = -2d_{33}(\mathbf{x})$. The velocity vector field is axisymmetric about the r_1-axis, Fig. 5, and it consists of four vortex rings. The two smaller rings create the axisymmetric strain at **x**, and the two outer rings are consequences of the negative tail of f(r).

Plain strain at x with the principal axis of positive extension laying in the $r_1 - r_2$ plane at 45° to the r_1-direction and the principal axis of compression lying at 135° is given by u(**x**) = **0** and $d_{12}(\mathbf{x}) = d_{21}(\mathbf{x})$ with all other components of **d**(**x**) vanishing. In this case the nature of **d** precludes axisymmetry but implies reflectional symmetry with respect to the r_1-r_2, $r_2 - r_3$ and $r_1 - r_3$ planes, and with respect to the planes at 45° and 135° to the r_1-axis. The flow consists of four inner vortical regions and four outer regions connected as shown by the vortex lines in Fig. 5.6.

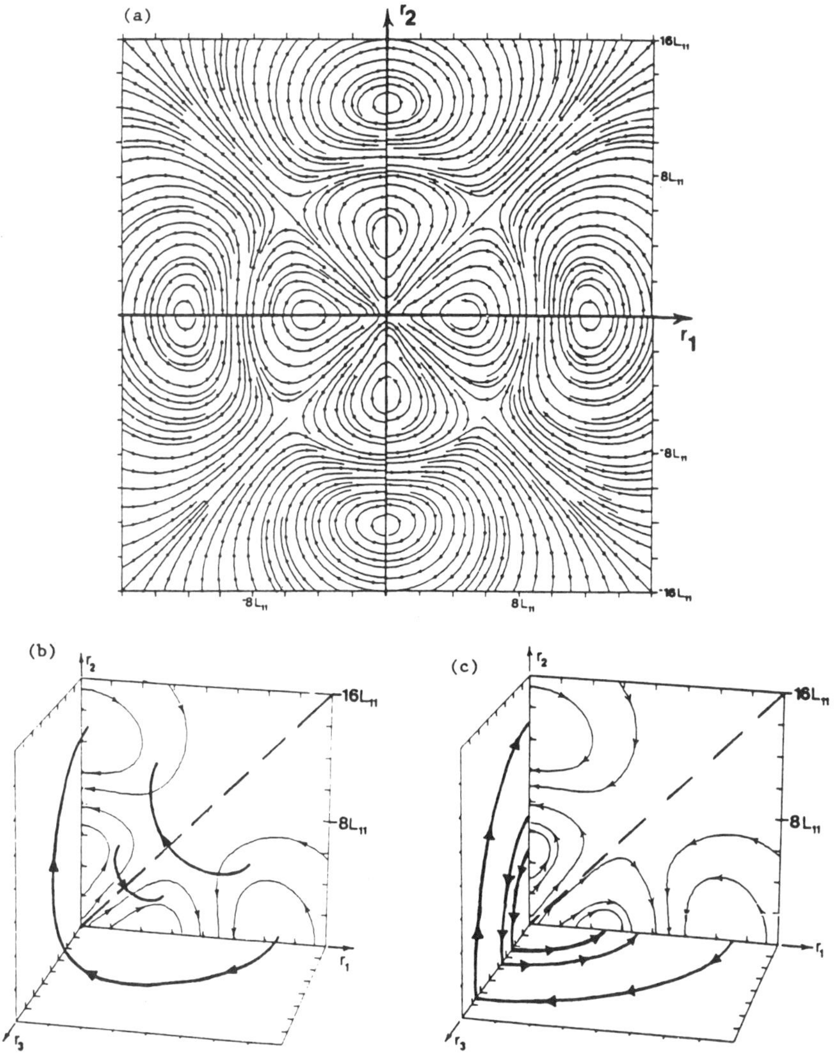

Figure 5.6 Conditional eddy given plane strain. a) streamlines in the $r_1 - r_2$ plane; b) streamlines (light ink) in the $r_1 - r_2$ plane and out-of-plane vortex lines (heavy ink); c) streamlines in the $r_1 - r_2$ plane and vortex lines in the $r_1 - r_3$ and $r_2 - r_3$ planes.*[from Ditter &Adrian 1988]*.

6. CONDITIONAL EDDIES IN ANISOTROPIC TURBULENCE

In anisotropic turbulence, the correlation tensor and, hence, the linear stochastic estimate become more complicated and reflect the asymmetries of the flow. For example, in homogeneous shear flow the form of the conditional eddy is a hairpin vortex similar to that shown in Fig. 6.1 (Adrian and Moin 1988; Adrian 1990). If the fluctuating vorticity is given one of its most probable values, which is inclined at approximately 45° to the direction of the mean flow, the plane of the hairpin lies in the same direction. This is the direction of maximum straining motion, as expected. The hairpin pumps fluid by vortex induction in such a way that if the hairpin is inclined forward, it produces a second quadrant event, i.e., negative u--component and positive v-component. The maximum induced velocity is approximately perpendicular to the plane of the hairpin, and being large, it contributes substantially to the mean Reynolds shear stress. The hairpins may also be inclined negatively, in which case the vortex induction produces a fourth quadrant event. Hairpins such as these are observed in the instantaneous realizations of direct numerical simulations of turbulence by Rogers and Moin (1987), and their scale and geometry are mimicked well by the linear stochastic estimates computed by Adrian and Moin (1988). This comparison between the linear stochastic estimate and the hairpins observed in random realizations is one of the best pieces of evidence that the stochastic estimates, given a proper condition, are able to faithfully reproduce all of the salient aspects of a coherent structure.

The structure of conditional eddies in inhomogeneous anisotropic shear flow has been studied using direct numerical simulations of low Reynolds number turbulent channel flow (Adrian, Moin and Moser 1987; Moin, Adrian and Kim 1987; Kendall 1992) The three-dimensional database makes it possible to obtain the full two-point spatial correlation tensor.

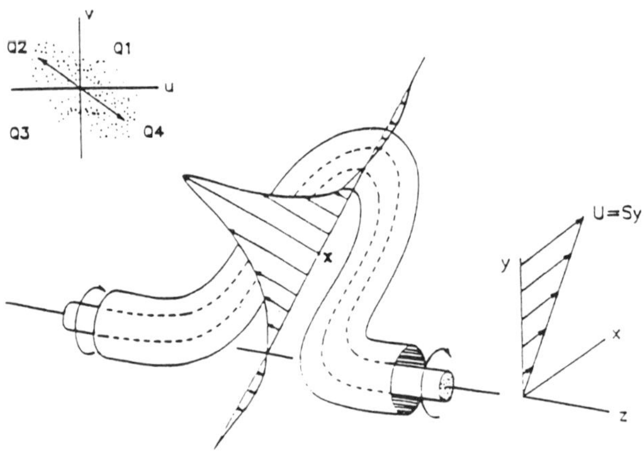

Figure 6.1 Inclined, hairpin-shaped conditional eddies in homogeneous shear flow induce flows that create strong Reynolds stress events in the second and fourth quadrants.*[from Adrian 1990].*

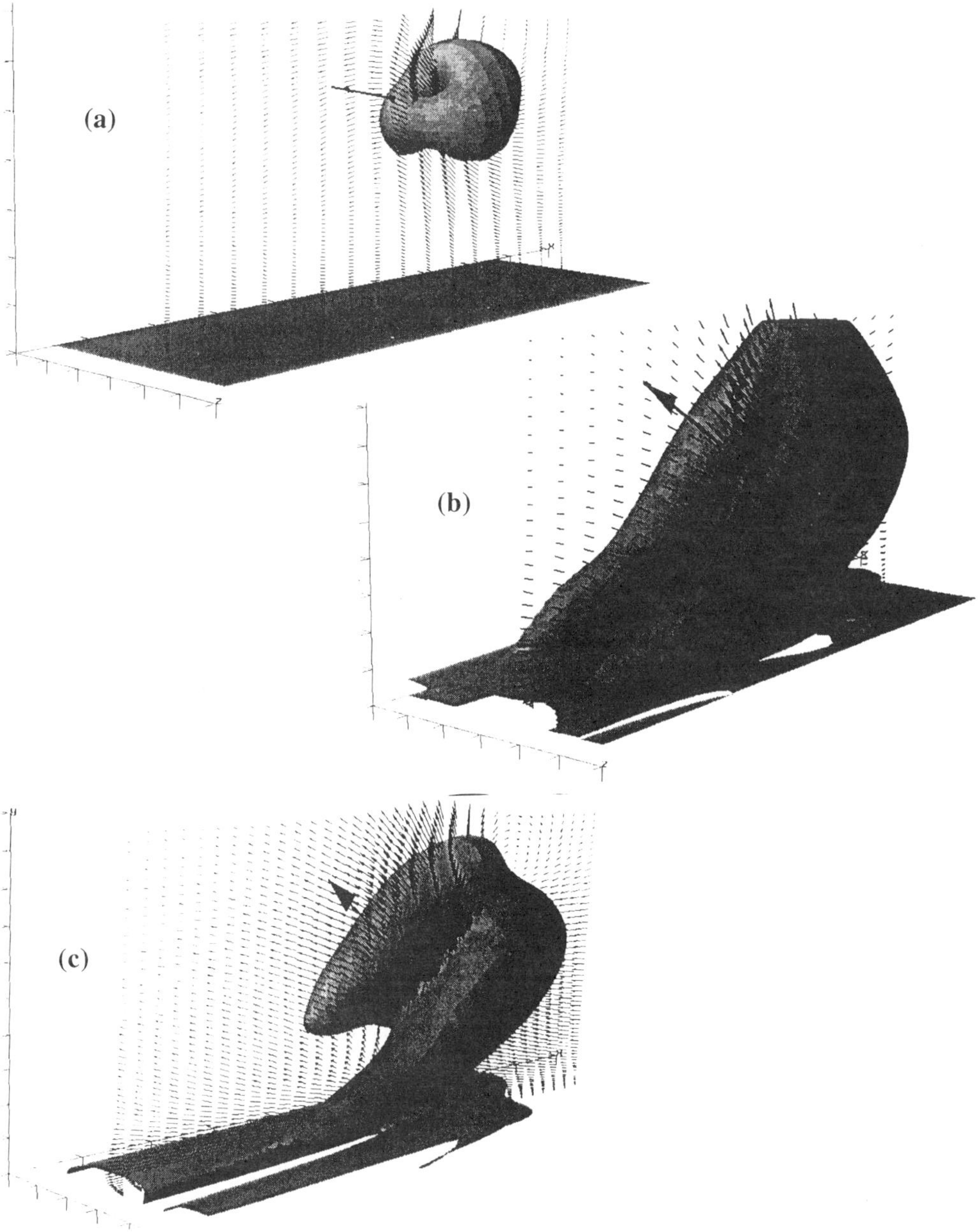

Figure 6.2 Conditional eddies in low Reynolds number turbulent channel flow estimated given a single-point conditional event $\mathbf{u}(\mathbf{x})$. Surfaces correspond to constant enstrophy fluctuation. Given vectors are the second quadrant events that correspond to the maximum contribution to the mean Reynolds shear stress. (a) $y^+= 180$ (channel centerline); $\mathbf{u}^+=(-0.81, 0.07, 0)$; (b) $y^+=103$, $\mathbf{u}^+=(-1.76, 1.1, 0.0)$; (c) $y^+=103$, $\mathbf{u}^+=(-1.76, 1.1, 2.0)$.

The results of the stochastic estimation based on a second-quadrant velocity vector event at a single point indicate an inclined hairpin structure with long quasi-streamwise vortical legs extending upstream from the inclined hairpin, such as those shown in Fig. 1.4 ($y^+=49$) and Fig.6.2(b) ($y^+=103$). In these figures the fluctuating magnitude of the vorticity is constant on the surfaces. The vectors denote velocity, and the heavy vector is the given velocity. The values of the velocity given in the caption to Fig. 6.2(b) correspond to the maximum probable contribution to the mean Reynolds stress. Very close to the wall the hairpin is not evident, and the primary feature of the conditional eddy is two long, counter-rotating, quasi-streamwise wall vortices. Farther away from the wall, the hairpin inclines upwards at a greater angle, and ultimately, on the centerline of the channel, where the mean shear is zero, the hairpin transforms into a simple vortex ring as shown in Fig. 6.2(a) ($y^+=180$). The inclined hairpin vortices on the wall induce low momentum flow to move upwards as do the long streamwise vortices behind the hairpin. This low momentum flow away from the wall creates strong Q2 events which are a dominant contributor to the mean Reynolds shear stress, $-\langle uv \rangle$ in the near wall region. The similarity between the hairpin in the wall layer and the hairpin in the homogeneous shear flow is noteworthy. However, in the wall flows, the quasi-streamwise vortices behind the hairpins are also very important in the creation of Reynolds shear stress.

Within the accuracy of linear estimation, specifying a Q4 event instead of a Q2 event simply reverses the sign of the velocity vector field of the conditional eddy. Thus, given a sweep event, the form of the conditional eddy is just an inclined hairpin with down-flow toward the wall. Although the fluctuating field is modified in a trivial way by such a sign change, the sum of the mean field plus the fluctuating conditional eddy field is different for the Q2 event relative to the Q4 event. This is particularly important when the fluctuation is strong enough for the non-linear interaction between the mean flow and the fluctuation to be important, for then sign difference will matter.

One of the criticisms of the stochastically estimated hairpin structure found from a Q2 event is that it is symmetric with respect to x-y plane through the center of the event. Robinson (1991) has noted that symmetric hairpin vortices are seldom observed in realizations of direct numerical simulations of turbulent channel flow and boundary layer flow, and statistical analysis by Guezennec, Piomelli and Kim (1987) has shown that hairpins are often asymmetric in nature, having circulation around one leg that is stronger than around another. The question is whether this sort of behavior can be obtained from stochastic estimation which is based on an inherently symmetric two-point spatial correlation tensor. In fact, one-legged hairpins are readily created from stochastic estimation by including a non-zero w-component of velocity in the conditional vector event. The effect of the non-zero w-component of velocity is to create a flow in the z-direction which generates a shear that has an x-component of vorticity. This x-component of vorticity intensifies one leg of the hairpin and weakens the other. The result is as shown in Fig. 6.2(c), which has been computed from the turbulent channel flow direct numerical simulation.

As well as their kinematics, the dynamics of conditional eddies are also important. While the conditional eddy is extracted from the physically based spatial correlation tensor, there is no guarantee that this structure is a stable structure, or that it does not quickly evolve into some other type of structure. For the case of turbulent channel flow at low Reynolds number discussed above, the dynamics of the eddies have been studied by Kendall (1992) and Zhou, Adrian and Balachandar (1995), by using the linear stochastically estimated vortex

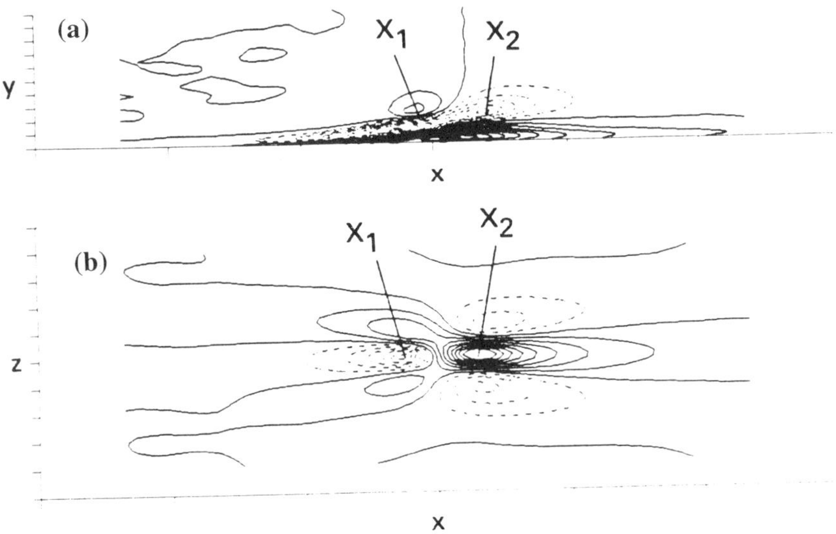

Figure 6.3 Linear estimation of channel flow given a Q2 event at $\mathbf{x_2}$ and a Q4 event at $\mathbf{x_1}$. (a) Contours of ω_z in the plane passing through $\mathbf{x_1}$. (b) Contours of the v-component of velocity in the x-z plane passing through $\mathbf{x_1}$. Solid line denote positive values.*[from Adrian, Moser and Moin 1987].*

as an initial condition and computing its evolution with time in a turbulent channel flow. The computations take the eddy in combination with the mean turbulent velocity profile. The results indicate that eddies whose conditional vector is less than approximately one rms velocity fluctuation in strength evolve very slowly after a brief transient from the initial conditional eddy into a slight sharper hairpin. After the transient, the eddy remains stable for several hundred viscous time units, which is more than long enough to contribute substantially to the mean time statistics. During this time the eddies sometimes grow slightly, but then generally decay slowly, while continuing to hold the shape of an initial conditional hairpin. More recent work by Zhou, Adrian, and Balachandar (1995) indicates that at sufficiently high intensity the initial hairpin vortex undergoes nonlinear evolution and begins to spawn further hairpin vortices.

One of the deficiencies of the hairpins estimated by specifying a single point Q2 vector event is that they do not exhibit the shear layer that is observed in direct numerical simulations and in experimental work to lie above and to the back of the hairpin vortex. This shear layer is an interaction between the slow, upward-moving fluid induced to flow between the legs of the hairpin, and faster, sweep-like fluid motions that approach the hairpin from behind. In order to observe this sort of flow field in the stochastic estimate, Adrian, Moser, and Moin (1987) and Guezennec (1989) imposed a two-point event which was a Q2 event at a point $\mathbf{x_2}$, as in Fig. 6.2, with a Q4 event slightly upstream of it at a point $\mathbf{x_1}$. The net effect was to create a shear layer very similar to that observed in the direct numerical simulations. We note that this combination Q2-Q4 event could also have been specified in effect by specifying a stagnation-point flow for the deformation tensor with a velocity vector of zero fluctuation intensity. An example of the Q2-Q4 combination event is shown in Fig. 6.3.

(a)

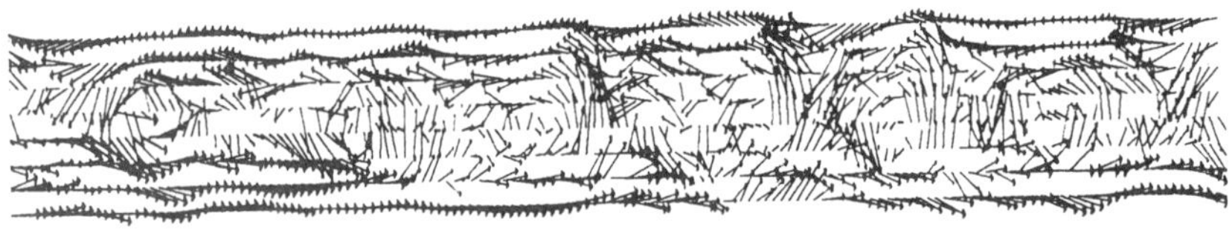

(b)

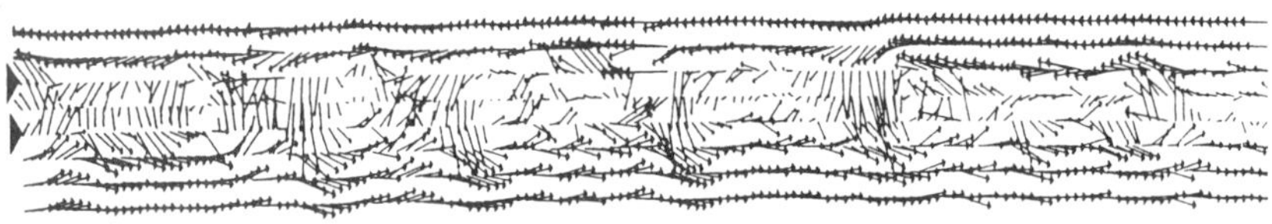

Figure 6.4 (a) Velocity vectors along horizontal lines in the shear layer of a round jet; (b) stochastically estimated field given data along the two line indicated by the arrows.*[From Ukeiley, et al. 1993].*

In turbulent shear layers, the linear stochastic estimate, given a velocity vector in the streamwise cross-normal plane at a single point, indicates spanwise roll cells in the fully turbulent zone of the turbulent shear layer (Tung 1980). Similar but much richer results are found in the shear layer of a turbulent jet by Cole, Glauser and Guezennec (1992). In this latter study the conditional event consisted of the time series of sampled velocity vectors at two radial locations bracketing the shear layer. Within the accuracy of Taylor's hypothesis the time series corresponded to sampling along two streamwise lines, which are indicated in Fig. 6.4. Comparison of the instantaneous realization in Fig.6.4(a) to the linear stochastic estimate in Fig.6.4(b) shows broad agreement. The stochastic estimate is somewhat smoother, indicating a low pass filtering effect, and the estimated fields decrease more rapidly than the realizations in the normal direction to the shear layer.

Roll cells in a turbulent wake have been studied by Gieseke and Guezennec(1993), using stochastic estimation given a single vector, Fig. 6.5. The flow pattern takes the form of two pairs of inclined vortices, which agrees rather well with the double roller model inferred from the correlation functions (Townsend 1976). Another way to view the conditional eddy is as a vortex ring that has been folded about a line in the streamwise-spanwise (x-z) plane. combined with time series of velocity vectors. Gieseke and Guezennec(1993) also used a time series along a line in the x-z plane to form a time-varying estimate of the velocity field given instantaneous data along the lines. The creation of a time varying depiction of the flow field in this way is similar to the work of Hassan (1980) in pipe flow, except that the latter estimates used only time varying data at one-point. From the line samples in time, Gieseke and Guezennec(1993) were able to predict a time-varying stochastic estimate which showed the translation and evolution of coherent structures in the wake. The results indicate inclined roller vortices. The movies show how the double roller structures interact in as they move. It is remarkable that fields of this complexity, c.f. Fig. 6.6, can be inferred from a single set of

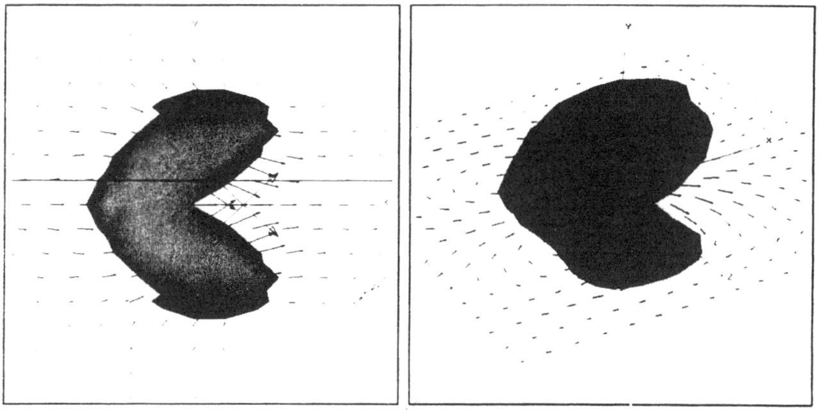

Figure 6.5 Stochastically estimated velocity field and surface of constant vorticity magnitude given Q2 event below the centerline and a Q4 event above the centerline of a turbulent wake. *[From Gieseke and Guezennec 1993}.*

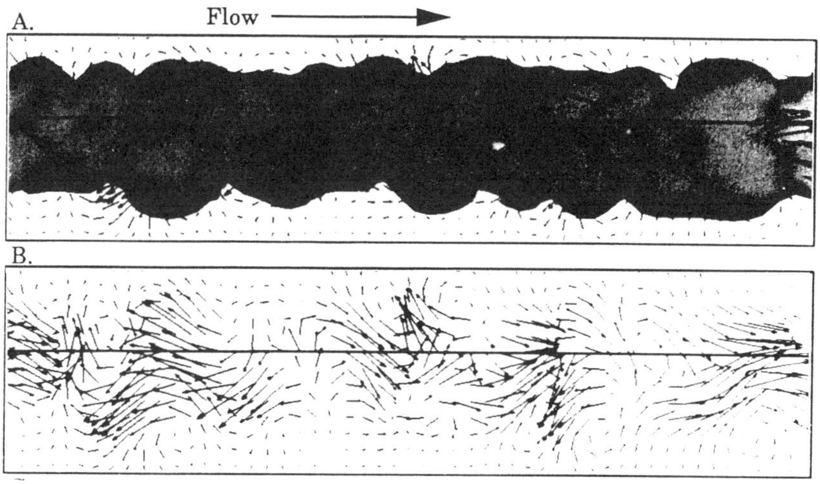

Figure 6.6 Estimated field constructed from a time series. Side view of surface of constant enstrophy.*[From Gieseke and Guezennec 1993}.*

correlation function data almost as easily as fields given the velocity at say, one or two points.

Turbulent thermal convection between uniform, horizontal planes should have statistics that are symmetric with respect to rotations about the vertical direction of the gravitational vector. This symmetry occurs in the presence of strong inhomogeneity in the vertical direction. It is known from physical observations that the flows in thermal convection often take the form of plumes hot (or cold) fluid that rise(or fall) across the horizontal layer, or unsteady, mushroom shaped thermals that occur in transient fashion as fluid moves away from one of the planes. At low Rayleigh numbers it is also known that convective roll cells form which are nearly two-dimensional. Interestingly, all of these phenomena can be observed by stochastic estimation, but only if the conditional event is chosen carefully (Balachandar and Adrian 1993). For example, stochastic estimation given a positive vertical velocity and a warm temperature fluctuation should give a roll cell or a thermal. In fact, unless the point for the event is very close to the surface, the flow is a large vortex ring that spans the fluid layer with its axis pointing vertically. After a little reflection it is clear that this shape is almost inevitable because an event that is axisymmetric about the vertical coupled with a two-point spatial correlation that is axisymmetric about the vertical must yield an estimated flow field with the same axisymmetry. How then can stochastic estimation ever predict two-dimensional roll-cells in such a flow? The answer is that one must use multi-point events to impose the two-dimensionality. Alternatively, but equivalently, one can work in both physical space and wavenumber space, using the latter to specify a two-dimensional event.

The stochastic estimate given a warm, rising event at one point is essentially an axisymmetric plume for all locations of the event except those very close to the horizontal boundaries, where the conditional eddy takes the form of an axisymmetric thermal with the characteristic mushroom-shaped head. This weaker flow field can also be found further away from the boundaries if one first estimates the large-scale roll-cell using a datum far from the wall, then subtracts the large scale field to obtain the smaller scale thermal field. This process is effective in extracting small eddies embedded in the fields of larger, stronger eddies.

In summary, for inhomogeneous and anisotropic flow fields, the linear stochastic estimate yields structures that are very similar to the flow structures observed in reality, provided that the given data reflect the important characteristics of the structures. The results are not very sensitive to the data, if they are reasonably chosen, but neither are they insensitive.

7. NEW DIRECTIONS IN STOCHASTIC ESTIMATION

New developments in the theory of stochastic estimation are being made on several fronts, most of them having to do with the intelligent selection of given events and/or optimization criteria. Developments are also being made in the application of stochastic estimation to a variety of turbulent flow problems.

Compared to modal decomposition methods that have global basis functions, like proper orthogonal decomposition, one of the strengths of stochastic estimation is that it yields fields that are compact in space if the events are located in a compact region (certainly this is true for single point events). By virtue of their compactness such fields can be viewed as representing eddies. However, from the viewpoint of interpolation and extrapolation, the fact that the fields decay with distance from the conditional events implies a loss of accuracy. The reduced energy is evident in Fig.6.4(b) compared to Fig.6.4(a). Cole, Glauser and Guezennec (1992) view this as a loss of signal energy and suggest that proper location of the event data can reduce the effect. Leboeuf and Mehta (1994) go further and recast the mean square error minimization problem as one in which the loss of energy of the signal is minimized. The resulting estimation problem is no longer linear, but the non-linear algebraic equations are not difficult to solve. The solution effectively amplifies the decaying correlation functions to enhance their magnitudes at large distances from the event data. The process can enhance the signal, but there is clearly a limit to this process, as the loss of correlation that occurs with increasing separation between the event data and the estimated fields is real, and cannot be replaced by amplifying small numbers.

Closure of the pdf equations requires closure of conditional averages of the velocity and the Reynolds stress. Linear estimation can be used for the conditional eddy terms, but a different sort of approximation is needed for the conditional Reynolds stress. This problem is much like the closure for the unconditional Reynolds stress, except that the only place that the conditional Reynolds stress appears in the pdf equations is in the non-linear part of the pressure term, i.e. the slow pressure term. Efforts to approximate this term and to experimentally evaluate their success have been made by Tung (1980) and Elam (1987). Chang (1985) also measured the conditional pressure term and its stochastic estimate directly in the shear layer of an axisymmetric jet.

The problems of large eddy simulation are naturally phrased in terms of stochastic estimation. In Adrian (1977) and later in (Adrian 1990) the sub-grid scale closure problem was formulated as one of estimating the subgrid scale stresses given data in the form of the resolved scale field. It was shown that an 'optimal" algorithm could be developed that was optimal in the sense that it minimized the mean square error involved in the estimate of the time-rate-of-change derivative at each time step given the available grid data. For the usual large eddy formulation in which the resolved scales field is defined by a low pass filtering operation, the best estimate of the sub-grid scale Reynolds stresses was shown to be the conditional average of the SGS stresses given the large scale field. In this sense it is not

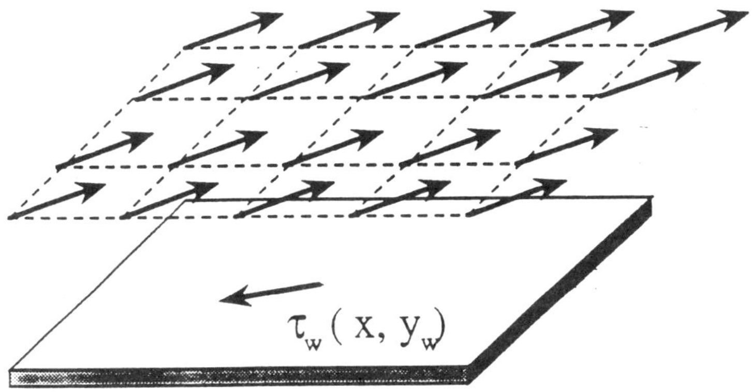

Figure 7.1 Stochastic estimation of the wall shear stress given velocity data on a neighboring numerical grid plane.

necessary to develop a relationship between the instantaneous SGS stresses and the instantaneous resolved scale field, explaining, perhaps, why the well known lack of correlation between these quantities does not have graver consequences for the closure approximations that are used in large eddy simulation.

A related problem in stochastic estimation is that of formulating a boundary condition for the resolved scale field in the wall region of a large eddy simulation. The boundary conditions that have been used are often ad hoc, and usually have little physical motivation. The work by Piomelli (1987) and Schumann (1975) expresses the wall shear stress in terms of the velocity of the resolved scales at the grid point closest to the wall. Schumann (1975) takes the point immediately above the wall location, while Piomelli (1987) takes a downstream point located so as to maximize the correlation between the wall shear stress and the grid velocity. Either boundary condition can be viewed as a linear estimate of the wall shear stress given the grid velocity datum.

These efforts suggest a generalized formulation of the problem in which all of the velocities computed on the first grid plane above the wall are used to estimate the wall shear stress as a function of position on the wall. c.f. Fig. 7.1. Note that the estimated field of wall shear stress is fully two-dimensional and time-dependent since the resolved scale data are evolving in time. This approach has been evaluated by Bagwell, Adrian, Moser and Kim (1993) using data from direct numerical simulation of channel flow to provide the correlation

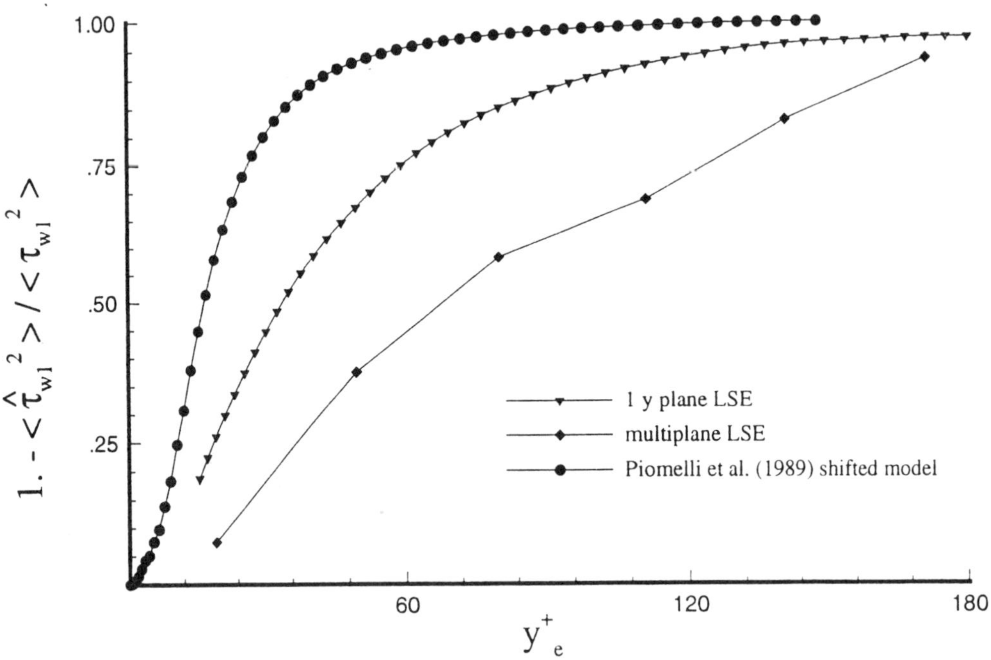

Figure 7.2 Relative mean square error in the estimate of the wall shear stress. *[from Bagwell, Adrian, Moser and Kim 1993]*

function used in the estimate and the data used to make a evaluation of the mean square error of the linear estimate. The results, shown in Fig. 7.2, indicate the mean square error for the wall shear stress as a function of the location y_e of the grid plane used for the data. Clearly the error must vanish as the grid plane approaches the wall. The error of the linear stochastic estimate is uniformly less than that of Piomelli's estimate, as should be the case because much more data is used in forming the estimate. When the value of y_e^+ exceeds about 30, the error approaches one-hundred per cent because the grid plane data become uncorrelated with the wall shear stresses. Further simulations were conducted by Bagwell (1994) to ascertain the effectiveness of this scheme when the computation of the resolved scale flow was coupled dynamically to the linear stochastic boundary condition. Results were stable and gave improved computations relative to earlier boundary conditions. Extensions to different Reynolds numbers require improved understanding of the scaling of the two-point spatial correlation function in the near-wall region, and extension to complex flow conditions requires an even greater generalization of the empirical representation of the correlation function.

The relationship between linear stochastic estimation and proper orthogonal decomposition (POD) continues to be an interesting topic of investigation. Both methods take the two-point correlation function as input, and it therefore appears at first glance as though they must be related. This is, of course, true, and under certain simplifying assumptions relations can be obtained. For example Berkooz, et al. (1993) consider joint normal random processes. Since the stochastic estimate is exactly equal to the conditional average in this case, they are able to develop several interesting relationships. However, in the general case the relationships between POD and LSE are not simple. For example, there is no simple relationship between a POD eigenmode and a conditional eddy. The eigenmode has a region of support that extends over the entire space of the decomposition, while the conditional eddy is compact. This is particularly obvious in homogenous turbulence where the eigenmodes from POD are simply trigonometric functions that individually carry no special information about the structure of turbulence. The LSE, on the other hand is capable of extracting structure regardless of the homogeneous state of the turbulence.

Since they form an orthogonal basis the correct way to use the POD eigenmodes in structural studies is to construct the structure from superposition of groups of the modes. The structure can be exactly represented by the superposition, and this is the strength of POD. In contrast LSE only approximates a structure locally. If greater accuracy is needed from a linear estimate, it is necessary to add more event data. It can be shown that there is a sampling theorem for linear stochastic estimation which shows that the linear stochastic estimate of a random function becomes exact if the data are sampled on a uniformly spaced grid and the sample spacing approaches the Nyquist sampling frequency. In fact, in the limit of Nyquist sampling the linear estimation coefficients exactly equal the sinc function specified in the Shannon Sampling Theorem. Thus, both linear estimation and POD become uninteresting when applied to fully sampled stationary random function.

The more interesting applications occur when the number of POD modes is limited so as to represent only a part of the signal or the number of event data are limited, corresponding to partial information for the stochastic estimate. In this regard Bonnet, et al. (1994) have recently combined POD and stochastic estimation by performing estimates of partial sums of POD modes and by examining estimation of the POD Fourier coefficients.

8. Summary and Conclusions.

Stochastic estimation is the approximation of a random variable in terms of given information. The best stochastic estimate is the conditional average. It has been shown that the conditional average itself can be estimated in terms of the data and the correlation functions between the data and the variable being estimated. The power of this result lies in the fact that once the estimation coefficients are found, the values of the event can be varied at will, making it possible to fully explore the space of events. A related property is that it is relatively easy to construct very complex events from sets of relatively simple data, again making it possible to explore a large class of events without having to perform conditional sampling

Acknowledgement

This work was supported by grants from the Office of Naval Research and the Air Force Office of Scientific Research. The assistance of R. Moser is greatly appreciated.

References

Adrian, R. J. (1977) On the role of conditional averages in turbulence theory. In J. Zakin and G. Patterson, Eds. *Turbulence in Liquids*. Princeton, NJ: Science Press, 323-332.

Adrian, R. J. (1979) Conditional eddies in isotropic turbulence. *Phys. Fluids* **22**, 2065-2070.

Adrian, R. J., Moin, P., Moser, R. D. (1987) Stochastic estimation of conditional eddies in turbulent channel flow, in: Moin, P., Reynolds, W. R., Eds. *Proc. of the 1987 Summer Program of the Center for Turbulence Research*, CTR-S87. NASA Ames Research Center, Moffett Field, CA, 7-20..

Adrian, R. J., Moin, P. (1988) Stochastic estimation of organized turbulent structure: homogeneous shear flow. *J. Fluid Mech.* **190**, 531-559.

Adrian, R. J., Jones, B. G., Chung, M. K., Hassan, Y., Nithianandan, C. K., Tung, A. T.-C. (1989) Approximation of turbulent conditional averages by stochastic estimation. *Phys. Fluids A.* **1**, 992-998.

Adrian, R. J. (1990) Linking correlations and structure: stochastic estimation and conditional averaging. in: Kline, S. J., Afgan, N. H., Eds. *Near Wall Turbulence, Proc. Zaric Memorial Intl. Symp.*, Washington D.C.: Hemisphere, 430-436.

Adrian, R. J. (1990) Stochastic estimation of sub-grid scale motions. *Appl. Mech. Rev.* **43**, 5214-5218.

Bagwell, T. G., Adrian, R. J., Moser, R. D., Kim, J. (1993) Improved approximation of wall shear stress boundary conditions for large eddy simulations. In: So, R., Launder, B., Eds. *Near Wall Turbulent Flows*. Amsterdam: Elsevier Science, 265-275.

Bagwell, T. G. (1994) Stochastic Estimation of Near Wall Closure in Turbulence Models. Ph. D. thesis, Univ. Illinois, Urbana, Illinois.

Balachandar, S., Adrian, R. J. (1993) Structure extraction by stochastic estimation with adaptive events. *J. Theoret. Comput. Fluid Dyn.* **5**, 243-257.

Batchelor, G. K. (1986) *The Theory of Homogeneous Turbulence*, Cambridge Univ. Press, Cambridge.

Berkooz, G.,Elezgaray, J., Holmes, P. Lumley, J. and Poje, A. (1993) The proper orthogonal decomposition, wavelets and modal approaches to the dynamics of coherent structures, In: Bonnet, J. P., Glauser, M. N., Eds. *Eddy Structure Identification in Free Turbulent Shear Flows.* Dordrecht: Kluwer. 325-336.

Brereton, G. J. (1992) Stochastic estimation as a statistical tool for approximating turbulent conditional averages. *Phys. Fluids A* **4**, 2046-2054.

Bonnet, J. P., Cole, D. R., Delville, J., Glauser, M. N., Ukeiley, L. S. (1994) Stochastic estimation and proper orthogonal decomposition: Complementary techniques for identifying structure. *Exp. Fluids* **17**, 307-314.

Chang, P. J. (1985) Fluctuating Pressure and Velocity Fields in the Near Field of a Round Jet. Ph.D. thesis, Univ. Illinois, Urbana, Illinois.

Chang, P. J., Adrian, R. J., Jones, B. G. (1985) Fluctuating pressure and velocity fields in the near field of a round jet. Univ. Illinois, Urbana, Illinois, Theo. & Appl. Mech. Report No. 475, UILU-ENG 85-6006.

Cole, D. R., Glauser, M. N., Guezennec, Y. G. (1992) An application of the stochastic estimation to the jet mixing layer. *Phys. Fluids A* **4**, 192-194.

Ditter, J. L. (1987) Stochastic Estimation of Eddies Conditioned on Local Kinematics: Isotropic Turbulence. M.S. thesis, Univ. Illinois, Urbana, IL.

Ditter, J.L. and Adrian, R.J. (1988). "Local flow structures around kinematic events in isotropic turbulence," presented at 11th Symp. on Turbulence, October 17-19, 1988, Univ. of Missouri-Rolla.

Deutsch, R. (1965) *Estimation Theory*. New York: Prentice-Hall, 1965.

Elam, K. S. (1987) Conditional Reynolds Stresses in the Near Field of a Round Jet. M.S. Thesis, Univ. Illinois, Urbana, Illinois.

Gieseke, T. J., Guezennec, Y. G. (1993) Stochastic estimation of multipoint conditional averages and their spatio-temporal evolution. in: Bonnet, J. P., Glauser, M. N., Eds. Eddy *Structure Identification in Free Turbulent Shear Flows*. Dordrecht: Kluwer, 281-292. ISBN: 0-7923-2449-8.

Guezennec, Y., Piomelli, U., and Kim, J. (1987) in *Proc. 1987 Summer Program of Center for Turbulence Research*, NASA Ames/Stanford Univ., Stanford, CA 263.

Guezennec, Y. (1989) Stochastic estimation of coherent structure in turbulent boundary layers. *Phys. Fluids A* **1**, 1054.

Hassan, Y. A. (1980) Experimental and Modeling Studies of Two-Point Stochastic Structure in Turbulent Pipe Flow. Ph.D. thesis, Urbana, IL: Univ. Illinois.

Johnsen, H., Pecseli, H. L., Trulsen, J. C. (1985) Conditional eddies, or clumps, in ion-beam-generated turbulence. *Phys. Rev. Lett.* **55**, 2297-2300.

Johnsen, H., Pecseli, H. L., Trulsen, J. C. (1986) Conditional eddies in plasma turbulence. *Plasma Physics and Controlled Fusion* **28**, 1519-1523.

Kendall, T. M. (1992) Dynamics of Conditional Vortices in Turbulent Channel Flow: A Direct Numerical Simulation. M.S. thesis, Univ. Illinois, Urbana, Illinois.

LeBoeuf R. L., Mehta, R. Improved methods for linear estimation of velocity records. *Exp. Fluids* **17**, 32-38.

Lundgren, T.S. (1967) Distribution functions in the statistical theory of turbulence. *Phys. Fluids* **10**, 969.

Moin, P., Adrian, R. J., Kim, J. (1987) Stochastic estimation of organized structures in turbulent channel flow, in: *Proc. 6th Turbulent Shear Flow Symp.*, Toulouse, 1987, 16.9.1-16.9.8.

Monin, A. S. (1967a) Equation for finite-dimensional probability distributions of turbulent field. *Dokl. Akad. Nauk SSSR* **177**, 1036-1038.

Nithianandan, C. K. (1980) Fluctuating velocity pressure field structure in a round jet turbulent mixing region. Ph.D. thesis, University of Illinois, Urbana, Illinois.

Novikov, E. A. (1967) Kinetic equations for vorticity field. *Dokl. Akad. Nauk. SSSR* **177**, 299-301.

Novikov, E. A. (1993) A new approach to the problem of turbulence, based on the conditionally averaged Navier-Stokes equations. *Fluid Dynamics Res.* **12**, 107-126.

Papoulis, A. (1984) *Probability, Random Variables and Stochastic Theory*. New York: McGraw-Hill.

Perry, A. E., and Chong, M. A. (1987) A description of eddying motions in flow patterns using critical-point concepts. *Ann. Rev. Fluid Mech.* **19**, 125-155.

Piomelli, U. (1987) Models for Large Eddy Simulations of Turbulent Channel Flows Including Transpiration. Ph.D. thesis, Stanford Univ., Stanford, California.

Robinson, S. K. (1991) Coherent motions in the turbulent boundary layer, *Ann. Rev. Fluid Mech.* **23**, 601-640.

Rogers, M. M., Moin, P. (1987) The structure of the vorticity field in homogeneous turbulent flows, *J. Fluid Mech.* **176**, 33.

Schumann, U. (1975) Subgrid scale model for finite difference simulations of turbulent flows in plant channels and annuli. J. *Comput. Phys.* **18**, 376-404.

Townsend, A. A. (1976) *The Structure of Turbulent Shear Flow*, 2nd Ed. Cambridge University Press, Cambridge.

Tung, A.T.C. and Adrian, R. J. (1980) Higher-order estimates of conditional eddies in isotropic turbulence. *Phys. Fluids* **23**, 1469-1470.

Tung, A. T. C. (1982) Properties of Conditional Eddies in Free Shear Flows. Ph.D. thesis, Univ. Illinois, Urbana, Illinois.

Ukeiley, L., Cole, D. R., Glauser, M. (1993) An examination of the axisymmetric jet mixing layer using coherent structure detection techniques. In: Bonnet, J. P., Glauser, M. N., Eds. *Eddy Structure Identification in Free Turbulent Shear Flows.* Dordrecht: Kluwer. 325-336.

Van Atta, C. W. and Chen, W. Y. (1968) Correlation measurements in grid turbulence using digital harmonic analysis. *J. Fluid Mech.* **34**, 497-515.

Willmarth, W. W., and Wooldridge, C. E. (1963) Measurements of the correlation between the fluctuating velocities and the fluctuating wall pressure in a thick turbulent boundary layer. AGARD Rep. 456.

Zhou, Z., Adrian, R. J., and Balachandar, S. (1995) Autogeneration of near-wall vortical structures in channel flow. Submitted to *Phys. Fluids.*

TOPOLOGICAL CLASSIFICATION AND IDENTIFICATION OF SMALL-SCALE TURBULENCE STRUCTURE

J.C. Vassilicos
DAMTP, Cambridge, UK

1. Lecture I: classification of turbulence singularities — fractals and wavelets

We seek ways to characterise and identify small-scale turbulence structure without ad hoc preconceived ideas of what a turbulent eddy, for example, should be, or what properties it should have. The turbulence field is a complex web of chaotic, random and ordered phenomena, and a numerical high Reynolds number solution of the Navier-Stokes equation cannot provide much insight into the kinematics and dynamics of the turbulence by itself. Direct Numerical Simulation (DNS) of turbulent flows generates large data sets, and gives us the satisfaction of having transformed a cryptic partial differential equation into a large set of inarticulate numbers. The problem of characterisation and identification of small-scale turbulence structure is a problem of transforming this inarticulate set of numbers into meaningful qualitative and quantitative statements about the kinematics and dynamics of the turbulence. In these lectures we introduce a recent attempt to characterise and identify *topological and geometrical* features of the turbulence based on a *classification of turbulence singularities* which follows as a *consequence* of Kolmogorov's 1941 similarity theory of small-scale turbulence. A classification, characterisation and identification of turbulence singularities is important because this is where the highest velocity gradients are to be found in the turbulence.

In 1941 Kolmogorov wrote two seminal papers on *small-scale isotropic* and *homogeneous equilibrium* turbulence *at the limit where the Reynolds number* $Re \to \infty$. In the first paper he derived the scaling of the second order structure function of turbulence fluctuations from simple dimensional arguments in the so-called inertial range of length scales (Kolmogorov 1941a). In the second paper he derived from first principles the form of the third order structure function in the same range of length scales (Kolmogorov 1941b).

So little is understood about the kinematics and dynamics of turbulence, that the simplest and apparently obvious insight can be of significant importance. Kolmogorov chose to work with velocity *differences* $\delta\mathbf{u}(\mathbf{x},\mathbf{r})$ between two different points $\mathbf{x}$ and $\mathbf{x}+\mathbf{r}$ in the flow, that is

$$\delta\mathbf{u}(\mathbf{x},\mathbf{r}) = \mathbf{u}(\mathbf{x}+\mathbf{r}) - \mathbf{u}(\mathbf{x}) \qquad (I1)$$

where $\mathbf{u}(\mathbf{x})$ is the flow velocity at point $\mathbf{x}$. A setting of the problem in such terms has the double advantage that relative velocities between fluid elements are Galilean invariant and a direct measure of turbulence *fluctuations*. The large scale mean flow is subtracted out in (I1) and does not influence the relative velocities $\delta\mathbf{u}(\mathbf{x},\mathbf{r})$. It is the relative velocity fluctuations $\delta\mathbf{u}(\mathbf{x},\mathbf{r},t)$ which are assumed to be in homogeneous and isotropic statistical equilibrium, not the absolute velocities $\mathbf{u}(\mathbf{x},t)$. Therefore, the statistics of $\delta\mathbf{u}(\mathbf{x},\mathbf{r},t)$ do not depend on time t, position $\mathbf{x}$ and unit vector $\hat{\mathbf{r}} = \frac{\mathbf{r}}{r}$ ($r \equiv |\mathbf{r}|$) in the range of length scales where Kolmogorov's theory applies. Kolmogorov's results are the following:

(i) The second order structure function

$$< \delta \mathbf{u}^2(r) > \sim (\epsilon r)^{2/3} \tag{I2}$$

in the inertial range of length scales r, i.e. provided that $L \gg r \gg \eta_K$, where L is an integral length scale characteristic of the energy containing eddies, $\eta_K \sim (\nu^3/\epsilon)^{1/4}$ is the viscous microscale below which turbulence fluctuations are smoothed out by viscous decay, ν is the kinematic viscosity of the fluid and ϵ is the average rate of kinetic energy dissipation per unit mass. The inertial range of length scales exists in the limit where* $Re \sim (L/\eta_K)^{4/3} \to \infty$. The brackets $<>$ denote an average over either space or realisations of the turbulence, and because of homogeneity and isotropy, $< \delta \mathbf{u}^2(r) >$ is neither a function of $\mathbf{x}$ nor of $\hat{\mathbf{r}}$ but only of the distance r.

(ii) Defining $\delta u_{||} \equiv \hat{\mathbf{r}} \cdot \delta \mathbf{u}$,

$$< \delta u_{||}^3(r) > \approx -\frac{4}{5}\epsilon r \tag{I3}$$

in the inertial range of length scales r. This is the only result in turbulence which stems from first principles (from the Navier-Stokes equation) and is remarkable because it not only provides the scaling with r of the third order structure function, but also the *universal* constant $-\frac{4}{5}$.

Kolmogorov's formula (I3) has two immediate and important consequences: that turbulence velocity fluctuations are *not* Gaussian and *not* mirror symmetric. We start these lectures by deriving a classification of turbulence singularities from Kolmogorov's result (I2).

If $< \delta \mathbf{u}^2(r) >^{1/2} /r$ can be considered to approximate characteristic gradients of the turbulent velocity field, then it follows from (I2) that singularities (or rather near-singularities) must exist in the turbulent flow where the velocity gradients tend to infinity as $r/L \to 0$ and $r/\eta_K >> 1$. This conclusion is made more definite and more specific by considering the equivalent of (I2) in Fourier space (Hunt & Vassilicos 1991). An account of how the structure function $< \delta \mathbf{u}^2(r) >$ relates to the energy spectrum $E(k)$ of the turbulence can be found in Batchelor (1953), and I leave it for the reader to show that $< \delta \mathbf{u}^2(r) > \sim r^{\xi_2}$ for small r is equivalent to $E(k) \sim k^{-1-\xi_2}$ for large k (k is the wavenumber). In the Fourier representation, Kolmogorov's scaling law (I2) becomes

$$E(k) \sim \epsilon^{2/3} k^{-5/3}, \tag{I4}$$

and is valid in the inertial range of wavenumbers, that is where $\frac{2\pi}{L} \ll k \ll \frac{2\pi}{\eta_K}$.

---

* Kolmogorov's dimensional analysis leading to (I2) is based on the additional assumption that ϵ does *not* depend on ν, and therefore $\epsilon \sim \frac{u'^3}{L}$, where u' is the r.m.s. of the turbulent fluctuations $\delta \mathbf{u}$. Hence, it follows from $Re \equiv \frac{u'L}{\nu}$ that $Re \sim (L/\eta_K)^{4/3}$.

Two fundamental results of Fourier analysis are applicable to (I4) (see Courant & Hilbert 1953). The first is that in the limit where $k \to \infty$, $E(k)$ decays faster than any power law if and only if the velocity field is everywhere regular. The second result states that $E(k) \sim k^{-2p}$ as $k \to \infty$ with an *integral* p if the velocity field carries isolated discontinuities in its $(p-1)$th derivatives, and no discontinuities in any derivative of order lower than $p-1$. Kolmogorov's energy spectrum (I4) is of the form $E(k) \sim k^{-2p}$ but with a *non-integral* $p = 5/6$. The turbulent velocity field can therefore be neither everywhere regular nor have isolated discontinuities. There must be singularities (or near-singularities) in the turbulence, and these singularities must be 'worse' than isolated discontinuities. We now *classify* the different types of discontinuity that are possible under the constraint (I4). There is no qualitative loss of generality if we draw this classification in 1-D rather than 3-D. The formulas and figures are easier to write and draw in 1-D.

There are three essential types of singularity that have a high wavenumber spectrum of the form $E(k) \sim k^{-2p}$ where p is a *non-integer*:

(i) $u(x) = x^{\sigma}$ (see Figure 1i); the energy spectrum of such a singularity exists in 1-D when $-1 \leq 2\sigma < 0$, and $E(k) \sim k^{-2-2\sigma}$. An example of a *flow* with a near-singularity of this kind is the Burgers vortex which is a strained vortex tube and is discussed in lecture II.

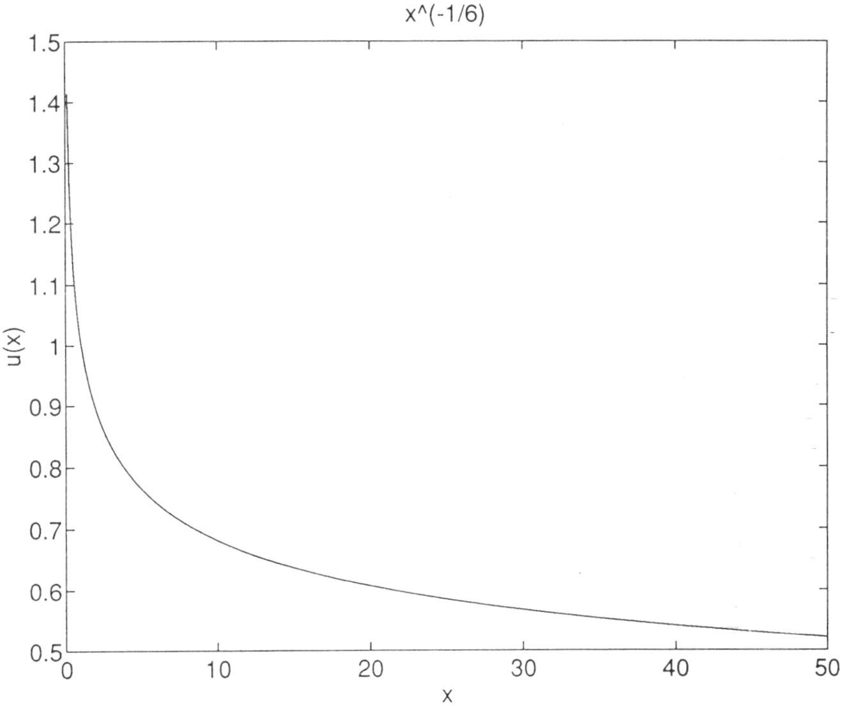

Fig. 1i, in 1-D, the function $u(x) = x^{-1/6}$ *has an energy spectrum* $E(k) \sim k^{-5/3}$.

(ii) $u(x) = e^{ix^{-\tau}}$ (see Figure 1ii); different types of singularities may be combined, as in $u(x) = x^{\sigma}e^{ix^{-\tau}}$, the high-wavenumber spectrum of which exists in 1-D when $-1 \leq 2\sigma \leq \tau$ and $\tau \geq 0$, and $E(k) \sim k^{-2p}$ where $p = \frac{2\sigma+\tau+2}{2(\tau+1)}$ (Hunt et al 1993). An example of a *flow* with a near-singularity of this kind is the Lundgren vortex which is also discussed in lecture II. The Lundgren vortex is a vortex tube with an internal spiral vortex sheet structure that is subjected to strain and differential rotation.

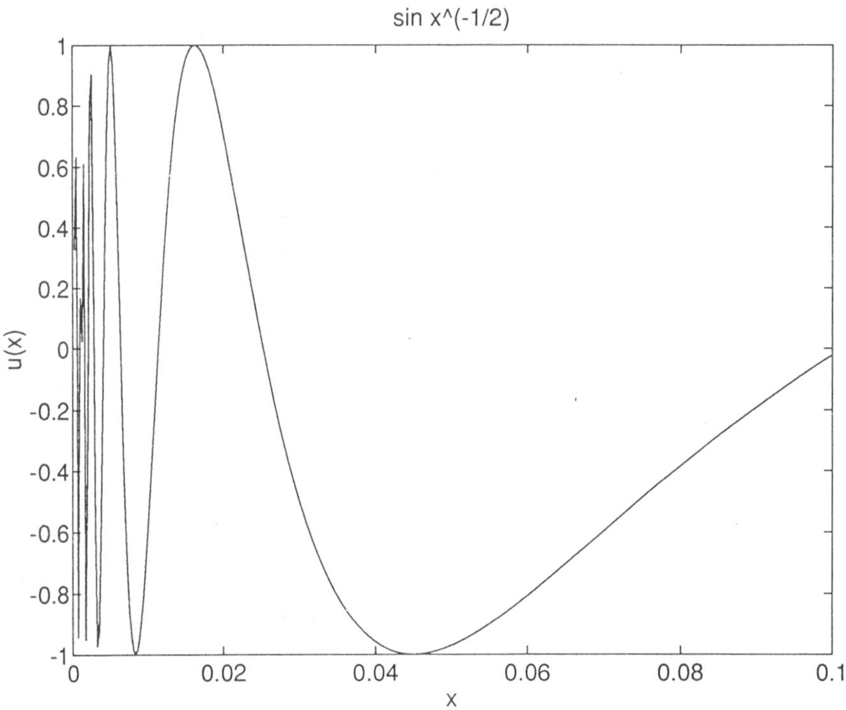

Fig. 1ii, in 1-D, the function $u(x) = \sin x^{-1/2}$ *has an energy spectrum* $E(k) \sim k^{-5/3}$.

(iii) $u(x)$ is a *fractal* function, such as a fractional Brownian motion (fBm) or a Weierstrass function. Turbulence fluctuations cannot be a fBm because they are not Gaussian, but they can, a priori, have a Weierstrass structure. Weierstrass functions $u(x)$ are defined by $u(x) = \sum_{n=0}^{\infty} k_n^{D-2} \sin(k_n x)$ where $k_n = y^n$ with $y > 1$ and $1 < D < 2$ (see Figure 1iii). The high wavenumber spectrum is $E(k_n) \sim k_n^{-2p}$ where $p = 2 - D$ and is a non-integer. I know of no example of a flow with a Weierstrass structure of near-singularities, but chaotic advection (Aref 1984) does generate a fractal topology similar to the one of the Hénon attractor by repeated stretching and folding (Fung & Vassilicos 1991).

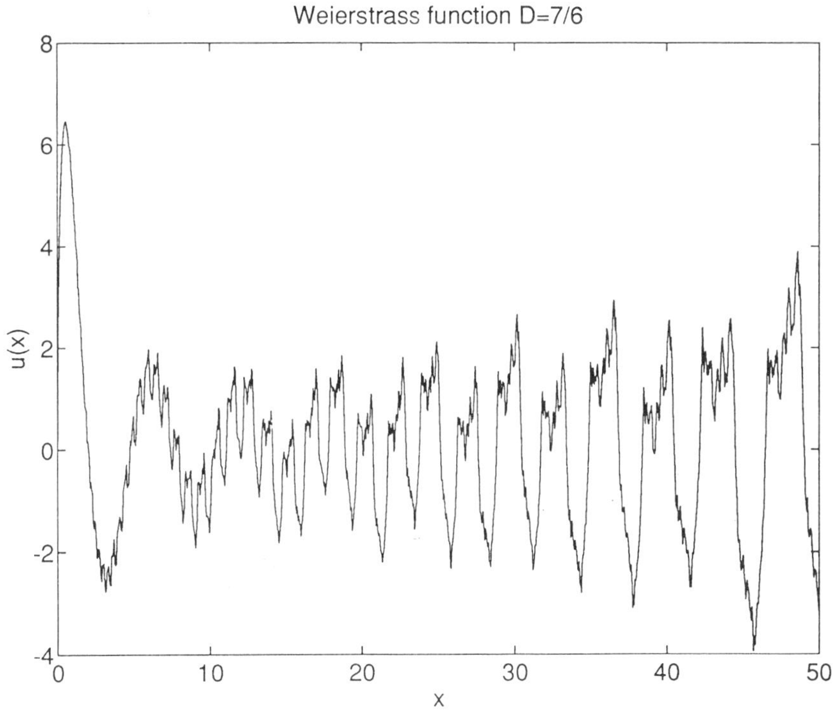

Fig. 1iii, the 1-D Weierstrass function for $D = 7/6$; $E(k) \sim k^{-5/3}$.

The same classification of singularities applies to turbulent passive scalar fluctuations. Following Kolmogorov's analysis, Obukhov, Corrsin and Batchelor (see Batchelor 1959) developed a universal statistical equilibrium theory of passive scalar fluctuations in small-scale isotropic and homogeneous high Reynolds number turbulence. A scalar $\theta(\mathbf{x},t)$, which may be the temperature or concentration carried by a fluid element at point $\mathbf{x}$ in time t, evolves through the combined actions of passive advection by the turbulent velocity field $\mathbf{u}(\mathbf{x},t)$ and molecular diffusion:

$$\frac{\partial\theta}{\partial t} + \mathbf{u}\cdot\boldsymbol{\nabla}\theta = \kappa\nabla^2\theta, \qquad (I5)$$

where κ is the molecular diffusivity of the scalar θ. The statistics of the turbulent passive scalar fluctuations $\delta\theta(\mathbf{x},\mathbf{r},t) \equiv \theta(\mathbf{x}+\mathbf{r},t) - \theta(\mathbf{x},t)$ are homogeneous and isotropic and therefore independent of $\mathbf{x}$ and $\hat{\mathbf{r}}$. Because the scalar fluctuations $\delta\theta(\mathbf{x},\mathbf{r},t)$ are assumed to be in equilibrium, the statistics of $\delta\theta(\mathbf{x},\mathbf{r},t)$ are independent of time. In the limit where $Pr \equiv \frac{\nu}{\kappa} \to \infty$, the equilibrium hypothesis may be applied to the following two ranges of length scales with the following results:

(i) In the inertial range of length scales r where $L \gg r \gg \eta_K$,

$$< \delta\theta^2(r) > \sim \frac{\chi}{\epsilon}(\epsilon r)^{2/3}, \tag{I6}$$

where χ is the mean rate of scalar power or 'energy' dissipation.

(ii) In the convective-diffusive range of length scales r where $\eta_K \gg r \gg \eta_B$,

$$< \delta\theta^2(r) > \sim \frac{\chi}{\gamma}, \tag{I7}$$

where $\eta_B \sim (\nu\kappa^2/\epsilon)^{1/4}$ is the Batchelor diffusive length scale, and $\gamma \sim (\epsilon/\nu)^{1/2}$ is a characteristic rate of strain. The convective-diffusive range exists in the limit where $Pr \to \infty$ since $Pr \sim (\eta_K/\eta_B)^2$.

In the Fourier representation, the scaling laws (I6) and (I7) can be written in terms of the power spectrum $\Gamma(k)$ of the turbulent passive scalar as follows:

(i) In the inertial range of length scales k where $\frac{2\pi}{L} \ll k \ll \frac{2\pi}{\eta_K}$,

$$\Gamma(k) \sim \chi\epsilon^{-1/3}k^{-5/3}. \tag{I8}$$

(ii) In the convective-diffusive range of wavenumbers k where $\frac{2\pi}{\eta_K} \ll k \ll \frac{2\pi}{\eta_B}$,

$$\Gamma(k) \sim \frac{\chi}{\gamma}k^{-1}. \tag{I9}$$

Because $\Gamma(k) \sim k^{-2p}$ and p is non-integral (in both the inertial and the convective-diffusive ranges of length scales), the passive scalar field $\theta(\mathbf{x})$ must carry singularities that are either simple as in Figure 1i, complex but isolated as in Figure 1ii, or non-isolated as in Figure 1iii. Hence, the same classification of singularities which applies to turbulent velocity fluctuations also applies to turbulent passive scalar fluctuations. This classification of singularities is partly based on the difference between *local* and *global* self-similarity and scaling. This distinction marks the essential *topological* difference between isolated complex singularities of type (ii) and non-isolated singularities of type (iii). As shown in the sequel, these singularities may be advantageously defined and studied by fractal analysis and wavelet transforms. We introduce the distinction between *local* and *global* self-similarity in the context of the simpler problem of the *geometry* and *topology* of passive *interfaces* in turbulent flows.

Consider the case where the passive scalar is black dye carried by the fluid, so that $\theta(\mathbf{x}) = 1$ where the fluid is black and $\theta(\mathbf{x}) = 0$ where the fluid is transparent. Black and transparent fluid are separated by a thin 'grey' interface, the thickness of which tends to zero as $\kappa \to 0$. In the limit where $Pr = \frac{\nu}{\kappa} \to \infty$ and $\nu \to 0$ (i.e. $Re \to \infty$), the thickness of that interface tends to 0. Because the power spectrum of the on-off passive scalar $\theta(\mathbf{x})$ defining the interface is given by (I8) and (I9), the interface necessarily has singularities that are either isolated complex singularities such as spirals or non-isolated singularities such as fractals (Figure 2). The classification of interfacial singularities is a subset of the classification of velocity and scalar singularities because interfaces are defined by on-off scalar fields which take only one of two values (e.g. 0 and 1) and cannot have singularities of type (i) like in Figure 1i.

Fig. 2, blob of black dye in the turbulence at time $t = 0$ and at a subsequent time when the turbulence has distorted the interface of the blob and brought its 'turbulent geometry' to a state with interfacial singularities. The scalar $\theta(\mathbf{x}) = 1$ in the black region, and $\theta(\mathbf{x}) = 0$ outside the black region.

We may expect the turbulent interfacial singularities to have a self-similar geometry since the power spectrum $\Gamma(k)$ follows a self-similar power law in wavenumber space. In fact, a number of experimental results in a variety of turbulent flows (see Sreenivasan (1991) for a review) indicate that the Kolmogorov capacity D_K (often called fractal or box dimension) of turbulent scalar interfaces exists, is well defined and is non-integral. The Kolmogorov capacity is one measure of geometrical self-similar scaling. Both isolated complex singularities and non-isolated singularities have well-defined and possibly non-integral Kolmogorov capacities if these singularities are self-similar. Hence, these singularities are responsible for both the spectral and the fractal properties of passive interfaces in turbulent flows. The Kolmogorov capacity and other fractal exponents, such as the Hausdorff dimension, provide quantitative definitions and measures of locally and globally self-similar singularities. In particular, the Kolmogorov capacity can be used to identify singular flow structure. In lecture II we apply fractal identification techniques to the investigation of singular eddy structure in small-scale DNS turbulence. We now introduce Kolmogorov capacities, Hausdorff dimensions, and *local* and *global* self-similarity.

Fractals

The Kolmogorov capacity is defined by the box-counting algorithm which was introduced by Kolmogorov in 1958. This algorithm forms the basis of most practical applications of fractal analysis. Consider an interface embedded in a space of euclidean dimension d (a line-interface is embedded in a plane of $d = 2$ and a surface-interface is embedded in a space of $d = 3$). Choose a length scale δ and cover the interface with boxes of size δ and 'volume' δ^d. $N_K(\delta)$ is the minimum number of such boxes required to cover the interface completely (see Figure 3). The Kolmogorov capacity D_K exists if

$$N_K(\delta) \sim \delta^{-D_K} \tag{I10}$$

for $\delta \to 0$. In practise the Kolmogorov capacity exists and is well defined if (I10) is valid in a range of length scales δ bounded from above by δ_{max} and below by δ_{min} and such that D_K retains the same value over that range as $\delta_{min} \to 0$ (see Figure 4). The lower cutoff scale δ_{min} may be either a viscous length scale such as η_K or a diffusive length scale such as η_B and may be taken to 0 by taking the viscosity or, respectively, the molecular diffusivity to 0.

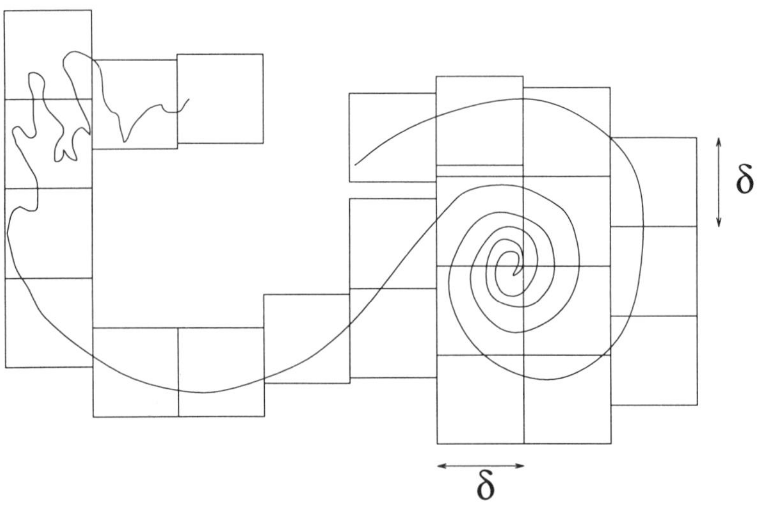

Fig. 3, covering of an interface with boxes of size δ. That is the first stage of the box-counting algorithm.

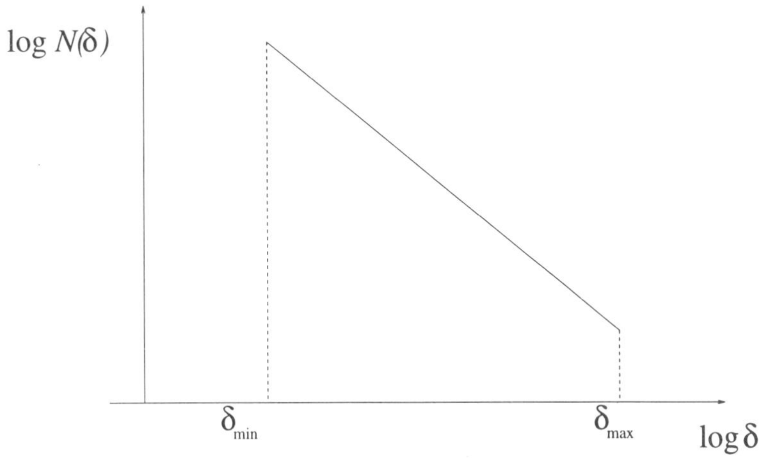

Fig. 4, log-log plot of the minimum number of boxes $N_K(\delta)$ needed to cover the interface in Figure 3 against the size δ of those boxes. In practise, a power law is only observed in a restricted range of length scales between δ_{max} and δ_{min}.

The Kolmogorov capacity D_K is such that $d-1 \leq D_K \leq d$. D_K cannot be strictly larger than d because the number of boxes of size δ needed to fully cover the embedding space scales as δ^{-d} and is larger than $N_K(\delta)$ for all small enough values of δ. D_K cannot be strictly smaller than $d-1$ because $\delta^{d-1}N_K(\delta) \sim \delta^{d-1-D_K}$ approximates the interfacial area as $\delta \to \delta_{min}$, and since that area does not vanish, $\delta_{min}^{d-1-D_K}$ cannot tend to 0 as $\delta_{min} \to 0$. For interfaces which are everywhere regular such as hyper-planes and hyper-spheres, $D_K = d-1$. But D_K can also be a non-integer between $d-1$ and d.

There are two generic classes of interfaces where $D_K > d-1$: *locally* and *globally* self-similar interfaces with, respectively, *isolated* and *non-isolated* irregularities. As well as defining the scaling of the interfacial geometry, D_K is also a measure of how locally or globally *space-filling* the interface is.

An important example of a locally self-similar interface is the spiral $r(\phi) \sim \phi^{-\alpha}$ where $\alpha > 0$ and (r, ϕ) are polar coordinates in the embedding plane of $d = 2$. The spiral has an isolated irregularity at its centre and is regular everywhere else. The covering of a spiral with boxes of size δ is sequential up to the point where the radial distance between successive coils of the spiral is approximately δ. The rest of the covering around the *accumulation* at the spiral's centre is space-filling (see Figure 5).

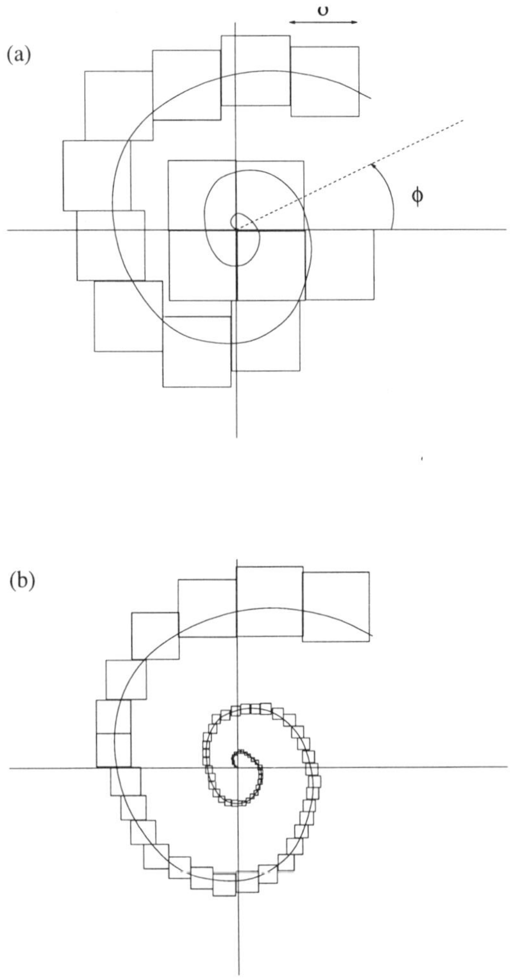

Fig. 5, (a) a spiral defined in the plane in polar coordinates r (distance from the centre) and angle ϕ (equal to a number of turns multiplied by 2π). The covering with boxes of size δ is first sequential, then space-filling. (b) The covering of the same spiral with boxes of decreasing sizes $\delta_i \leq \delta$ is sequential.

If L_δ is the length of the section of the spiral covered sequentially and A_δ the area of the inner space-filling covering, then

$$N_K(\delta) \approx L_\delta/\delta + A_\delta/\delta^2. \tag{I11}$$

Equation (I11) suggests the possibility of an intermediate scaling $N_K \sim \delta^{-D_K}$ with $1 < D_K < 2$. In the case of the algebraic spiral $r(\phi) \sim \phi^{-\alpha}$, it is possible to show that

$$D_K = max(1, \frac{2}{1+\alpha}), \qquad (I12)$$

and D_K is therefore non-integral when $\alpha < 1$ (Vassilicos & Hunt 1991). The non-integral value of D_K reflects the self-similar scaling of the spiral's *accumulation* at its centre. The spiral is more or less space-filling when D_K is more or less close to 2 since $D_K \to 2$ as $\alpha \to 0$ (see Figure 6). The spiral's self-similarity and space-fillingness are therefore *local*, i.e. only as the resolution is increased around the *centre* of the spiral is it possible to see more of the *spiral* at smaller scales. Such a blowing-up procedure at any other point of the spiral reveals nothing but a regular asymptotically straight segment of curve.

(a) (b)

Fig. 6, (a) a space-filling algebraic spiral with exponent α close to 0 and D_K close to 2 to be compared with (b) a less space-filling algebraic spiral with larger α and smaller D_K.

'Fractal' interfaces are *globally* self-similar interfaces. What is often meant by 'fractal interface' is an interface with fractional Hausdorff dimension D_H (see Mandelbrot 1982). The box-counting algorithm that defines D_H is different from Kolmogorov's and is illustrated in Figure 5b. The interface is covered with boxes labelled i of size $\delta_i \leq \delta$ and 'volume' δ_i^d. $N_H(\delta)$ is the minimum number of such boxes required to cover the interface completely. The Hausdorff dimension D_H exists if

$$N_H(\delta) \sim \delta^{-D_H} \qquad (I13)$$

as $\delta \to 0$. Note that the Hausdorff dimension of spiral interfaces is $D_H = d - 1$ ($D_H = 1$ in Figure 5b) because the Hausdorff covering of a spiral is sequential. However, the Hausdorff covering of an interface with non-isolated irregularities such as the Koch curve (Figure 7) cannot be sequential. If such an interface is self-similar and D_H exists, D_H is a non-integer reflecting the degree to which the Hausdorff covering is space-filling (Figure 8). The interface is then *globally* self-similar, because one can magnify the interface around nearly *any* point of the interface and resolve the fractal irregularities at smaller scales. Many such interfaces have also well-defined

Kolmogorov capacities D_K (e.g. the Koch and the Peano curves in Figures 7 and 8) which must be greater or equal to D_H because $N_H(\delta) \leq N_K(\delta)$ as $\delta \to 0$.

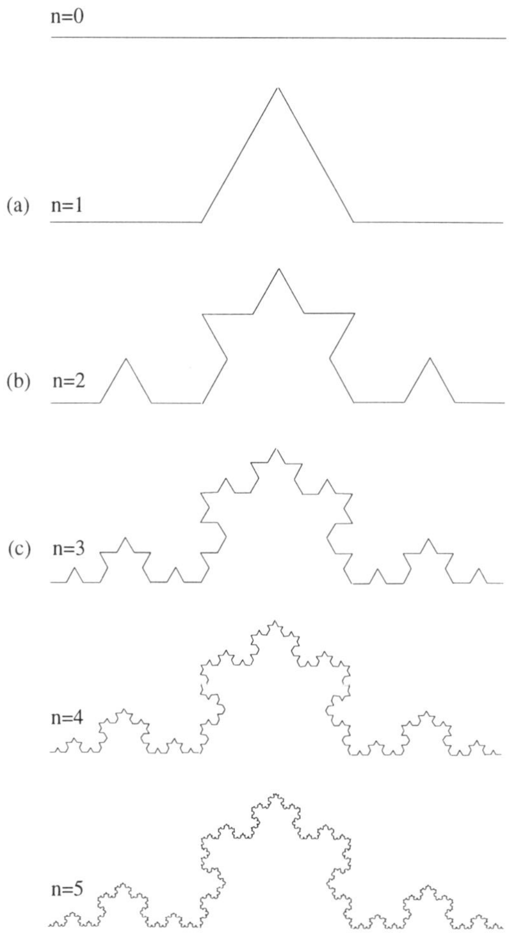

Fig. 7, (a) the generating pattern of the Koch curve. The length of the 4 straight line segments is 1/3. (b) Replace all straight line segments in the generating pattern by a rescaled generating pattern with 4 straight line segments of length 1/9. (c) Continue that process at infinitum and obtain the Koch curve. 4 boxes of size 1/3 are needed to cover the curve in (a). 4^2 boxes of size $(1/3)^2$ are needed in (b). In general, 4^n boxes of size $(1/3)^n$ are needed to cover the Koch curve at the nth resolution, and therefore $D_K = \ln 4^n / \ln 3^n \approx 1.27$.

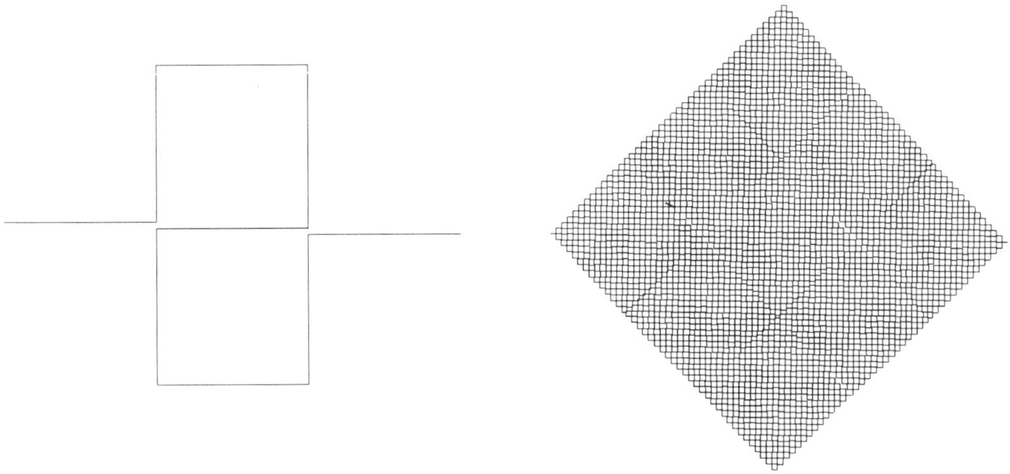

Fig. 8, the Peano curve. Its generating pattern has 9 *straight line segments of length* 1/3 *each. After enough iterations the Peano curve comes arbitrarily close to any point inside a square of side-length* 1/2 *and diagonal length* 1*. Thus, the Peano curve is globally space-filling, and* $D_H = D_K = \ln 9/\ln 3 = 2$*. The Hausdorff dimension of the Koch curve is also equal to its Kolmogorov capacity but is therefore smaller than 2. Hence the Koch curve is not as globally space-filling as the Peano curve.*

The difference between *locally* and *globally* self-similar interfaces is a difference in *topology*. An interface can tend to fill the space either locally or globally. However, unlike the Hausdorff dimension, the Kolmogorov capacity is insensitive to this topological distinction and provides a way to study interfacial singularities in both cases of isolated and non-isolated singularities. This topological distinction is identical to the distinction between accumulating oscillations and fractal functions in the classification of velocity singularities that are a priori possible in small-scale turbulence (see Figures 1ii and 1iii). Figure 9 illustrates this topological distinction for the scalar $\theta(\mathbf{x})$ on a 1-D cross-section through irregularities of an interface with non-integral D_K. The set of point intersections of such interfaces with the 1-D straight line cross-section has a non-integral Kolmogorov capacity D'_K where $0 < D'_K < 1$, whether this set of point-intersections is locally (Figure 9a) or globally (Figure 9b) self-similar.

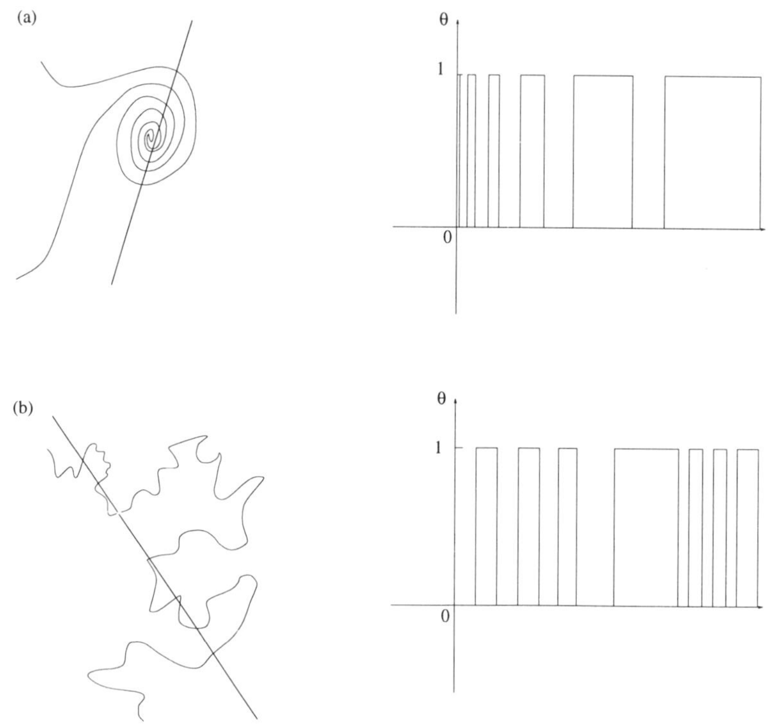

Fig. 9, (a) a 1-D cross-section cutting across the centre of a spiral interface and the scalar θ along this cross-section. The origin on the horizontal axis is the spiral's centre. (b) A 1-D cross-section through a fractal interface and the scalar θ along this cross-section. In both (a) and (b) the set of point-intersections where θ jumps from 1 to 0 or 0 to 1 may have a non-integral Kolmogorov capacity D'_K if this set of points is self-similar. However, the self-similarity is local in (a) and global in (b).

Kolmogorov capacities and Hausdorff dimensions can also be used to define quantitatively local and global self-similarity in signals, functions and fields such as turbulent velocity and scalar fluctuations. One can cover the functions $u(x)$ in Figures 1ii and 1iii like in the example of Figure 10a and derive D_H and D_K respectively from Hausdorff and Kolmogorov box-counting operations. A function with globally self-similar singularities such as in Figure 1iii is such that $D_H > 1$ and $D_K > 1$, whereas a function with a locally self-similar singularity such as in Figure 1ii is such that $D_K > 1$ but $D_H = 1$.

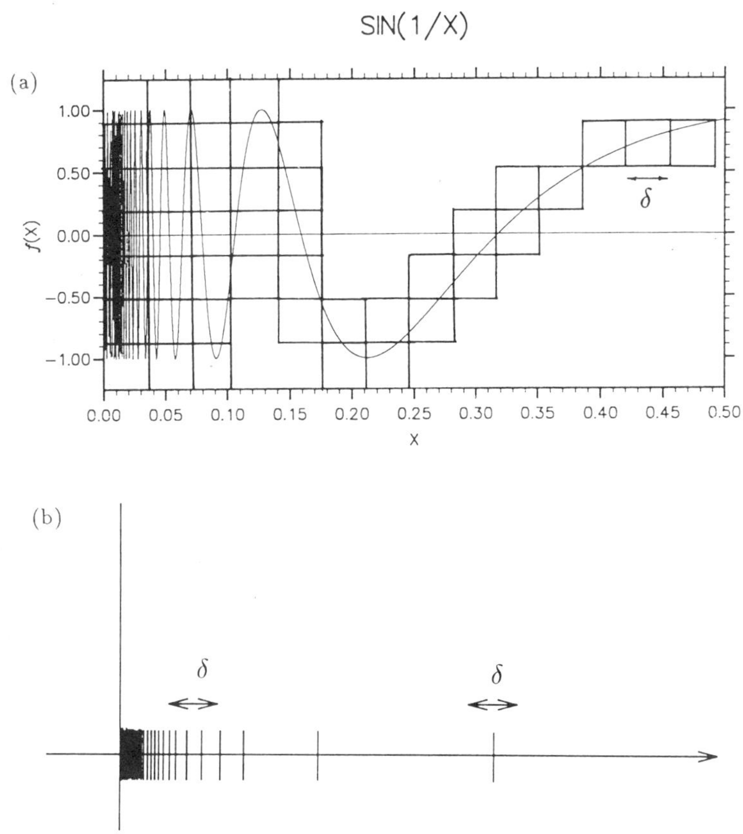

Fig. 10, calculating the Kolmogorov capacity of the function $sin(1/x)$ by Kolmogorov box-counting. (a) Covering the curve with boxes in the 2-D plane. (b) Covering the intersections of $sin(1/x)$ with the x-axis with 1-D line segments.

The characterisation of self-similar singularities in terms of Kolmogorov capacities and Hausdorff dimensions isolates important, if not defining, properties of self-similar singularities. Further detail concerning the singularity can be irrelevant. Consider, for example, the function $u(x) = x^{\sigma} e^{ix^{-\tau}}$ and its zero-crossings x_n which are such that $x_n \sim n^{-1/\tau}$. This set of zero-crossings has a Hausdorff dimension $D'_H = 0$ and a Kolmogorov capacity $D'_K = \frac{\tau}{\tau+1} > 0$ (see Figure 10b and Vassilicos & Hunt 1991). An essential property of the locally self-similar singularity $u(x) = x^{\sigma} e^{ix^{-\tau}}$ is its accumulation which is quantitatively characterised by the parameter τ. The Kolmogorov capacity D'_K is a one-to-one function of τ and is, unlike τ, a robust practical measure of self-similarity. A direct measurement of τ entails fitting $x^{\sigma} e^{ix^{-\tau}}$ and $x_n \sim n^{-1/\tau}$ against the data, which is both more difficult and besides the main point.

A measurement of D_K and/or D'_K by box-counting captures the essential quantitative feature of the self-similar singular accumulation easily and directly. In the next lecture we apply this property of capacities to the streamlines of a turbulent flow in order to identify singularities in that turbulent flow.

Kolmogorov capacities are not only natural and easy measures of the self-similar accumulating character of singularities, they can also be *more accurate* measures of self-similarity than the more traditional Fourier power spectra. Whilst interfaces in a variety of turbulent flows are found experimentally to have very well defined Kolmogorov capacities even at comparatively low Reynolds numbers (Sreenivasan 1991a), the scalar field defining these interfaces has a Fourier power spectrum $\Gamma(k)$ which approaches its asymptotic $\Gamma(k) \sim k^{-5/3}$ self-similar form very slowly as the Reynolds number increases (Sreenivasan 1991b). One explanation is that many terms are needed in a Fourier series for it to converge to its asymptotic form when the Fourier expanded field has self-similar properties. For example, a practical measurement of the power spectrum of a scalar field with a sharp spiral interface cannot reveal the spectrum self-similar scaling form and the local self-similarity of the spiral interface when the spiral has not as much as 50 turns (Vassilicos & Hunt 1991). It is shown in Vassilicos & Hunt (1991) that if the Kolmogorov capacities D_K and D'_K of the spiral interface are well defined in a range of length scales between, say, δ_{min} and 1, then the power spectrum $\Gamma(k)$ can only be measured accurately in a much smaller range of length scales between $\delta_{min}^{1-D'_K}$ and 1 ($0 \leq D'_K \leq 1$). On the other hand, as we shall see in the next lecture on specific numerical examples, only two turns of a spiral interface are enough to measure its Kolmogorov capacity accurately over a very restricted range of length scales. In Figure 11 we show a scalar interface in a turbulent jet. Notice the many spiral structures and their small number of turns.

Fig. 11, black and white rendering of $|\nabla\theta|^2$ *where* θ *is the scalar concentration of dye in a turbulent jet.* $Re = 4000$. *From Sreenivasan (1990, personal communication).*

We have defined quantitatively the difference between locally and globally self-similar singularities in terms of Hausdorff dimensions and Kolmogorov capacities. Kolmogorov capacities are easily accessible in practise, but Hausdorff dimensions are not. Wavelet transforms are mathematical microscopes that can be used to discriminate between isolated and non-isolated singularities both in theory and in practise. Unlike Fourier transforms, which decompose a signal into delocalised sinusoidal oscillations, wavelet transforms can *localise* oscillations. We now briefly introduce wavelet transforms and apply them in lecture III to the analysis of a one-dimensional turbulence signal with a goal to identify locally and globally self-similar singularities in high Reynolds number turbulence experimental data.

Wavelet transforms

All transforms $Tu(p_n)$ of functions or signals $u(x)$ with respect to a basis or an 'analysing' function $\Theta(x,p_n)$ are calculated by means of an integral of the form

$$Tu(p_n) = \int u(x)\Theta^*(x,p_n)\,dx. \tag{I14}$$

By virtue of the power theorem of Fourier transforms, one can also compute $Tu(p_n)$ in terms of the Fourier transforms $\hat{u}(k)$ and $\hat{\Theta}(k,p_n$ of $u(x)$ and $\Theta(x,p_n)$ by integrating over Fourier k-space:

$$Tu(p_n) = \int \hat{u}(k)\hat{\Theta}^*(k,p_n)\,dk. \tag{I15}$$

An example of a transform is Shannon sampling, in which case $\Theta(x,x_0) = \delta(x-x_0)$ where δ is the Dirac delta function and x_0 is the unique parameter of the transform. The Fourier transform of $\Theta(x,x_0)$ is $\hat{\Theta}(k,x_0) = Const.$ It is clear that the integration over x in (I14) need only be carried out in an infinitesimal region around x_0, whereas the integration over k in (I15) must be carried out over the entire range of k to yield the same results as with (I14).

Shannon sampling organises the information contained in $u(x)$ in such a way that $Tu(x_0)$ is a very good estimate of $u(x)$ at $x = x_0$; on the other hand, the entire Fourier transform $\hat{u}(k)$ –for all k– is needed to calculate $Tu(x_0)$. We may say that the transformed information about $u(x)$ is very well localised in x-space, but also very widely spread in k-space.

The basis function of Fourier transforms is a sinusoid: $\Theta(x,k_0) = \frac{1}{\sqrt{2\pi}}e^{-ik_0x}$ in which case the parameter is a wavenumber k_0, and $\hat{\Theta}(k,k_0) = \delta(k-k_0)$. The Fourier transform reorganises the information in the 'opposite' way from what Shannon sampling does. The information is now very well localised in k-space, but is widely spread in x-space.

No transform provides perfect localisation in both x- and k- spaces. This is Heisenberg's uncertainty principle: the intervals Δx and Δk over which the integrals in (I14) and (I15) respectively need to be carried out for an accurate estimation of $Tu(x)$ are always such that $\Delta x\Delta k \geq 1$. Shannon sampling is the limit where $\Delta k \to \infty$ and $\Delta x \to 0$; the Fourier transform corresponds to the opposite limit where $\Delta x \to \infty$ and $\Delta k \to 0$.

A good compromise for localisation in both the x- and k-spaces is achieved if one chooses $\Theta(x,p_n)$ to be a wave-packet (see Figure 12), so that, unlike sine waves, it is sufficiently well localised in x-space, and its Fourier transform $\hat{\Theta}(k,p_n)$, though not a delta function, nevertheless has a significant magnitude around the value $k = k_0$. This is the wave-packet or Gabor transform, which is a function of *two* parameters, a location x_0 and a wavenumber k_0: $\Theta(x,x_0,k_0) = \frac{1}{\sqrt{2\pi}}e^{-ik_0x}e^{-\frac{(x-x_0)^2}{2s^2}}$ and $\hat{\Theta}(k,x_0,k_0) = se^{-\frac{(k-k_0)^2s^2}{2}}$ where s is a constant width. If we interpret Δx and Δk to be the widths of the analysing function and of its Fourier transform respectively, then $\Delta x = s$ and $\Delta k = \frac{1}{s}$ $(s>0)$. When $s=1$, the Gabor transform concentrates the information in x-k space in squares of constant size $\Delta x = \Delta k = 1$ (see Figure 13).

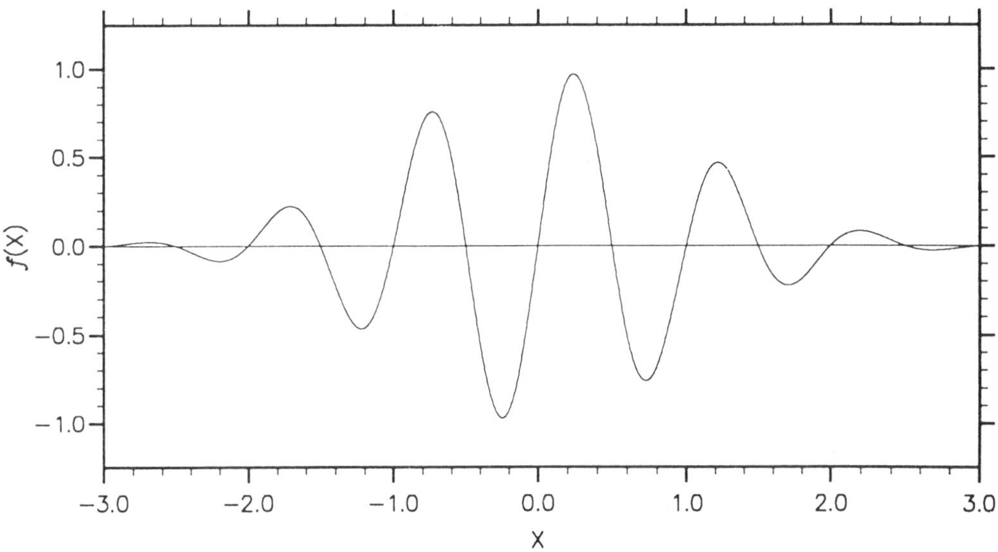

Fig. 12, The Gabor wave-packet $\sin(kx)\exp(-x^2/2s^2)$, $k = 2\pi$, $s = 1$.

A trade off between spatial and wavenumber localisations does not imply that the transform is a good mathematical microscope. A microscope is a device which increases spatial resolution when attention is concentrated on increasingly small-scale variations. It also 'magnifies' the amplitude of the event, in the sense that it effectively measures that amplitude with a smaller unit at smaller scales.

A feature of the analysing function used in the Gabor transform is that, as $k_0 \to \infty$, the number of oscillations in the wavepacket inside a constant portion of x-space increases. In order to have a mathematical microscope, the extent of the analysing function in x-space should *decrease* with scale, thus increasing its spatial resolution. In particular, if the typical wavelength (scale) of the oscillation decreases in proportion to the decrease of the size Δx, the analysing function should exhibit a constant number of oscillations at all sizes. A way to implement this with a straightforward departure from the Gabor transform is to set $s = a$, where the wavelength $a = \frac{2\pi}{k_0}$. The parameters are x_0 and a, essentially as with the Gabor transform, but $\Theta(x, x_0, k_0) =$

$\frac{1}{\sqrt{2\pi}} e^{-i\frac{2\pi x}{a}} e^{-\frac{(x-x_0)^2}{2a^2}}$ and $\hat{\Theta}(k, x_0, k_0) = a e^{-\frac{(k-k_0)^2 a^2}{2}}$. Clearly, $\Delta x \to 0$ and $\Delta k \to \infty$ as $a \to 0$, and $\Delta k \to 0$, $\Delta x \to \infty$ as $a \to \infty$. The transform we have just described is therefore asymptotically a Shannon sampling at small scales and a Fourier transform at large scales (see Figure 13). At intermediate scales it is comparable to a Gabor transform. For this transform to be a good mathematical microscope, it also needs to increasingly 'magnify' the amplitude of the small-scale event as $a \to 0$; this is achieved by including in the definition of the analysing function a multiplication with, say, a^{-1}. The transform whose analysing function is

$$\Theta(x, x_0, a) = \frac{a^{-1}}{\sqrt{2\pi}} e^{-i\frac{2\pi x}{a}} e^{-\frac{(x-x_0)^2}{2a^2}} \tag{I16}$$

is a mathematical microscope. But it has the disadvantage that the signal $u(x)$ cannot be reconstructed from this transform (see Farge 1992 and references therein), as it can from its Fourier transform.

The wavelet transform is both a mathematical microscope and also admits a reconstruction formula. In fact, as we shall now see, the wavelet transform is a mathematical microscope because its analysing function is defined to be *self-similar* as a varies.

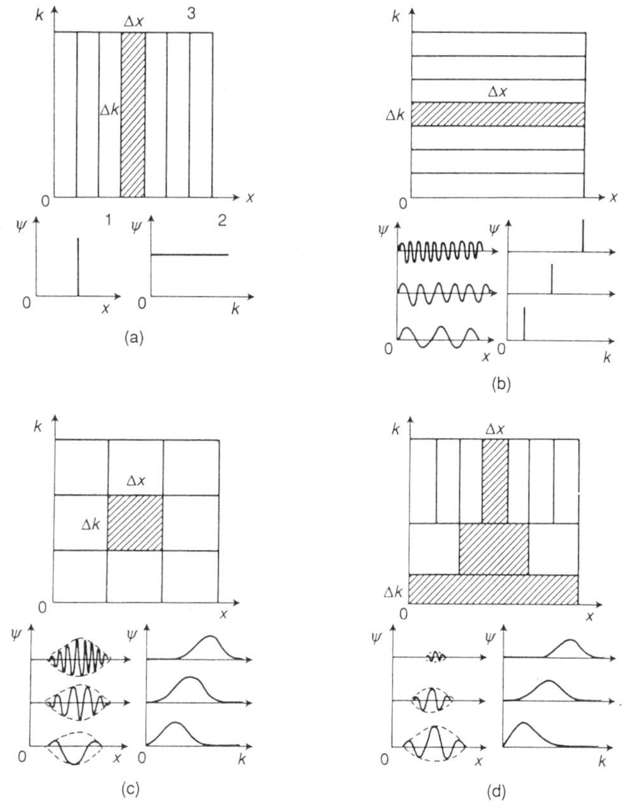

Fig. 13, a comparison of the way different transforms balance physical-space and wavenumber-space resolution (from Farge 1992). (a) Shannon sampling. Perfect physical-space resolution, no wavenumber-space resolution. (b) Fourier transform. Perfect wavenumber-space resolution, no physical-space resolution. (c) Gabor transform. Balance between wavenumber-space and physical-space resolution: the same at all length scales. (d) Wavelet transform. The balance between wavenumber-space and physical-space resolution varies with length-scale. Smaller length-scales are more finely resolved: mathematical microscope.

The analysing function of a wavelet transform is called a wavelet. Wavelets are generated from a 'mother' wavelet $\theta(x) \in L_r(R)$ by translations and dilations, in such a way that they are self-similar with respect to the norm of $L_r(R)$. The parameters are a location x_0 for the translation and a scale a for the dilation, and the wavelet is defined by

$$\Theta(x, x_0, a) = a^{-\frac{1}{r}}\theta(\frac{x - x_0}{a}). \qquad (I17)$$

Note that $|\Theta(x, x_0, a)|^r$ and the argument $(x - x_0)/a$ both scale in proportion to a^{-1}; this is the self-similarity of $\Theta(x, x_0, a)$ with respect to the norm of $L_r(R)$.

There are two requirements on the function $\theta(x)$ for it to be a 'mother' wavelet. The first is that it must be well localised in both physical and Fourier space, i.e. that its spread Δx in x-space is given by a finite constant s and its spread in k-space by $\Delta k = \frac{1}{s}$. As a consequence of the wavelet's self-similarity, its spread is sa in x-space, $1/sa$ in k-space, and its norm scales like a^{-1}; the structure of the wavelet in the x-a plane is described by its contours which point in a cone towards x_0 as $a \to 0$ (see Figure 13). This is why the wavelet transform is a mathematical microscope.

The second requirement is less important for identification purposes and more important for modelling and data compression. The 'mother' wavelet must be *admissible*, i.e. must satisfy the condition

$$\int_{-\infty}^{+\infty} |\hat{\theta}(k)|^2 \frac{dk}{k} < \infty, \qquad (I18)$$

where $\hat{\theta}(k)$ is the Fourier transform of $\theta(x)$. This requirement guarantees the existence of a reconstruction formula (see Farge 1992 and references therein). When $\theta(x)$ is integrable, (I18) implies that $\hat{\theta}(k)$ is rapidly decreasing as $k \to \infty$, and also that $\hat{\theta}(k = 0) = 0$. $\theta(x)$ must therefore have zero mean (i.e. $\int_{-\infty}^{+\infty} \theta(x), dx = 0$).

In practise these two requirements usually mean that one chooses a 'mother' wavelet which is an oscillating function about 0 with a rapidly decaying envelope. The one-dimensional continuous wavelet transform $\tilde{u}(a, x_0)$ of a signal $u(x)$ is then defined as follows:

$$\tilde{u}(a, x_0) = a^{-\frac{1}{r}} \int_{-\infty}^{+\infty} u(x)\theta^*(\frac{x - x_0}{a})\, dx. \qquad (I19)$$

The wavelet transform makes it possible to analyse the structure of a signal of at different length scales a and different positions x_0 simultaneously.

Because of the rapid decay of θ both near $k = 0$ and as $k \to \infty$, the Fourier transform of the family of wavelets $a^{-1/r}\theta(\frac{x-x_0}{a})$ constitutes a bank of passband filters. Indeed, the continuous wavelet transform (I19) can also be written in the form

$$\tilde{u}(a, x_0) = a^{-\frac{1}{r}} \int_{-\infty}^{+\infty} \hat{u}(k)\hat{\theta}^*(ak)e^{-ikx_0}\, dk. \qquad (I20)$$

The wavelet transform decomposes a simple singularity $u(x) \sim x^\sigma$ with $\sigma < 1$ (see Figure 1i) by positioning the large scale wavelets (large a) far away from $x = 0$ and the smaller scale wavelets close to $x = 0$. Indeed the derivatives $\frac{du}{dx} \to \infty$ as $x \to 0$ when $\sigma < 1$. An isolated complex singularity (e.g. $u(x) = e^{ix^{-\tau}}$ with $\tau > 0$, see Figure 1ii) is decomposed into wavelets in a similar way: the wavelets $a^{-1/r}\theta(\frac{x-x_0}{a})$ are placed at positions x_0 far away from $x = 0$ when a is large ($x = 0$ is the centre of the singularity where scales accumulate), and closer to $x = 0$ as $a \to 0$. The wavelet transform of a fractal function (the generic example of non-isolated singularities, see Figure 1iii) breaks up the function into wavelets that are distributed in space according to their size in such a way that they reproduce the globally self-similar generating pattern of the fractal function (see Holschneider, 1988 and Arneodo *et al.* 1988).

To summarise, the family of analysing wavelets $a^{-1/r}\theta(\frac{x-x_0}{a})$ may be compared to a mathematical microscope, for which the 'optical' characteristics are defined by θ, a^{-1} is the magnification and x_0 is the point of focus. Different correlations between the magnification and the point of focus can characterise different types of singularity, isolated or non-isolated, for example. The continuous wavelet transform has been extended to more than one dimensions by Murenzi using rotation in addition to dilation and translation (Murenzi 1990), in which case the mathematical microscope is also a polariser.

Conclusions

Kolmogorov's and Batchelor's -5/3 laws have been repeatedly confirmed by experimental evidence since the early 1960's. The first strong confirming evidence for Kolmogorov's $-5/3$ law was provided by measurements with hot film anemometers on a towed body in an oceanic tidal channel by Grant *et al.* (1962). Batchelor's (1959) theory was verified experimentally by Gibson (1963). Kolmogorov's non-integral scaling of the energy spectrum and Batchelor's non-integral scaling of scalar power spectra imply that the turbulent velocities, scalars and interfaces must hold singularities that are either topologically simple with a spatial dependence of the x^σ ($\sigma < 0$) type (Figure 1i) or topologically complex. Topologically complex singularities involve many scales in spatial topology (Figures 1ii and 1iii), and can be classified in terms of the distinction between *local* and *global* scaling and self-similarity. This distinction can be defined quantitatively by the difference between Kolmogorov capacities and Hausdorff dimensions, and wavelet transforms can provide practical ways to identify topologically different singularities. Kolmogorov capacities can be measured easily and accurately even at low resolutions. We exploit this property of the Kolmogorov capacity in lecture II where we introduce a fractal analysis of turbulent *streamlines* to identify singularities in DNS turbulent flows. In lecture III we introduce and develop wavelet analysis of singular flow structure which we apply to experimental turbulent velocity signals.

REFERENCES

1. Aref, H. 1984 Stirring by chaotic advection. J. Fluid Mech. **143**, 1.

2. Arneodo, A., Grasseau, G. & Holschneider, M. 1988 Phys. Rev. Lett. **61**, 2281.

3. Batchelor, G.K. 1953 *The theory of homogeneous turbulence.* Cambridge University Press.

4. Batchelor, G.K. 1959 Small-scale variation of convected quantities like temperature in turbulent fluid. Part 1. General discussion and the case of small conductivity. J. Fluid Mech. **5**, 113.

5. Courant, G. & Hilbert, D. 1953 *Methods of mathematical physics.* Interscience.

6. Farge, M. 1992 The wavelet transform. Ann. Rev. Fluid Mech. **24**, 395.

7. Fung, J.C.H. & Vassilicos, J.C. 1991 Fractal dimensions of lines in chaotic advection. Phys. Fluids A **3** (11), 2725.

8. Gibson, M.M. 1963 Spectra of turbulence in a round jet. J. Fluid Mech. **15**, 161.

9. Grant, H.L., Stewart, R.W. & Moilliet, A. 1962 Turbulence spectra from a tidal channel. J. Fluid Mech. **12**, 241.

10. Holschneider, M. 1988 J. Stat. Phys. **50**, 963.

11. Hunt, J.C.R., Kevlahan, N.-K.R., Vassilicos, J.C. & Farge, M. 1993 Wavelets, fractals and Fourier transforms: detection and analysis of structure. In *Wavelets, fractals and Fourier transforms*, ed. M. Farge, J.C.R. Hunt & J.C. Vassilicos (Oxford University Press), 1.

12. Hunt, J.C.R. & Vassilicos, J.C. 1991 Kolmogorov's contributions to the physical and geometrical understanding of small-scale turbulence and recent developments. Proc. R. Soc. Lond. A **434**, 183.

13. Kolmogorov, A.N. 1941a The local structure of turbulence in incompressible viscous fluid for very large Reynolds numbers. Dokl. Akad. Nauk. SSSR **30**(4). English translation in Proc. R. Soc. Lond. A (1991) **434**, 9.

14. Kolmogorov, A.N. 1941b Dissipation of energy in the locally isotropic turbulence. Dokl. Akad. Nauk. SSSR **32**(1). English translation in Proc. R. Soc. Lond. A (1991) **434**, 15.

15. Kolmogorov, A.N. 1958 A new invariant for transitive dynamical systems. Dokl. Akad. Nauk. SSSR **119**, 861.

16. Mandelbrot, B.B. 1982 *The fractal geometry of nature.* Freeman.

17. Murenzi, R. 1990 Ondelettes multidimensionelles et application à l' analyse d'images. Theése de Doctorat de l'Université Catholique de Louvain, 1348, Louvain-la-Neuve, Belgium.

18. Sreenivasan, K.R. 1991a Fractals and multifractals in fluid turbulence. Ann. Rev. Fluid Mech. **23**, 539.

19. Sreenivasan, K.R. 1991b On local isotropy of passive scalars in turbulent shear flows. Proc. R. Soc. Lond. A **434**, 165.

20. Vassilicos, J.C. & Hunt, J.C.R. 1991 Fractal dimensions and spectra of interfaces with application to turbulence. Proc. R. Soc. Lond. A **435**, 505.

2. Lecture II: fractal analysis of singular flow structure

In the two previous chapters we showed how different singularities can account for the non-trivial scaling properties of turbulent scalar and velocity structure functions and for the fractal properties of turbulent scalar and velocity topologies. In this lecture we set out to investigate what the singularity structure of the small-scale turbulence actually is.

Recent Direct Numerical Simulation (DNS) of a variety of turbulent flows (see Jiménez *et al.* 1993 and references therein) has revealed that small-scale turbulent vorticity is concentrated in long and slender tube-like structures (Figures 1 and 2), at least within the range of Reynolds numbers that is currently computationally attainable. Do these intense vortex tubes carry any of the singularities of the turbulence? If they do, are these singularities simple singularities of type (i) (Figure 1i in lecture I) or complex isolated singularities of type (ii) (Figure 1ii in lecture I)? Vortex tubes with simple singularities of type (i) could be created and sustained by turbulent strain, and vortex tubes with complex singularities of type (ii), such as rolled-up spiral vortex sheets, could be the outcome of shear instabilities. The nature of turbulence singularities is intimately linked to the dynamical process which generates them.

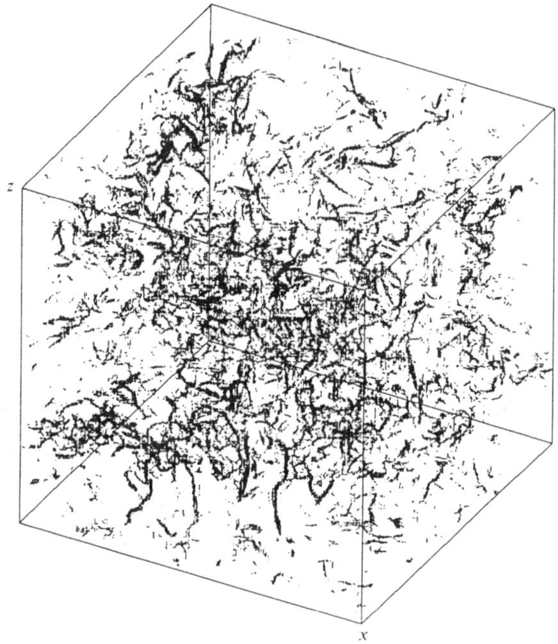

Fig. 1, vorticity field in homogeneous isotropic turbulence. The vorticity is represented by a vector proportional to the vorticity amplitude, and only vector amplitudes larger than a given threshold are shown. The Reynolds number $Re_\lambda \equiv \frac{u'\lambda}{\nu}$ based on the Taylor microscale λ is approximately 150 (u' is the r.m.s. value of the turbulent fluctuating velocity). (From Vincent & Meneguzzi 1991.)

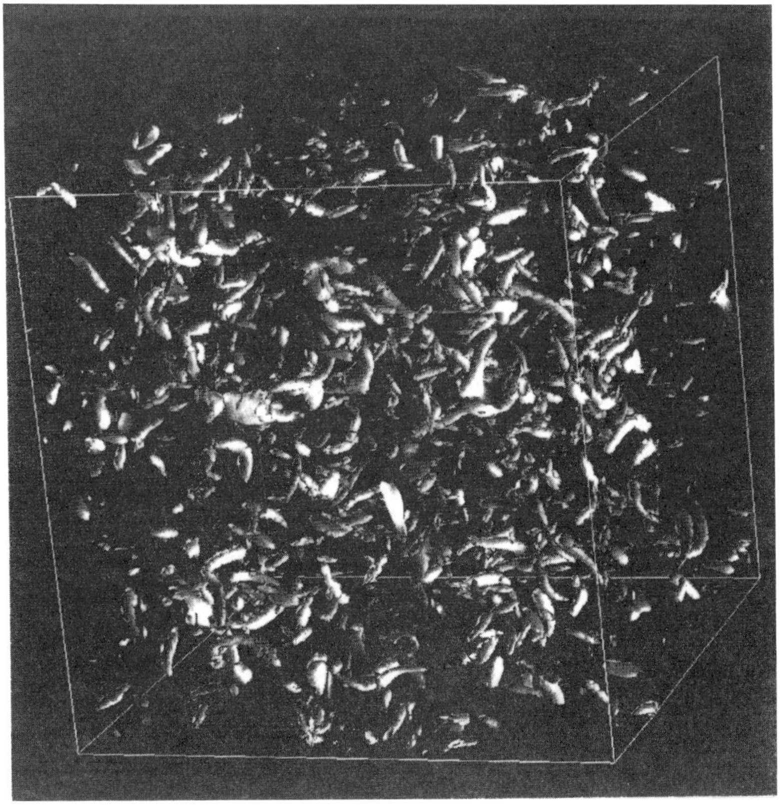

Fig. 2, iso-enstrophy contours in homogeneous isotropic turbulence. The enstrophy of these contours is 6.3 times the enstrophy's r.m.s. and $Re_\lambda \approx 19$*. (From Vassilicos et al. 1994.)*

The singularity structure of the velocity field can be related to scaling and self-similarity properties of the velocity field. We shall now see that the singularity structure of the velocity field can also be related to the scaling and self-similarity properties of *streamlines* and a fractal analysis of the topology of streamlines can provide information about the singularity structure of the flow. It is possible to prove that *in an axisymmetric incompressible flow with bounded vorticity at infinity, when the Kolmogorov capacity* D_K *of a spiral streamline is strictly larger than 1, there must be a velocity singularity on the axis of symmetry of the flow.* This result is a corollary of the *theorem of self-similar streamlines* which we prove in the sequel.

An example of an axisymmetric Navier-Stokes solution where a singularity of type (i) is sustained by strain is the Burgers vortex (Batchelor 1967, p. 271-273), and an example of a nearly axisymmetric asymptotic solution of the Navier-Stokes equation with a spiral singularity structure that could be the outcome of a shear instability is Lundgren's rolled-up spiral vortex sheet (Lundgren 1982, 1993). We first prove the theorem of self-similar streamlines. We then study the 'fractal' scaling properties of streamlines in the Burgers and Lundgren vortices, and show how the 'fractal' properties of streamlines relate to the velocity and vorticity singularity structures of these specific model vortices. Because Kolmogorov capacities need little resolution in practise to be determined accurately (Vassilicos & Hunt 1991), we are led to the development of a sensitive new technique, *the fractal analysis of streamlines for the identification of flow singularities* which has been applied to the study of the singularity structure of vortex tubes in DNS turbulence data (Vassilicos *et al.* 1994).

The theorem of self-similar streamlines

Let us start by stating a claim: that *in an axisymmetric, incompressible flow with bounded vorticity at infinity, if the velocity field is regular (no singularities) at the axis of symmetry, then the Kolmogorov capacity D_K of spiral streamlines is necessarily equal to 1.* How do we go about proving such a claim?

A function $u(r)$ is regular around $r = 0$ if it can be Taylor expanded around $r = 0$, i.e. $u(r) = u(0) + ru'(0) + \frac{r^2}{2}u''(0) + ...$, where u' and u'' are, respectively, the first and second derivatives of u. In an axisymmetric flow, the velocity field is conveniently decomposed in azimuthal, radial and axial components u_ϕ, u_r and u_z and the spatial coordinates are r, the distance from the axis of symmetry, ϕ, the angle around the axis of symmetry, and z, the distance along the axis of symmetry. Our basic premises of axisymmetry and incompressibility imply the existence of a Stokes stream function $\psi(r, z)$, and

$$u_r(r, z) = -\frac{1}{r}\frac{\partial}{\partial z}\psi(r, z), \tag{II1}$$

$$u_z(r, z) = \frac{1}{r}\frac{\partial}{\partial r}\psi(r, z). \tag{II2}$$

The velocity field's regularity at the axis of symmetry means that, as $r \to 0$,

$$u_\phi(r, z) \approx r^m G(z), \tag{II3}$$

$$\psi(r, z) \approx r^l F(z), \tag{II4}$$

where $G(z)$ and $F(z)$ are regular functions of z for all values of z, m is a non-negative integer, and l is an integer which must be greater or equal to 2 in order for u_z to be regular at the axis. In effect, *regularity is too strong a requirement for the streamlines' D_K to be equal to* 1; *we prove that* $D_K = 1$ *provided that m and l are real and such that $m > -1$ and $l > 0$.* Hence, there also exist some singular flows where $D_K = 1$. But there is no regular flow where $D_K > 1$.

The vorticity's boundedness at infinity and the regularity of $G(z)$ imply that $G(z)$ is a bounded function of z. By examining the spiral streamline in a (r, z) plane (Figure 3a) we find that $F(z)$ is a monotonically increasing function of z. Indeed, as $r \to 0$ along the spiral streamline,

$$\frac{dr}{dz} = \frac{u_r}{u_z} \approx -\frac{r}{lF(z)}\frac{dF}{dz}(z) \qquad (II5)$$

which can be integrated to yield

$$r^l F(z) \approx Const \qquad (II6)$$

as $r \to 0$ and r and z are streamline coordinates (see more detailed discussion in Vassilicos *et al.* 1994). It follows that $F(z) \to \infty$ as $r \to 0$ on the spiral streamline, and since $F(z)$ is regular, $F(z)$ must be a monotonically increasing function of z.

Examining the spiral streamline in the azimuthal plane (r, ϕ) (Figure 3b), we see that

$$\frac{d\phi}{dr} = \frac{u_\phi}{ru_r} \approx -r^{m-1}\frac{G(z)}{\frac{dF}{dz}(z)} \qquad (II7)$$

and

$$2\pi = \int_{n2\pi}^{(n+1)2\pi} d\phi \approx -\int_{r_n}^{r_{n+1}} dr r^{m-1}\frac{G[z(r)]}{\frac{dF}{dz}[z(r)]} \sim \int_{z_n}^{z_{n+1}} dz \frac{G(z)}{F^\beta(z)}, \qquad (II8)$$

where r_n and z_n are radial and axial coordinates of the spiral streamline at the nth turn of the spiral, and $\beta = \frac{m+1}{l}$. Since β is strictly positive, $\frac{G(z)}{F^\beta(z)}$ is monotonic decreasing as $z \to \infty$, and therefore $\Delta z_n \equiv z_{n+1} - z_n \to \infty$ as $n \to \infty$.

The distance Δ_n between successive turns after the nth turn of the spiral streamline in 3-D space is given by $\Delta_n^2 = \Delta z_n^2 + \Delta r_n^2$ where $\Delta r_n \equiv r_n - r_{n+1}$. Because $\Delta z_n \to \infty$ as $n \to \infty$, $\Delta_n \to \infty$ as $n \to \infty$ and therefore, *no accumulation of length scales exists on the spiral streamline, which implies that* $D_K = 1$. Indeed the Kolmogorov capacity D_K of streamlines is measured by covering the 3-D space with boxes, and even though the spiral's projection on the azimuthal plane does accumulate ($\Delta r_n \to 0$ as $n \to \infty$), the spiral streamline in 3-D space does not. It converges towards the central axis of symmetry, but in doing so, the distance Δ_n between successive turns increases indefinitely. Hence, there is no accumulation that can be reflected in a non-integral D_K.

The *theorem of self-similar streamlines* that we just proved can be stated as follows:

In an axisymmetric incompressible flow with bounded vorticity at infinity, if there is no singularity on the axis of symmetry of the flow or if there is a singularity of the type (II3) and (II4) where m and/or l are non-integral and $m > -1$, $l > 0$, then the Kolmogorov capacity D_K of a spiral streamline is equal to 1

The *theorem of self-similar streamlines* can also be stated in the following contrapositive way:

In an axisymmetric incompressible flow with bounded vorticity at infinity, when the Kolmogorov capacity D_K of a spiral streamline is strictly larger than 1, there must be a singularity on the axis of symmetry of the flow which is not a singularity of the type (II3) and (II4) where m and/or l are non-integral and such that $m > -1$ and $l > 0$.

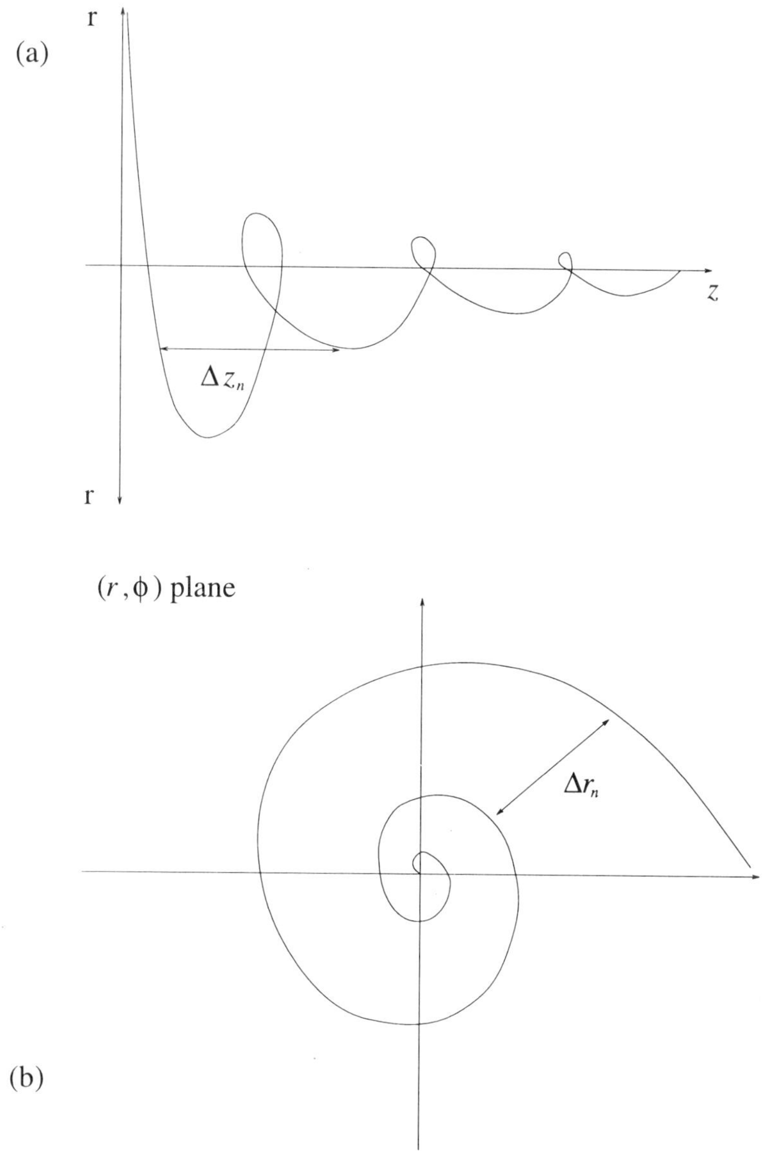

Fig. 3, (a) the spiral streamline as seen in a (r, z) plane and (b) as seen in the azimuthal plane (r, ϕ).

The Burgers and the Lundgren vortices and the effect of strain on D_K

The Burgers and the Lundgren vortices are examples of singular velocity fields where the Kolmogorov capacity D_K of streamlines can be strictly larger than 1. The Burgers vortex has a singularity of type (i) (Figure 1i in lecture I) and the Lundgren spiral vortex has a singularity of type (ii) (Figure 1ii in lecture I). Both vortices are sustained against viscous attrition by a persistent irrotational straining field

$$u_z = \gamma z \qquad (II9a)$$

and

$$u_r = -\frac{\gamma}{2} r, \qquad (II9b)$$

where γ is the strain rate which we assume to be constant in time. The Stokes streamfunction of this straining field is

$$\psi(r, z) = \frac{\gamma}{2} r^2 z, \qquad (II10)$$

and therefore, from (II4), $l = 2$.

The Burgers vortex is an exact solution of the Navier-Stokes equation and vorticity arises from the azimuthal velocity

$$u_\phi(r) = \frac{\Gamma}{2\pi r}(1 - e^{-\frac{r^2}{2R^2}}), \qquad (II11)$$

where Γ is the circulation, and the radius of the vortex core is $R = \sqrt{\frac{2\nu}{\gamma}}$, where ν is the kinematic viscosity. Outside the core,

$$u_\phi \approx \frac{\Gamma}{2\pi} r^{-1}. \qquad (II12)$$

Hence, comparing with (II3), the velocity field is singular with $m = -1$. The theorem of self-similar streamlines does not rule out spiralling streamlines with non-integral D_K when $m = -1$. As we shall now see, the Burgers vortex is an example of a flow singularity of type (i) (Figure 1i in lecture I) where $m = -1$ and where D_K is non-integral provided the Reynolds number $Re_\Gamma = \frac{\Gamma}{\nu}$ is large enough and the strain-rate γ is not too large.

The streamlines of the Burgers vortex are helical around the z-axis and their projection on the (r, ϕ) plane has the spiral form $r(\phi) \sim \phi^{-\frac{1}{2}}$. Applying formula (I12) of lecture I, we find that the Kolmogorov capacity D_{Kproj} of the streamlines' projection on the azimuthal plane is $D_{Kproj} = \frac{4}{3}$. A non-integral D_{Kproj} implies a non-integral D_K but only over a *restricted* range of length scales. The Kolmogorov box-coverings of streamlines in 3-D space and of their projection in the azimuthal plane are identical only when these coverings are restricted in the region of space where $\Delta z_n \ll \Delta r_n$. In that region of space, $D_K = D_{Kproj}$, and therefore $D_K = \frac{4}{3}$. However, where $\Delta z_n \gg \Delta r_n$, $D_K = 1$ because the distance between successive coils is $\Delta_n \approx \Delta z_n$ and grows indefinitely with n. From (II9), the condition $\Delta z_n \ll \Delta r_n$ is satisfied as long as $2z_n \ll r_n$. From (II9) and (II12),

$$r_n^2 = \frac{\Gamma}{4\pi^2\gamma} n^{-1} \qquad (II13),$$

and if we consider a streamline which passes through a point where $r = r_0$ and $z = z_0$ before completing its first turn,

$$z_n = z_0 \frac{4\pi^2 r_0^2}{\Gamma} \gamma n. \qquad (II14)$$

Hence, the number N of first turns of the spiral streamlines where $D_K = D_{Kproj}$ is

$$N \sim \frac{\Gamma}{4\pi^2\gamma} (2z_0 r_0^2)^{-2/3}. \qquad (II15)$$

A covering of the first N turns of a spiral streamline with boxes of size δ requires a minimum number of boxes of size δ that scales as $N_K(\delta) \sim \delta^{-D_K}$ with a well-defined $D_K > 1$ over a range of length scales such that $\Delta r_N < \delta < \Delta r_1$. A covering with boxes of size $\delta \leq \Delta r_N$ leads to $N_K(\delta) \sim \delta^{-D_K} = \delta^{-1}$. In particular, if the strain rate γ is so large that $N \approx 1$, then $D_K = 1$ over all length scales. Hence, the range of scales over which $D_K > 1$ shrinks as the strain rate γ increases.

It is remarkable that only two turns of a spiral can be sufficient for an estimation of D_K (Vassilicos & Hunt 1991). For example, if the box-counting algorithm is applied to a Burgers vortex streamline, we can expect to find $D_K > 1$ in the restricted range $\Delta r_N \leq \delta \leq \Delta r_1$ even when $N \approx 2$.

However, it follows from (II9b) and (II12) that a spiral streamline starting far outside the vortex core reaches the core after a number $Re_\Gamma/8\pi^2$ of turns. Hence, Re_Γ must be larger than about 200 for the streamline to have slightly more than two turns around the core. Inside the core, the velocity field is regular since it is dominated by strain and viscosity, and the theorem of self-similar streamlines implies that the Kolmogorov capacity of streamlines inside the core must be equal to 1. For D_{Kproj} to be well-defined and non-integral, Re_Γ must be larger than 200, and for D_K to be well-defined and non-integral, Re_Γ must be larger than 200 and the strain rate γ must be so small that $N \geq 2$.

Jiménez *et al.* (1993) calculated the distribution of Re_Γ values of intense vortex tubes in DNS homogeneous and isotropic turbulence, and found that Re_Γ is an increasing function of Re_λ, the Reynolds number of the turbulence based on the Taylor microscale λ. In particular, when $Re_\lambda = 35.8$, Jiménez *et al.* (1993) find that

$Re_\Gamma < 200$ for all the vortex tubes in the turbulence. If the intense vortex tubes in the small scales of a low Reynolds number turbulence are Burgers vortices, the streamlines around these vortex tubes should have trivial Kolmogorov capacities, i.e. $D_K = D_{Kproj} = 1$, because they shoot into the vortex core in less than two turns. Inside the vortex core, $D_K = 1$ by virtue of the theorem of self-similar streamlines. As we discuss towards the end of this lecture, Vassilicos *et al.* (1994) find streamlines with $D_K > 1$ around intense vortex tubes in isotropic turbulence of $Re_\lambda = 19$ and 21, thus indicating that these vortex tubes are not of the Burgers type and that the strain acting on them must be weak.

It has been suggested that the vortex tubes in the small-scale turbulence may have the Lundgren spiral vortex structure (Lundgren 1982, 1993). Lundgren vortices are a family of incompressible asymptotic solutions of the Navier-Stokes equation as time $t \to \infty$. The simplest example of such a vortex is the unstrained and inviscid 2-D spiral vortex sheet which is given by

$$\omega(r, \phi, t) = 2\pi f(r)\delta(\phi - \Omega(r)t), \tag{II16}$$

where the vorticity $\omega(r, \phi, t)$ is in the z-direction, and δ is the Dirac delta function. The 2-D vorticity profile (II16) is an incompressible long-time solution of the Euler equation provided that

$$rf(r) = \frac{d}{dr} r^2 \Omega(r) \tag{II17a}$$

and

$$\frac{d}{dr}\Omega(r) < 0. \tag{II17b}$$

The circulation around a circle of radius r is $\Gamma(r) = 2\pi r^2 \Omega(r)$, the ϕ-averaged vorticity is $f(r)$, and (II17a) is Kelvin's theorem. The differential rotation of the spiral vortex is represented by $\Omega(r)$; it should be noted that the asymptotic solution (II16) is fully determined when $\Omega(r)$ has been chosen under the constraint (II17b).

As noted by Lundgren (1982), when a solution of the 2-D Euler or Navier-Stokes equation is superposed and subjected to the straining field (II9) the resulting velocity field is a solution of the 3-D Euler or Navier-Stokes equation. Applying and superposing the irrotational straining field (II9) to (II16), the resulting velocity field takes the form

$$u_z = \gamma z, \tag{II18a}$$

and

$$u_r \approx -\frac{\gamma r}{2} + \frac{T(t)^{-2} f(\sqrt{S(t)}r)}{r(\frac{d\Omega}{dr}(\sqrt{S(t)}r))^2}[\phi - \Omega(\sqrt{S(t)}r)T(t) - 2\pi int(\frac{\phi - \Omega(\sqrt{S(t)}r)T(t)}{2\pi}) - \pi], \tag{II18b}$$

$$u_\phi \approx S(t)[r\Omega(\sqrt{S(t)}r) + O(T(t)^{-1})] \tag{II18c}$$

as $t \to \infty$, where the stretch ratio $S(t) = e^{\gamma t}$ and the strained time $T(t) = \int_0^t S(\tau)\,d\tau$ (see Vassilicos *et al.* 1994). The velocity field (II18) is an incompressible asymptotic solution of the 3-D Euler equation as $t \to \infty$. There are contributions to u_r from both the strain field and the spiral vortex sheet. Note the jump in u_r across the sheet of vorticity at $\phi = \Omega(\sqrt{S(t)}r)T(t)$. A similar jump can also be found in the $O(T(t)^{-1})$ term in u_ϕ. The contribution to u_r and u_ϕ of the induced velocity from the vortex sheet tends to 0 as $t \to \infty$. Thus, the azimuthal velocity is dominated by the differential rotation $\Omega(\sqrt{S(t)}r)$ which rotates the entire sheet structure. Note also that the leading order in (II18) is axisymmetric, so that the flow tends towards axisymmetry as $t \to \infty$.

The delta function in (II16) and the velocity field (II18) may be modified to take viscosity into account (Lundgren 1982). Here we only discuss the inviscid Lundgren solution which has a complex singularity of type (ii) (Figure 1ii in Lecture I) down to vanishing scales. Taking viscosity into account would only replace the singularity by a near-singularity, and the conclusions concerning the fractal properties of streamlines would remain essentially the same. The Lundgren singularity is complex of type (ii) because of the spiral accumulating structure of the jumps in velocity which extends all the way to the spiral's centre. In effect, the Lundgren singularity is two-fold: the spiral accumulation of the vortex sheet –see (II16), and the simple singularity in $\Omega(r)$, since $\Omega(r)$ is monotonically increasing as $r \to 0$ –see (II17b). Such complex singularities are not included in the domain of validity of the theorem of self-similar streamlines, and streamlines may exist with $D_K > 1$. The Lundgren vortex is an example of a flow singularity of type (ii) (Figure 1ii in Lecture I), and in some cases, D_K is non-integral provided the strain rate γ is not too large.

The equation $r = r(\phi)$ of a streamline's projection in the (r, ϕ) plane may be obtained by integrating

$$\frac{d\phi}{dr} = \frac{u_\phi}{r u_r} \approx -\frac{4\pi\sqrt{S(t)}}{\gamma}\frac{\Omega(\sqrt{S(t)}r)}{r} \tag{II19}$$

to leading order in time. If $\Omega(r) \sim r^{-\frac{1}{\alpha}}$ where $\alpha > 0$, then $r(\phi) \sim \phi^{-\alpha}$, which is *identical to the spiral form* $\phi = \Omega(\sqrt{S(t)}r)T(t) \sim r^{-\frac{1}{\alpha}}$ *of the vortex sheet.* Equation (I12) of lecture I is

$$D_{Kproj} = max(1, \frac{2}{1+\alpha}), \tag{II20}$$

and therefore $D_{Kproj} > 1$ when $\alpha < 1$. The Kolmogorov capacity of the spiral intersection of the vortex sheet with the azimuthal plane is also D_{Kproj}. Hence the accumulation pattern of the streamlines is a direct reflection of the accumulation structure of the vortex sheet.

The Kolmogorov capacity D_K of a Lundgren streamline in 3-D space is equal to D_{Kproj} if the box-counting operation is limited to the first N turns of the spiral streamline where $\Delta z_n \ll \Delta r_n$, and N turns out again to be inversely proportional to the strain rate γ as in (II15). Hence, the effect of strain on the value of D_K is identical for both the Lundgren and the Burgers vortices. The Kolmogorov capacity D_K of streamlines is well-defined and non-integral over a range of length scales δ between Δr_N and Δr_1 and $D_K = 1$ where $\delta \leq \Delta r_N$. *The range of scales over which the Kolmogorov capacity D_K of streamlines is strictly larger than 1 shrinks as the strain rate γ increases.*

The fractal analysis of streamlines around vortex tubes in low Reynolds number isotropic turbulence

A fractal analysis of streamlines can probe the internal structure of DNS turbulence vortex tubes and detect singularity and self-similarity properties even at surprisingly low Reynolds numbers where the traditional measure of self-similarity, the energy spectrum $E(k)$, fails to achieve its asymptotic ($Re_\lambda \to \infty$) form $E(k) \sim k^{-5/3}$. It is important that only two turns of a streamline are enough to measure D_K. The Kolmogorov capacity of streamlines may be a measure that is robust to variations in Re_λ; one may expect the number of turns of spiral streamlines and spiral vortex sheets to increase in some self-similar way as the Reynolds number increases, in which case the D_K measured at low values of Re_λ can be a good estimate of the asymptotic value of D_K as $Re_\lambda \to \infty$. Some parts of the scaling and singularity flow structure of the small-scale turbulence that is responsible for Kolmogorov's non-integral power law $E(k) \sim k^{-5/3}$ may already be present at values of Re_λ that are too low for measurements of $E(k)$ to give $k^{-5/3}$. However, these parts of the singular turbulence structure that are robust to variations in Re_λ may be *localised* in space and may not be the only features responsible for the asymptotic form of $E(k)$, since the energy spectrum is a global quantity integrated over the entire turbulent flow field.

By virtue of the theorem of self-similar streamlines, the fractal analysis of streamlines around individual axisymmetric vortex tubes can reveal the existence of a singularity with self-similar scaling in the vortex tube. This analysis has the added double advantage of being robust to variations in Reynolds number and of being applicable to selected localised flow regions within the turbulence, such as vortex tubes. The traditional measures of self-similarity, the power spectra $E(k)$ and $\Gamma(k)$, converge slowly with Reynolds number (Sreenivasan 1991, Vassilicos & Hunt 1991) and are average quantities over the entire flow and therefore blind to specific localised flow features of the turbulence.

Vassilicos *et al.* (1994) developed a *systematic* algorithm which, given an Eulerian velocity field on a grid of discrete points in space, interpolates velocities between grid nodes and calculates streamlines, then identifies streamline sections with persistence of strong curvature (such as in Figure 3), and finally calculates the Kolmogorov capacities of these streamline sections. Because the algorithm is automatic it can be applied to many regions of the flow at once, and a large number of streamlines can be studied simultaneously. Vassilicos *et al.* (1994) applied this systematic fractal analysis of streamlines to DNS isotropic and decaying turbulence with low Reynolds numbers $Re_\lambda = 19$ and 21. In some of their runs, they integrated approximately 4000 streamlines on a 128^3 grid, spaced uniformly one integral length scale apart from one another. In another run on a similar data set, they integrated approximately 23000 streamlines in the high enstrophy regions of the flow –inside the enstrophy contours in Figure 2. Vassilicos *et al.* (1994) applied the algorithm to data sets of DNS turbulence with a variety of resolutions (the same instance of isotropic decaying turbulence in 128^3, 256^3 and 512^3 grids). In all these cases it was possible to collect statistics of streamline capacities D_K around vortex tubes in the small-scale turbulence. A significant number of these capacities are well-defined and strictly larger than 1. This result is important because it suggests the existence of flow singularities in the small-scale vortex tubes, and because only a very restricted class of smooth curves have well-defined non-integral capacities. For example, only a restricted class of spirals, for example algebraic spirals, can have well-defined non-trivial Kolmogorov capacities (see (I12) in lecture I). The fact that streamlines exist with $D_K > 1$ around the vortex tubes of the small-scale turbulence sets tight constraints on what the flow structure of these vortex tubes can be.

The box-counting algorithm used in the systematic fractal analysis of streamlines gives the value of D_K automatically over a non-prescribed range of scales δ. Specifically, the lower bound δ_{min} of the range of box-sizes δ is not prescribed, but the upper bound δ_{max} is set equal to the Taylor microscale λ. The Taylor microscale $\lambda = 5u'^2/\omega'^2$ where u' and ω' are the r.m.s. values of the turbulent velocity and vorticity respectively, and so λ can be roughly interpreted to be a mean radius of streamline curvature in the small-scale turbulence. The output of the box-counting algorithm is obtained by calculating

$$D_K(\delta) = -\frac{log[N_K(\delta)/N_K(\delta_{min})]}{log(\delta/\delta_{min})} \qquad (II21)$$

directly as a function of box-size δ. Applied to an artificial spiral $r(\phi) \sim \phi^{-1/2}$ generated on the computer (Figure 4a), the output of this box-counting algorithm is shown in Figure 4b.

(a)

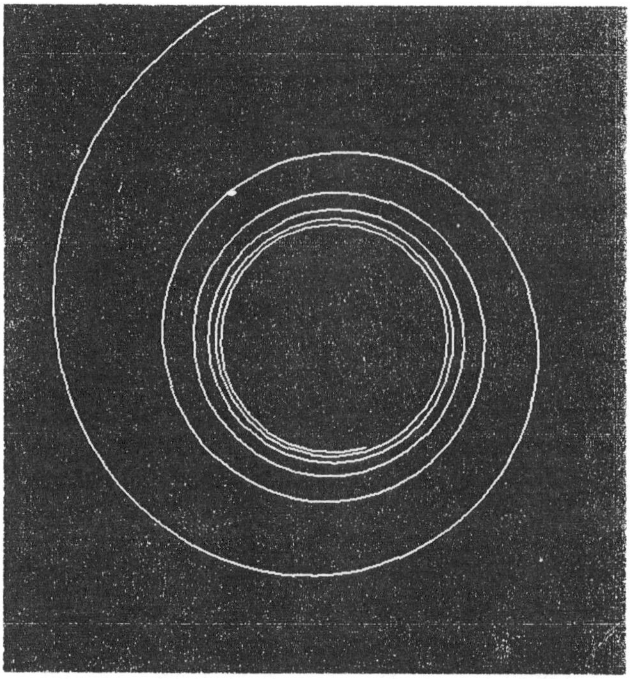

(b)

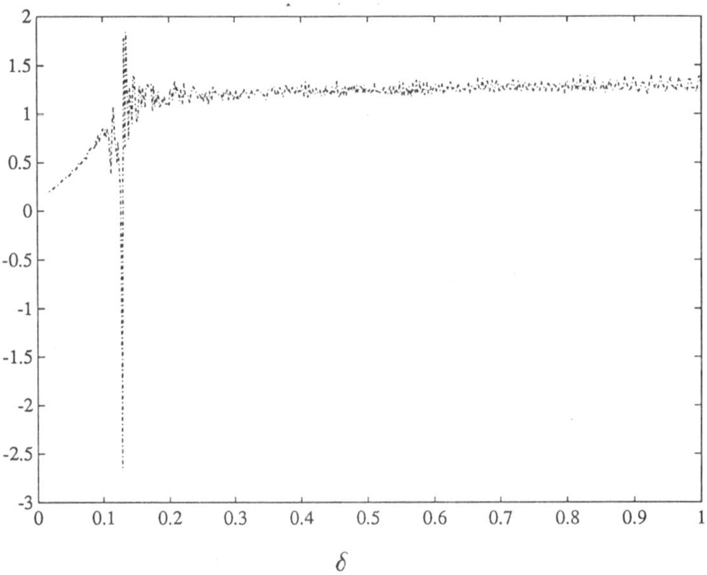

Fig. 4, (a) the spiral $r(\phi) \sim \phi^{-1/2}$ and (b) the measured function $D_K(\delta)$ for that spiral. Note how the $D_K(\delta)$ curve points at both δ_{min} and D_K.

The algorithm identified $\delta_{min} = 0.127$ as the lower bound of the range of scales where D_K is best defined in the sense that the standard deviation of $D_K(\delta)$ over values of δ between δ_{min} and δ_{max} is minimal when $\delta_{min} = 0.127$. The capacity D_K has a well defined constant value in that range (no upward or downwards trend), and the distance between the innermost consecutive coils of the spiral in Figure 4a is indeed close to 0.12. The measured capacity is $D_K = 1.3 \pm 0.08$, which is satisfactory, given that $D_K = 4/3$ for spirals $r(\phi) \sim \phi^{-1/2}$ (see (I12) in lecture I).

Figures 5a and 5b show two examples of box-counting results for streamlines around small-scale vortex tubes in an isotropic DNS turbulence of $Re_\lambda = 19$. Note that the smallest length scale δ_{min} of the range of self-similar box-counting scaling is slightly smaller than η for one streamline and slightly larger for the other. There is also a variation of D_K from streamline to streamline. In Figure 5a, for example, $D_K \approx 1.05$, and in Figure 5b, $D_K \approx 1.25$. It also turns out that different streamlines in the turbulence may have qualitatively different kinds of geometrical 'cascades' of length scales of motion. Examples of streamlines are shown in Figures 6 and 7. In Figure 6 the streamlines accumulate towards a central axis and the accumulation is primordially radial, whereas in Figure 7a the streamline accumulates towards a limit-circle and the accumulation of length scales is *along* the vortex tube. The limit-circle lies in a region of the vortex tube where the magnitude of the vorticity attains a local minimum (Figure 8). These two types of streamline accumulating topology appear to be quite generic in the small-scale turbulence field. Figure 7b shows an example of a streamline with only two turns which proved enough for the box-counting algorithm to give $D_K \approx 1.3$ over the appropriate range of scales.

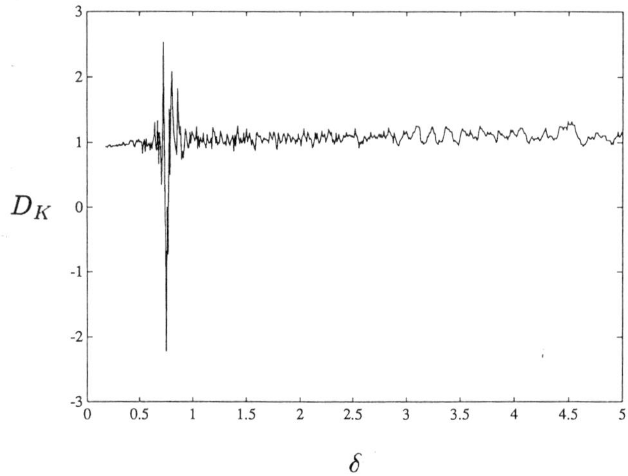

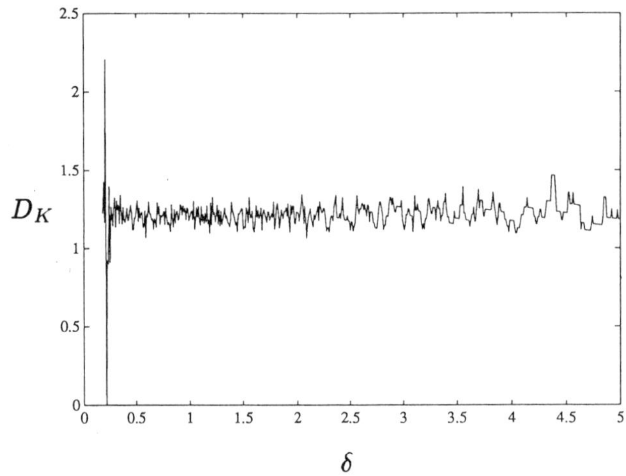

Fig. 5, Examples of box-counting results for two spiral streamlines in low Reynolds number DNS turbulence from Vassilicos et al. (1994). The Taylor microscale $\lambda \approx 5\Delta x$, and the Kolmogorov scale $\eta \approx 0.55\Delta x$, where Δx is the grid-size (the δ axes are in units Δx). Both D_K and δ_{min} are unambiguously obtained by the algorithm.

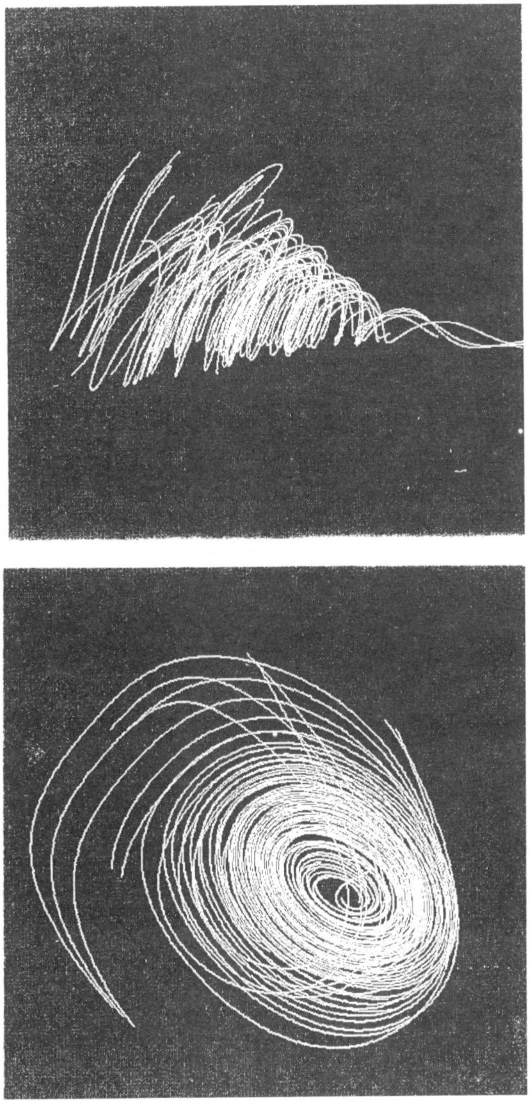

(a)

(b)

Fig. 6, From an isotropic DNS turbulence of $Re_\lambda \approx 21$ *(Vassilicos et al. 1994). (a) Projection of streamlines on a plane transversal to a vortex tube in the small-scale turbulence. (b) Projection of the same streamlines on the azimuthal plane of the vortex tube. The Kolmogorov capacities of these streamlines vary between 1.2 and 1.3.*

(a)

(b)

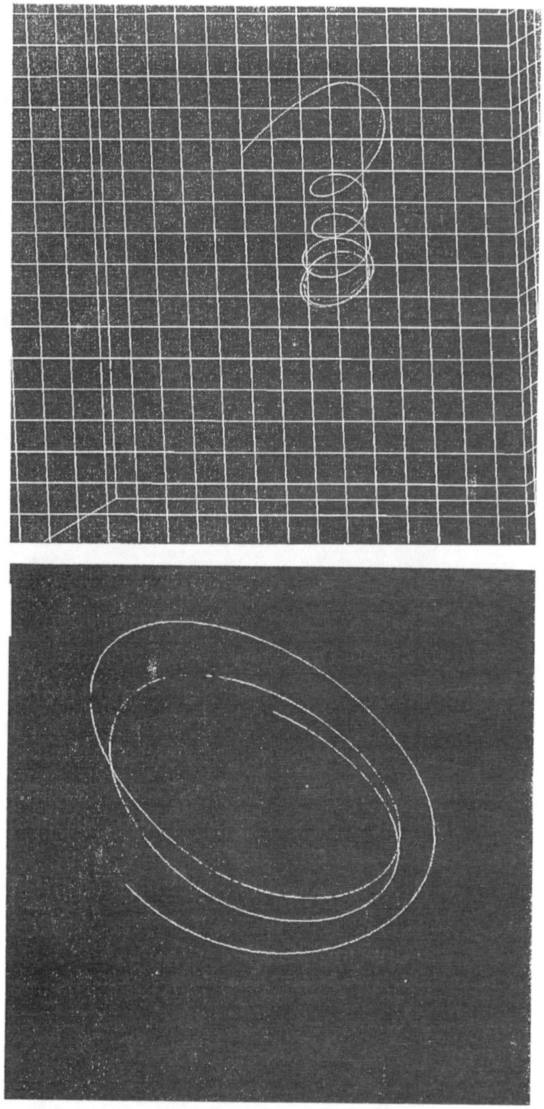

Fig. 7, From an isotropic DNS turbulence of $Re_\lambda = 19$. (a) A number of axially accumulating streamlines with $D_K \approx 1.33$ around the same vortex tube. The vorticity vectors are directed along the axis around which the streamlines coil, and point from left to right. The mesh size of the grid superimposed on the picture (not the DNS grid) is $2\Delta x \approx 4\eta \approx 0.4\lambda$ (reproduced with the permission of J.C. Vassilicos, B.P. Moquin & J.G. Brasseur). (b) Only two turns of a streamline are enough to measure D_K. Here, $D_K \approx 1.3$ (from Vassilicos et al. 1994).

(a) (b) (c) (d)

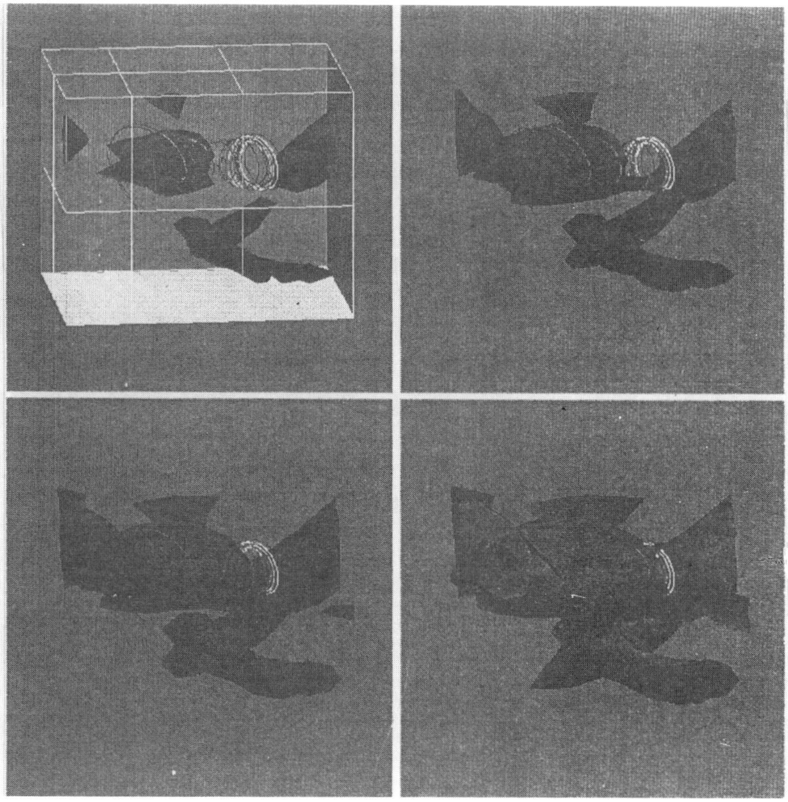

Fig. 8, Same axially accumulating streamlines as in Figure 7a plotted with iso-enstrophy surfaces where ω^2 is equal to (a) 15 r.m.s., (b) 12 r.m.s., (c) 10 r.m.s. an (d) 8 r.m.s. (the r.m.s. is of ω^2 over the entire flow simulation). Note that the streamlines approach limit circles at the region of minimum enstrophy along the axis. (From Vassilicos et al. 1994).

Figure 9 is an example of an histogram of streamline capacities D_K around small-scale vortex tubes of an isotropic unforced DNS turbulence where the distance between grid nodes is approximately half the Kolmogorov viscous length scale η and $Re_\lambda \approx 21$. This numerically generated data set of isotropic unforced turbulence was obtained by DNS on a 128^3 grid, but such histograms have also been confirmed by DNS simulations of identical turbulent flows on 256^3 and 512^3 grids (Vassilicos *et al.* (1994). The lower bound δ_{min} is always found to have a value which ranges between, approximately, $\frac{1}{2}\eta$ and 2η. In practise, we can safely state that $D_K > 1$ if $D_K - \sigma > 1$ where σ is the standard deviation of $D_K(\sigma)$ and is calculated over values of δ ranging between $\delta_{max} = \lambda$ and $\delta_{min} = O(\eta)$. One fifth of the streamlines in the histogram of Figure 9 are such that $D_K - \sigma > 1$ and such significant numbers of streamline sections with $D_K - \sigma > 1$ are also found in other isotropic decaying turbulence of similar Reynolds number but with far better spatial resolution (Vassilicos *et al.* 1994).

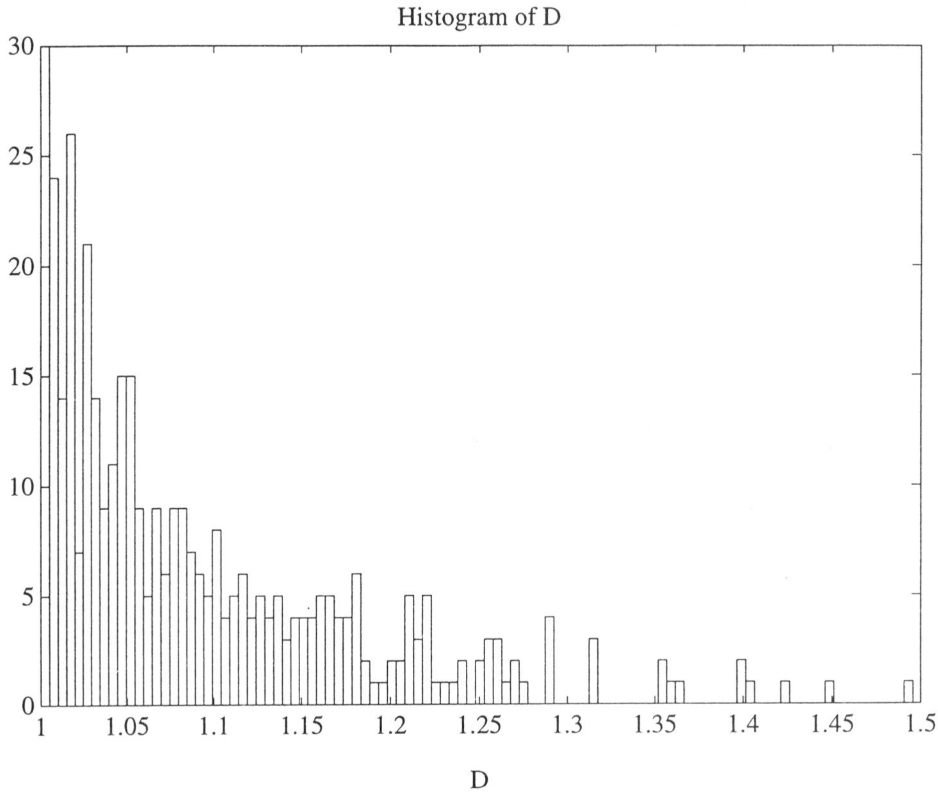

Fig. 9, Histogram of capacities D_K of streamlines around small-scale vortex tubes in a low Reynolds number DNS isotropic and unforced turbulence. 4000 streamlines were integrated spaced uniformly slightly less than one integral length scale apart from one another. 795 streamline sections with persistence of strong curvature were found around small-scale vortex tubes. Of these streamline sections, 180 were independent; independent streamlines coil around different vortex tubes. The histogram is based on such 180 streamline sections. (From Vassilicos et al. 1994).

Conclusion

There is enough numerical evidence that streamlines around small-scale turbulence vortex tubes have well-defined scaling geometrical structures with well-defined Kolmogorov capacities D_K over length scales ranging between the Taylor and the Kolmogorov microscales. The theorem of self-similar streamlines implies that if some of these vortex tubes are axisymmetric with streamlines spiralling towards the central axis, then these small-scale turbulence vortex tubes must carry flow singularities that are not as simple as (II3) and (II4) with non-integral m and/or l such that $m > -1$ and $l > 0$.

Because the evidence discussed above concerns DNS turbulence of Reynolds numbers Re_λ as low as 19 and 21, the vortex tubes where streamlines have non-integral Kolmogorov capacities cannot be Burgers vortices. Based on the results of Jiménez *et al.* (1993) we argued that a Reynolds number $Re_\lambda < 35.8$ is too small for streamlines to spiral around Burgers vortex tubes more than once or twice before reaching the viscous core. Hence, were the small-scale vortex tubes of a low Reynolds number turbulence Burgers vortices, their streamlines would have trivial Kolmogorov capacities $D_K = 1$.

These results are confirmed by a study of the spatial correlation of the enstrophy ω^2 with the square of the viscous force f^2 in the vortex tubes where streamlines with persistent curvature are identified, and in particular, where $D_K > 1$ (Vassilicos *et al.* 1994). If this spatial correlation is calculated over the space of vortex tubes where ω^2 is larger than some threshold value ω_{th}^2, and if the dependence of this spatial correlation C is plotted against the ratio $\omega_{th}^2/\omega_{max}^2$, where ω_{max}^2 is the local peak enstrophy of individual vortex tubes, then the result obtained from a low Reynolds number DNS isotropic turbulence is shown in Figure 10 and is compared with the corresponding $C(\omega_{th}^2/\omega_{max}^2)$ curve for Burgers vortices in Figure 11.

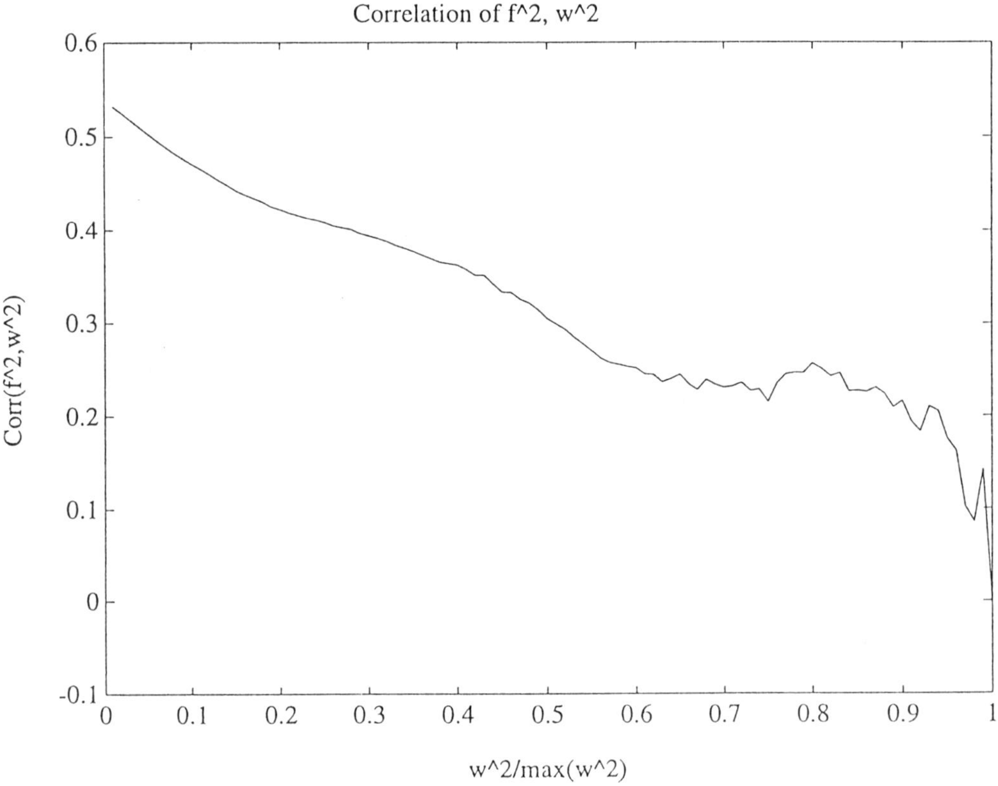

Fig. 10, Spatial correlation of enstrophy ω^2 *with the square of the viscous force* f^2 *as a function of the locally scaled threshold* $\omega_{th}^2/\omega_{max}^2$. *Only vortex tubes where streamlines have been detected with persistent curvature are included in the calculation of the correlation* C. *The DNS turbulence is unforced and isotropic and generated on a* 128^3 *grid, the Reynolds number* $Re_\lambda = 19$, *and the grid spacing is roughly equal to* 1.8η. *The part of the curve where* $\omega_{th}^2/\omega_{max}^2 > 0.8$ *is noise due to probing below the grid. (From Vassilicos et al. 1994).*

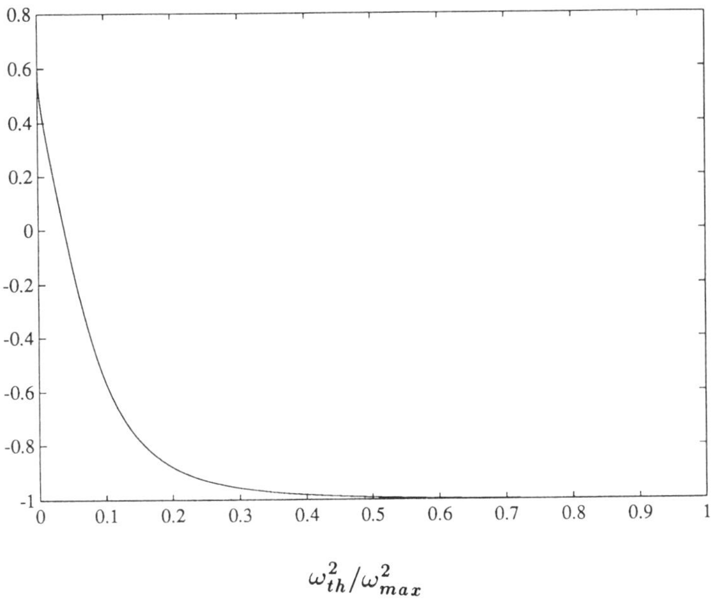

Fig. 11, Spatial correlation of enstrophy with the square of the viscous force as a function of $\omega_{th}^2/\omega_{max}^2$ for the Burgers vortex. Analytical result from Vassilicos et al. (1994).

The internal structure of the vortices associated with streamlines having non-trivial Kolmogorov capacities is different from that of a Burgers vortex. Because the viscous force $\mathbf{f}$ is essentially a spatial derivative of the vorticity ($\mathbf{f} = -\nu\rho\nabla \times \boldsymbol{\omega}$ where ρ is the density of the fluid) the significant non-negative correlation between f^2 and ω^2 is indicative of a singular vortex tube structure down to scales of order η. The Burgers dependence of C on $\omega_{th}^2/\omega_{max}^2$ is accounted for by the schematic profile of radial variations of enstrophy and viscous force in Figure 12. The viscous force is small where the enstrophy is large and vice-versa.

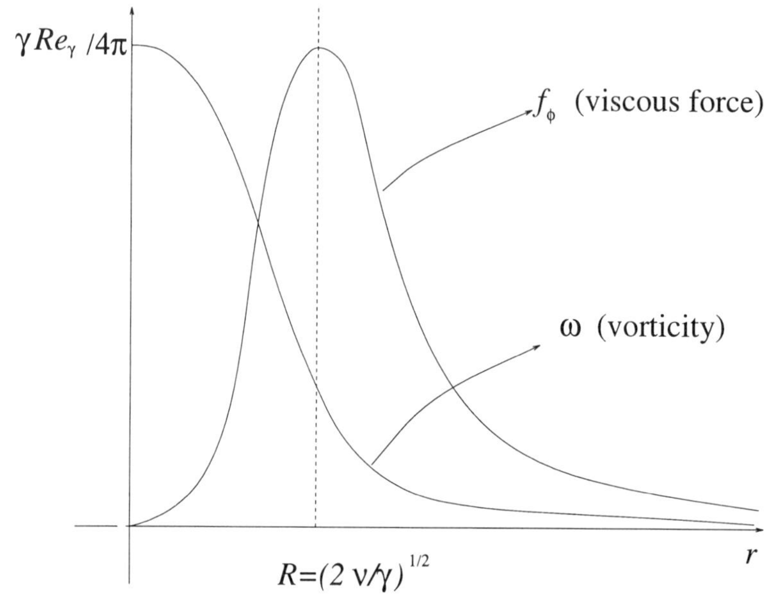

Fig. 12, Schematic profile of radial variations of the enstrophy and the viscous force in the Burgers vortex. Note that the viscous force peaks at $r = R = \sqrt{\frac{2\nu}{\gamma}}$ where the vorticity has reached small-to-intermediate values, and note the vanishing viscous force at the centre of the vortex where the motion is asymptotically one of solid body rotation.

Let me close this lecture with a few speculations. Recent evidence by Brasseur & Lin (1991) (see Figure 13) and Vincent & Meneguzzi (1994) suggests that the vortex tubes in the small-scale turbulence are produced by shear instabilities of vortex sheets which appear at first and then roll up to form vortex tubes with a singular spiral internal structure. Presumably, the non-integral capacities of vortex tube streamlines discussed in this lecture reflect the internal spiral vorticity structure of these tubes. However, it is questionable whether these vortex tubes remain stable as the Reynolds number increases. The self-similar scaling of vortex tube streamlines extends from the Kolmogorov viscous length scale to the Taylor microscale, and in the next lecture we present results which may suggest that a qualitative difference in turbulence structure exists above and below the Taylor microscale λ. Below the Taylor microscale, the turbulence singularity structure may be dominated by the outcome of vortex sheet spiral roll-ups. Above the Taylor microscale, the turbulence may be governed by vortex tube instabilities and by stretching and folding mechanisms which generate a fractal singularity structure similarly to chaotic advection. In fact, an increasing number of mechanisms may come into play as $Re_\lambda \to \infty$. It remains a mystery how and why all these different and complex mechanisms balance the 'turbulence accounts' to give a sum total that obeys simple dimensional similarity and leads to Kolmogorov's -5/3 law.

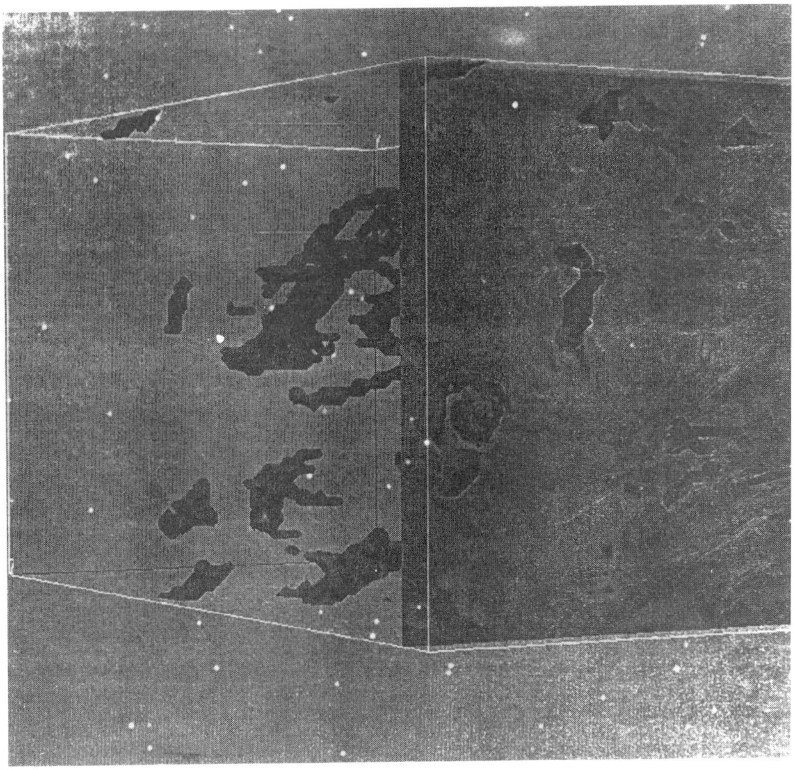

Fig. 13, Homogeneous turbulent shear flow. Iso-contours of ω^2. Note the spiral vortex sheet. (From J.G. Brasseur & W.-Q. Lin (19 November 1991, personal communication).

REFERENCES

21. Jiménez, J., Wray, A.A., Saffman, P.G. & Rogallo, R.S. 1993 The structure of intense vorticity in homogeneous isotropic turbulence. J. Fluid Mech. **255**, 65.

22. Kevlahan, N.K.-R. & Vassilicos, J.C. 1994 The space and scale dependencies of the self-similar structure of turbulence. Proc. R. Soc. Lond. A (to appear).

23. Lundgren, T.S. 1982 Strained spiral vortex model for turbulent fine structure. Phys. Fluids **25** (12), 2193.

Lundgren, T.S. 1993 A small-scale turbulence model. Phys. Fluids A **5** (6), 1472.

24. Sreenivasan, K.R. 1991 On local isotropy of passive scalars in turbulent shear flows. Proc. R. Soc. Lond. A **434**, 165.

25. Vassilicos, J.C. & Hunt, J.C.R. 1991 Fractal dimensions and spectra of interfaces with application to turbulence. Proc. R. Soc. Lond. A **435**, 505.

26. Vassilicos, J.C., Moquin, B.P. & Brasseur, J.G. 1994 Self-similar spiral flow structure in low Reynolds number isotropic and decaying turbulence. J. Fluid Mech. (under review).

27. Vincent, A. & Meneguzzi, M. 1991 The spatial structure and statistical properties of homogeneous turbulence. J. Fluid Mech. **225**, 1.

28. Vincent, A. & Meneguzzi, M. 1994 The dynamics of vorticity tubes in homogeneous turbulence. J. Fluid Mech. **258**, 245.

3. Lecture III: wavelet analysis of singular flow structure

Let us become familiar with the wavelet transform by way of examples.

(i) The wavelet transform of the Dirac delta function $u(x) = \delta(x - x_1)$ is

$$\tilde{u}(a, x_0) = a^{-1/r} \int_{\infty}^{+\infty} \delta(x - x_1)\theta^*(\frac{x - x_0}{a})\, dx = a^{-1/r}\theta^*(\frac{x_1 - x_0}{a}), \qquad (III1)$$

and therefore has the wavelet's cone like structure in the (x_0, a) plane (Figure 13 in lecture I) pointing towards $x_0 = x_1$ as $a \to 0$ (Figure 1). The wavelet transform can therefore be used to detect the location of sharp discontinuities in $u(x)$ and its derivatives, and sudden violent 'blips' like delta functions.

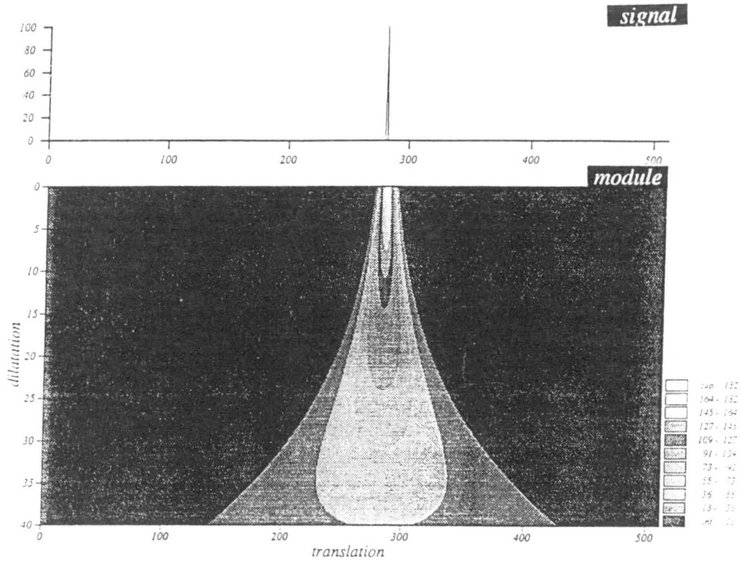

Fig. 1, the modulus of the wavelet transform of a delta function (reproduced with the permission of Marie Farge & Gabriel Rabreau). Light colouring indicates high values of the modulus.

(ii) The Fourier transform of a sinusoidal function $u(x)$ with an abrupt change in wavelength at a point x_1 (Figure 2 shows a related but different example; the difference is not essential) is the sum of two delta functions plus a high frequency contribution $\hat{u}_h$ from the discontinuity in the first derivative of $u(x)$ at $x = x_1$, that is $\hat{u}(k) = \delta(k-k_1) + \delta(k-k_2) + \hat{u}_h(k)$ where $2\pi/k_1$ is the wavelength at $x < x_1$ and $2\pi/k_2$ is the wavelength where $x > x_1$. We expect this discontinuity to appear in the wavelet transform of $u(x)$ as a cone structure pointing towards x_1 as $a \to 0$ in the plane (x_0, a). However, where a is larger than both $2\pi/k_1$ and $2\pi/k_2$, the wavelet transform is well approximated by

$$\tilde{u}(a, x_0) = a^{-1/r} \int_{-\infty}^{+\infty} \hat{u}(k)\hat{\theta}^*(ak)e^{-ikx_0}\, dk \approx a^{-1/r}\hat{\theta}^*(ak_1)e^{-ik_1x_0} + a^{-1/r}\hat{\theta}^*(ak_2)e^{-ik_2x_0}, \tag{III2}$$

and therefore has a periodic structure directly reproducing the periods $2\pi/k_1$ and $2\pi/k_2$ and even pointing towards $a = 2\pi/k_1$ and $a = 2\pi/k_2$ where $x_0 < x_1$ and $x_1 < x_0$ respectively (see Figure 2). Hence, the wavelet transform can also detect localisation in Fourier space, and thereby well-defined periodicities in a signal.

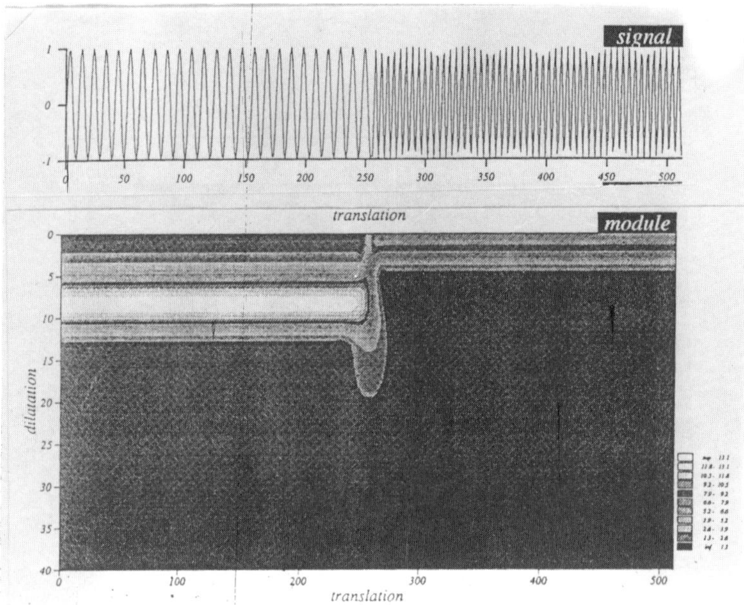

Fig. 2, the modulus of the wavelet transform of a period doubling function (reproduced with the permission of Marie Farge & Gabriel Rabreau). Light colouring indicates high values of the modulus.

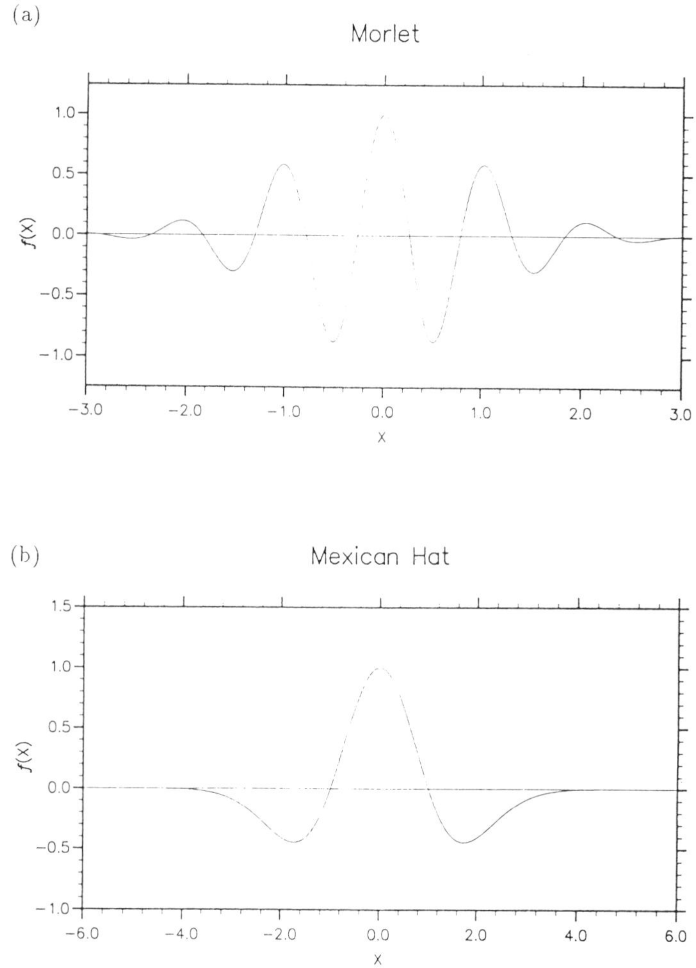

Fig. 3, examples of wavelets. (a) Imaginary part of Morlet wavelet. (b) Mexican hat (second derivative of the Gaussian).

(iii) The wavelet transform $\tilde{u}(a, x_0)$ of an isolated complex singularity $u(x) = x^\sigma e^{-ix^{-\tau}}$ can be estimated in the limit where $a \to 0$ by the method of stationary phases if the mother wavelet has the Morlet form $\theta(x) = e^{ix}e^{-\frac{x^2}{2}}$ (see Figure 3), in which case

$$\tilde{u}(a, x_0) = a^{-1/r} \int dx x^\sigma e^{ix^{-\tau}} e^{i\frac{x-x_0}{a}} e^{-\frac{(x-x_0)^2}{2a^2}}, \tag{III3}$$

and

$$\tilde{u}(a, x_0) \sim a^{-1/r} a^p e^{i\phi(a,x_0)} e^{-\frac{(x_0-\xi)^2}{2a^2}} \tag{III4}$$

as $a \to 0$, where the locations ξ of local wavelet transform maxima in the (x_0, a) plane depend on the length-scale a according to

$$\xi = (\tau a)^{\frac{1-D'_K}{2}}, \tag{III5}$$

D'_K being the Kolmogorov capacity of the zero-crossings of $u(x)$, i.e. $D'_K = \frac{\tau}{\tau+1}$, the phases $\phi(a, x_0)$ are given by

$$\phi(a, x_0) = (1+\tau)(\tau a)^{-D'_K} - \frac{x_0}{a}, \tag{III6}$$

and the power

$$p = \frac{2\sigma + \tau + 2}{2(\tau+1)} = 1 + \sigma - (\sigma + \frac{1}{2})D'_K. \tag{III7}$$

The wavelet transform of an accumulating oscillation is cone-shaped in the (x_0, a) plane pointing towards the centre of the accumulating singularity (Figure 4) and the shape $\xi \sim a^{\frac{1-D'_K}{2}}$ of the cone is a measure of the space-filling property and the Kolmogorov capacity D'_K of the accumulation. Note from lecture I that the Fourier energy spectrum of $u(x) = x^{\sigma} e^{-ix^{-\tau}}$ is $E(k) \sim k^{-2p}$ where the exponent p is given by (III7). The Fourier transform $\hat{u}(k) \sim k^{-p} e^{i\phi(k)}$ where the Fourier phases $\phi(k) \sim k^{D'_K}$ (Hunt *et al.* 1993). The dependence of $\phi(k)$ on the wavenumber k is similar to the dependence of the wavelet phases $\phi(a, x_0)$ on a^{-1}. Clearly, wavelet transforms can detect isolated singularities and even some of their properties, such as D'_K, the scaling exponent p of the Fourier power spectrum, and properties of the Fourier phase structure.

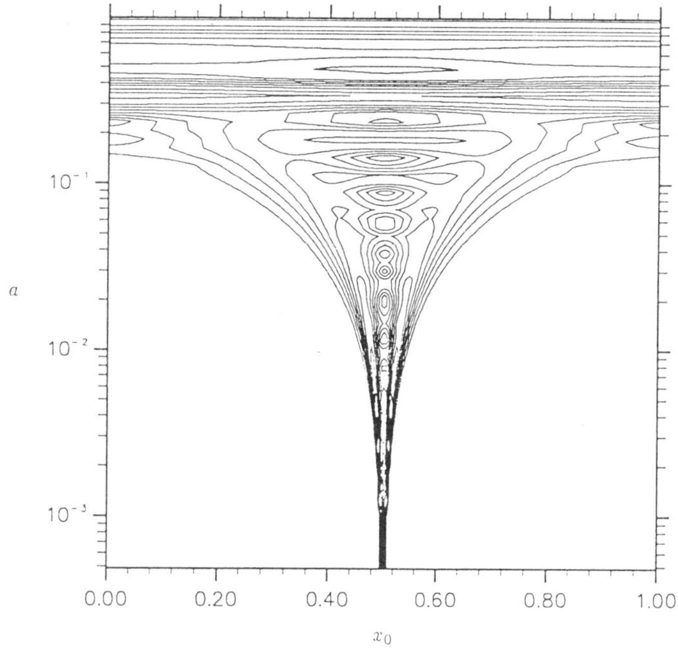

Fig. 4, Morlet wavelet transform of $sin(2\pi x^{-1/2})$. *(From Kevlahan & Vassilicos 1994).*

(iv) An isolated complex singularity can be transformed into a non-isolated singularity and a local self-similarity into a global self-similarity by randomising the Fourier phases. For example, by randomising the Fourier phase spectrum of an isolated accumulation such as $u(x) = x^{\sigma}e^{ix^{-\tau}}$, the isolated accumulation becomes a function with non-isolated singularities, but the energy spectrum $E(k) \sim k^{-2p}$ remains the same (see Figure 5).

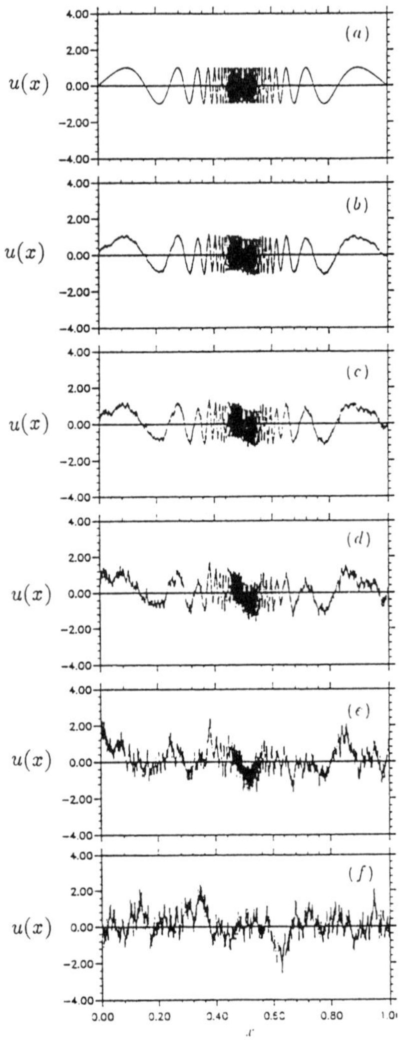

Fig. 5, The Fourier phase spectrum of $u(x) = x^{\sigma} e^{ix^{-\tau}}$ is $\phi_S \sim k^{D'_K}$ where $D'_K = \frac{\tau}{1+\tau}$. To investigate the effect of varying amounts of phase noise we replace $\phi_S(k)$ by $\phi(k) = (1-f)\phi_S(k) + f\phi_R(k)$ where $\phi_R(k)$ is a random function of k and $0 \leq f \leq 1$. (a) $sin(2\pi/x)$ with no phase noise ($f = 0$), (b) with 5% noise ($f = 0.05$), (c) 10% noise ($f = 0.1$), (d) 20% noise ($f = 0.2$), (e) 40% noise ($f = 0.4$) and (f) 100% noise ($f = 1$). (From Kevlahan & Vassilicos 1994).

The wavelet transform of a random fractal function similar to Figure 5f is shown in Figure 6. Wavelet transforms of fractal functions, not necessarily with random phases, also have a structure of non-isolated cones in the (x_0, a) plane. Pioneering works on the wavelet transform of fractal and random functions show how Holder exponents and Hausdorff dimensions of fractal functions can be obtained from the wavelet transform (Holschneider 1988, Arneodo *et al.* 1988, Flandrin 1993). However, little work has been done on the wavelet transforms of isolated singularities and their difference from wavelet transforms of non-isolated singularities such as fractal functions. Kevlahan & Vassilicos (1994) exploited the cone-like structure of wavelet transformed singularities to distinguish between isolated and non-isolated singularities in practice.

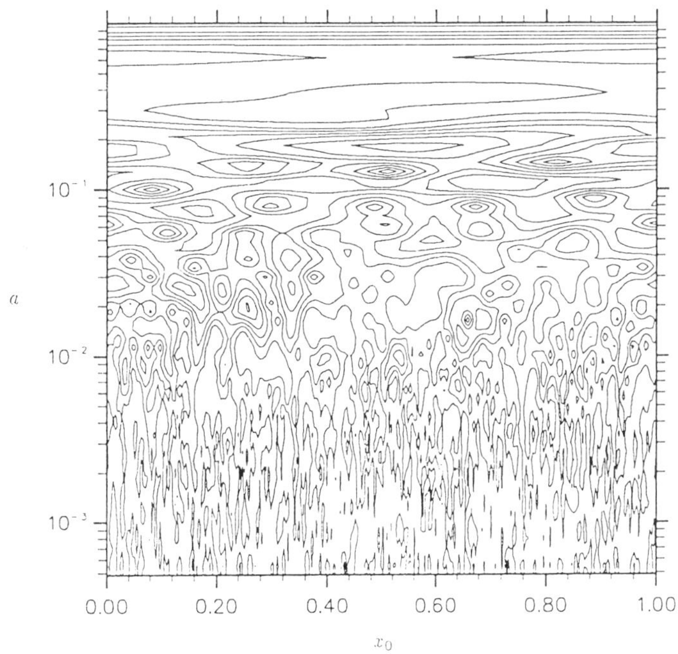

Fig. 6, Morlet wavelet transform of a fractal signal obtained by fully randomising the Fourier phases of $sin(2\pi x^{-1/2})$. *Compare with Figure 4. (From Kevlahan & Vassilicos 1994).*

Wavelet transforms of experimental turbulent velocity signals are shown in Figures 8, 9 and 10. Figures 8 and 9 show the wavelet transforms of transverse and streamwise turbulent velocity fluctuations in a turbulent mixing layer where the Reynolds number $Re = O(10^4)$ (see Figure 7 for details and Delville, Chahine & Bonnet (1987) for a full description of the experiment). Wavelet transforms capture the location of significant events either by concentration of wavelet energy around a particular position x and length-scale a as in Figure 8, or by more genuine cone-like structures as in Figure 9. The wavelet transform in Figure 9 also shows what could be (though not necessarily) part of a cascade process since large cones break up into smaller cones at smaller scales a. Cascade processes usually generate fractal fields and therefore non-isolated singularities. However, looking at the wavelet transform of an experimental high Reynolds number grid turbulence in Figure 10, it is difficult to judge whether the cones point to isolated or to non-isolated singularities. In the sequel we describe a first attempt by Kevlahan & Vassilicos (1994) to identify isolated versus non-isolated singularities in turbulent flows. Techniques based on the wavelet transform in conjunction with other identification criteria or event extraction algorithms may also be used to determine which flow structures are responsible for which part of the Fourier energy spectrum and in particular for Kolmogorov's $k^{-5/3}$ law. An example is provided by Kevlahan (1994) which, albeit rudimentary and inconclusive, does illustrate the point. In Figure 9b, Kevlahan (1994) arbitrarily marked two regions of activity and one region of quiescence. Regions such as (i) and (iii) could have been extracted, for example, by an eddy identification algorithm. The wavelet transform allows the definition of a *local* wavelet energy spectrum $E(a,x_0) = |\tilde{u}(a,x_0)|^2$ which, if integrated over the entire space x_0, can be related to the Fourier energy spectrum $E(k)$ (e.g. see Figure 7 in Meneveau 1991). Defining $E_w(a) = \int_{-\infty}^{+\infty} E(a,x_0)\,dx_0$ and using equation (I20) of lecture I, $E_w(a) = a^{-2/r}\int dk E(k)|\hat{\theta}(ak)|^2$, so that when $E(k) \sim k^{-2p}$, the scaling exponent p can be derived from $E_w(a) \sim a^{2p-1-2/r}$. However, if integrated only over one of the extracted regions or events (i), (ii) or (iii), $\int E(a,x_0)\,dx_0$ may reveal whether this flow event is responsible for the overall spectrum's scaling $E(k) \sim k^{-2p}$. The wavelet energy spectrum of the data in Figure 9b averaged over all locations appears to be roughly $E_w(a) \sim a^{4/3}$, and so are the wavelet spectra averaged *only* over the regions (i) or (iii). Over the quiescent region (ii), the wavelet spectrum seems to be $\int E(a,x_0)\,dx_0 \sim a^2$ (see Figure 12). The wavelet transform can be combined with other eddy structure identification techniques to identify the flow singularities responsible for the scaling of overall statistics.

Furthermore, spatially averaged wavelet energy spectra are much smoother than Fourier spectra because of a *double integration* in both physical and wavenumber spaces (see equation (I20) in lecture I). This smoothness is an advantage for analysing *individual* realisation of the flow and individual coherent structures within it. The Fourier transform is only smoothed out after averaging over many realisations of the turbulence.

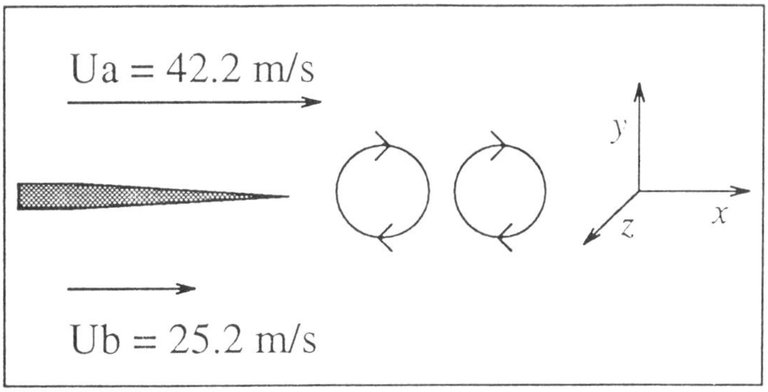

Fig. 7, Schematic diagram of an incompressible turbulent mixing layer. The splitter plate separates high and low speed (U_a and U_b) turbulent air streams which mix downstream of the plate. The spanwise direction is along z, transverse along y and streamwise along x.

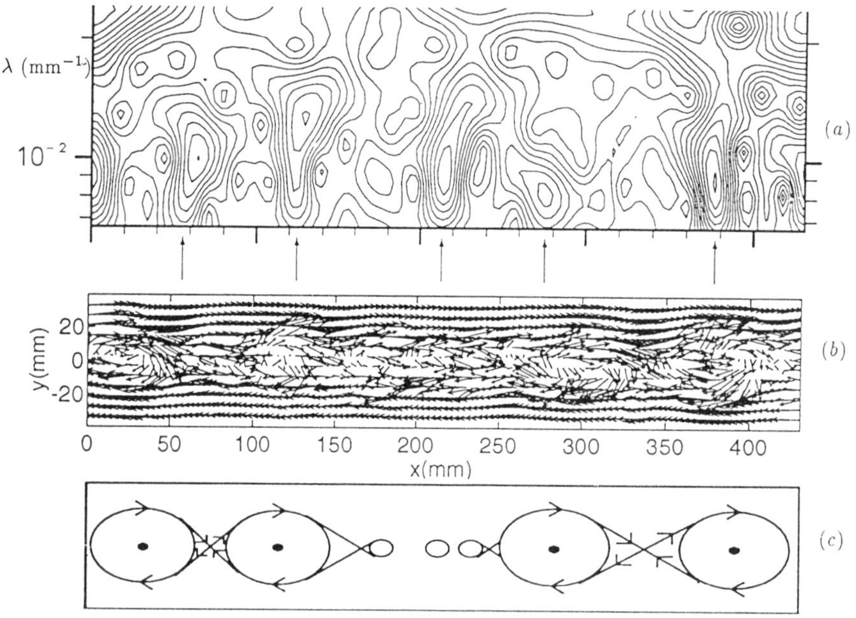

Fig. 8, (a) contour plot of the modulus of the Morlet wavelet transform of 128 points of transverse velocity along the streamwise coordinate x at $y = -3mm$. (b) Velocity vectors in mixing layer obtained from a y-aligned rake with 12 hot-wire probes. The arrows indicate that the same number of high energy eddy events are seen in both the wavelet transform and the mixing layer, approximately at the same locations. (c) Schematic diagram of the mixing layer showing structures. (From Kevlahan 1994).

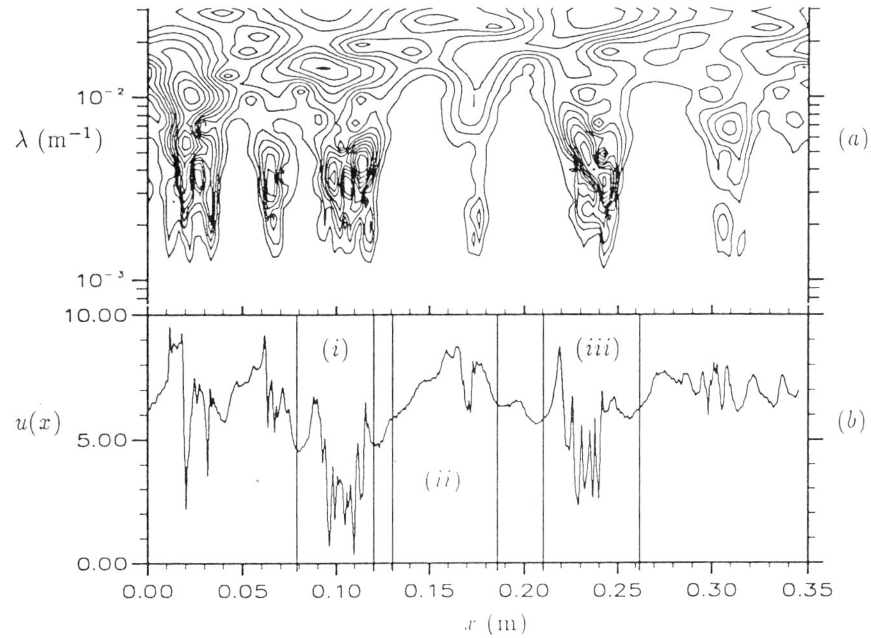

Fig. 9, (a) contour plot of the modulus of the Morlet wavelet transform of 1024 points of streamwise turbulent velocity along the streamwise coordinate. (b) Streamwise turbulent fluctuations as a function of the streamwise coordinate x. Vigorous events (i) and (iii) and quiescent event (ii) have been marked. (From Kevlahan 1994).

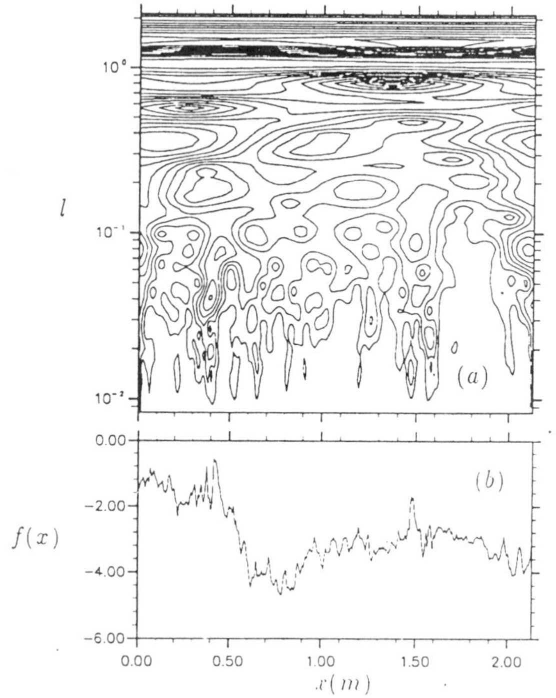

Fig. 10, (a) Morlet wavelet transform of the 512 points of the experimental longitudinal turbulent velocity component in (b). (b) Longitudinal turbulence data from the Modane wind tunnel by Gagne (1987). $Re_\lambda = 2720$, the Taylor microscale $\lambda = 0.036m$, the Kolmogorov viscous length-scale $\eta_K = 0.00035m$ and the integral scale is approximately $2500\eta_K$. This data set has a very high Reynolds number but the resolution only extends down to scales between λ and η_K. The Kolmogorov scale η_K is not adequately resolved. (From Kevlahan & Vassilicos 1994).

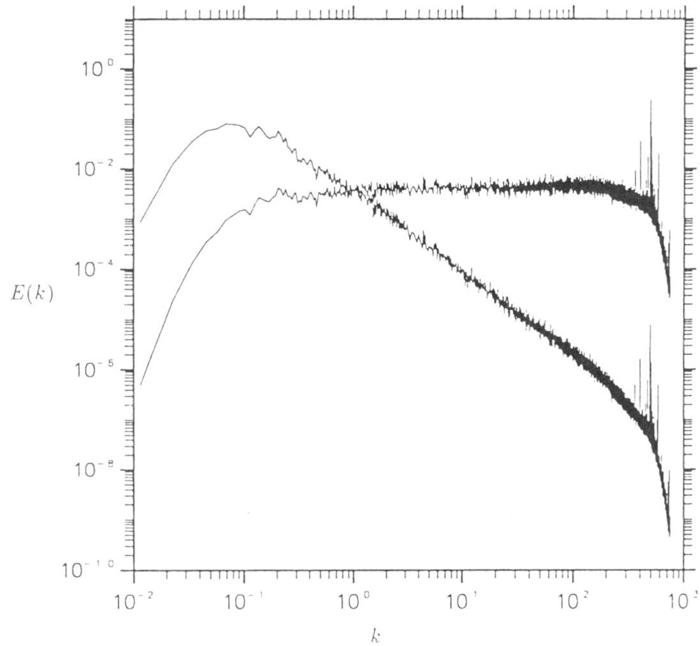

Fig. 11, Fourier energy spectrum of the turbulence data in Figure 10b. The horizontal curve is $E(k)k^{5/3}$. *The integral scale is at about* $k = 7.18m^{-1}$, *the Taylor microscale at* $k = 175m^{-1}$ *and the Kolmogorov viscous scale (not resolved) at* $k = 18000m^{-1}$. *(From Kevlahan & Vassilicos 1994).*

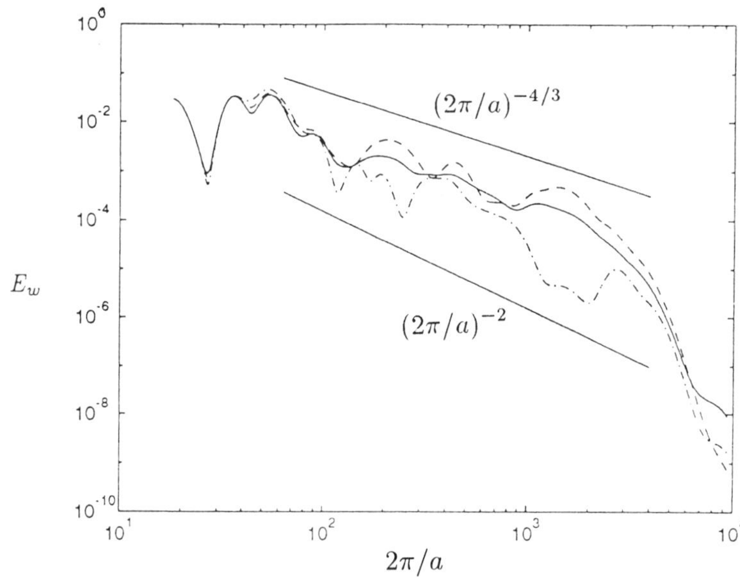

Fig. 12, spatially averaged wavelet energy spectra compared with $(2\pi/a)^{-4/3}$ *and* $(2\pi/a)^{-2}$. – *Over all locations,* - - *over the two regions marked (i) and (iii) in Figure 9b, over the region marked (ii) in Figure 9b. Note that (i) and (iii) are close to* $(2\pi/a)^{-4/3}$ *whilst the region (ii) is close to* $(2\pi/a)^{-2}$. *(From Kevlahan 1994).*

A singularity structure identification technique based on the wavelet transform

If a stochastic signal $u(x)$ has globally self-similar scaling, then $u(x)$ carries a *dense* set of (non-isolated) singularities and has accumulating length scales everywhere. The wavelet transform of such a signal is shown in Figure 6, and is characterised by long slender conical structures that point towards the x_0 axis in the direction where $a \to 0$. The wavelet transform of an isolated singularity has a single conical structure; the wavelet transform of a signal with a dense set of singularities should have a dense set of cones pointing towards the x_0 axis. The *average* wavelet transform $\tilde{U}(a, x_0)$ of the wavelet transforms $\tilde{u}_r(a, x_0)$ of N different realisations $u_r(x)$ ($r = 1, 2, ..., N$) of the stochastic signal should have no cones. A superposition of dense sets of conical structures of random sign should produce cancellations of the conical structures everywhere.

If, on the other hand, the stochastic signal $u(x)$ has a locally self-similar scaling, then $u(x)$ carries a *sparse* set of isolated singularities. Figure 4 shows the single cone produced by a single locally accumulating function. In this case, an averaging of the wavelet transforms $\tilde{u}_r(a, x_0)$ of N realisations $u_r(x)$ ($r = 1, 2, ..., N$) may not result in a cancellation of the cones. Instead, we expect $\tilde{U}(a, x_0)$ to be the collection of the cones in all realisations simply because the likelihood that two cones from two different realisations coincide is practically zero when the set of singularities is sparse. A sparse set of singularities implies a collection of cones, whereas a dense set of singularities implies a cancellation of cones in the wavelet transform.

Let us develop this distinction between locally and globally self-similar signals by way of examples.

(i) A single realisation $u_r(x)$ of a random phase signal can be written

$$u_r(x) = \int |\hat{u}(k)| exp(i\phi_r(k)) exp(-ikx)\, dk, \qquad (III8)$$

in terms of the modulus $|\hat{u}(k)|$ of its Fourier transform $\hat{u}(k)$ and of a realisation of the random phase spectrum $\phi_r(k)$. Thus, each realisation has the same distribution of energy in wavenumber space, but different phases. The wavelet transform $\tilde{u}_r(a, x_0)$ of $u_r(x)$ can be calculated in Fourier space as follows:

$$\tilde{u}_r(a, x_0) = a^{-1/r} \int |\hat{u}(k)| exp(i\phi_r(k)) \hat{\theta}^*(ak) exp(-ikx_0)\, dk. \qquad (III9)$$

The average $\tilde{U}(a, x_0)$ of the wavelet transforms of N realisations is

$$\tilde{U}(a, x_0) = \frac{1}{N} \sum_{r=1}^{N} \tilde{u}_r(a, x_0), \qquad (III10)$$

and it follows that

$$\tilde{U}(a, x_0) = a^{-1/r} \int dk |\hat{u}(k)| \hat{\theta}^*(ak) exp(-ikx_0) \frac{1}{N} \sum_{r=1}^{N} exp(i\phi_r(k)). \qquad (III11)$$

Because the phases $\phi_r(k)$ are random numbers between $-\pi$ and $+\pi$, the sum $\sum_{r=1}^{N} exp(i\phi_r(k))$ is essentially equivalent to a random walk, and there exists a random function $\phi'(k, N)$ such that

$$\sum_{r=1}^{N} exp(i\phi_r(k)) \approx \sqrt{N} exp(i\phi'(k, N)), \qquad (III12)$$

if N is large enough. Therefore,

$$\tilde{U}(a, x_0) \sim \frac{a^{-1/r}}{\sqrt{N}} \int dk |\hat{u}(k)| \hat{\theta}^*(ak) exp(-ikx_0) exp(i\phi'(k, N)), \qquad (III13)$$

which means that $\tilde{U}(a, x_0)$ is equal to one particular realisation of the wavelet transform $\tilde{u}_r(a, x_0)$ divided by $\sqrt{N}$.

The cones are the regions in (x_0, a) space where the most of the 'wavelet energy' $|\tilde{u}(a,x_0)|^2$ concentrates. The cone structure in $\tilde{U}(a,x_0)$ is qualitatively the same as in $\tilde{u}_r(a,x_0)$ but weaker by a factor $\sqrt{N}$. Random phase signals are globally self-similar and their singularities are dense. Indeed an averaging over many realisations of such signals removes, gradually, the cone structure in the wavelet transform.

We define σ_w, the integral over all locations of the modulus of the wavelet transform,

$$\sigma_w(a) = \int |\tilde{U}(a,x_0)|\, dx_0. \qquad (III14)$$

$\sigma_w(a)$ is large if either many or strong (or both many and strong) cones cross the coordinate a in the wavelet plane (x_0, a) (by 'strong' we mean energetic). From (III13) we see that $\sigma_w(a) = O(N^{-\frac{1}{2}})$ for random phase signals. This result is confirmed numerically at several different scales in Figure 13a.

These properties are not unique to random phase signals, but also appear in other globally self-similar signals such as the Weierstrass function (see lecture I). In Figure 13b, a Weierstrass function's $\sigma_w(a)$ is plotted as a function of the number of realisations N, and in spite of the non-random phase spectrum of Weierstrass functions, $\sigma_w(a)$ decreases like $O(N^{-1/2})$ at all scales a.

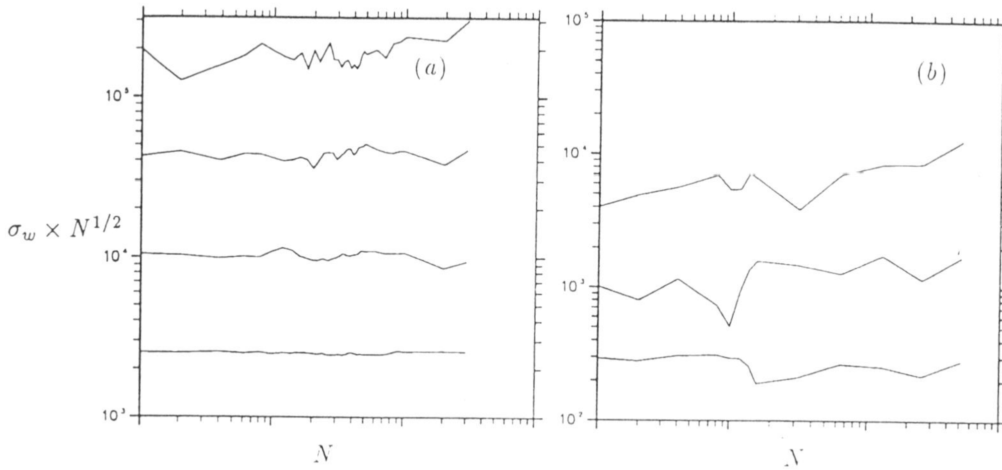

Fig. 13, (a) $\sigma_w(a)N^{\frac{1}{2}}$ of a random phase signal as a function of number of realisations at scales $a = 1.94E-2, 0.106, 0.575, 3.13$ (from bottom). (b) $\sigma_w(a)N^{\frac{1}{2}}$ of a Weierstrass function $u(x)\sum_{n=1}^{\infty} y^{(D-2)n} sin(y^n x)$, $y = 1.5$, $D = 1.5$ (the series is truncated after the first 50 terms) as a function of number of realisations at scales $a = 1.14E-3, 6.20E-3, 0.0337$ (from bottom). (From Kevlahan & Vassilicos 1994).

(ii) Now consider a locally accumulating signal. A single realisation of a 'spiral' function

$$x^{\sigma} exp(ix^{-\tau}) \qquad (III15)$$

is

$$u_r(x) = (x - c_r)^{\sigma} exp(i(x - c_r)^{-\tau}), \qquad (III16)$$

where c_r is the centre of the spiral (a random number between, say, 0 and 1). The 'spiral' function is an isolated (and therefore sparse) singularity and is locally self-similar since it has a local accumulation of length scales towards the centre, c_r, of the spiral. The Kolmogorov capacity of the zero-crossings of $u_r(x)$ is $D'_K = \frac{\tau}{1+\tau}$, but the Hausdorff dimension $D'_H = 0$.

If the Morlet 'mother' wavelet $\theta(x) = e^{ix}e^{-\frac{x^2}{2}}$ (Figure 3) is used, by a trivial change of variables the wavelet transform of (III16) is

$$\tilde{u}_r(a, x_0) = a^{-1/r} \int x^{\sigma} exp(i(x^{-\tau} + \frac{x - x_0 + c_r}{a}))exp(-\frac{(x - x_0 + c_r)^2}{2a^2})\, dx. \qquad (III17)$$

This integral is estimated in the limit where $a \to 0$ by the method of stationary phases (see (III4) to (III7)):

$$\tilde{u}_r(a, x_0) \sim a^{p-1/r} exp(-\frac{(x - c_r - (\tau a)^{\frac{1}{1+\tau}})^2}{2a^2})exp(i[(1+\tau)(\tau a)^{-\frac{\tau}{1+\tau}} - \frac{x_0 - c_r}{a}])\, dx \qquad (III18)$$

as $a \to 0$, where p is given by (III7). If we define $\tilde{u}_0(a, x_0)$ when $c_r = 0$, the average wavelet transform $\tilde{U}(a, x_0)$ is

$$\tilde{U}(a, x_0) \sim \tilde{u}_0(x_0, a)\frac{1}{N}\sum_{r=1}^{N} exp(i\frac{c_r}{a})exp(-\frac{c_r^2}{2a^2})exp(\frac{c_r}{a^2}(x_0 - (\tau a)^{\frac{1}{1+\tau}})^2), \qquad (III19)$$

and a numerical computation of the sum in (III19) leads to

$$\sum_{r=1}^{N} exp(i\frac{c_r}{a})exp(-\frac{c_r^2}{2a^2})exp(\frac{c_r}{a^2}(x_0 - (\tau a)^{\frac{1}{1+\tau}})^2) \approx NF(a, x_0) \qquad (III20)$$

for large enough values of N. It then follows that $\tilde{U}(a, x_0)$ is asymptotically independent of N as $N \to \infty$!

It is clear from Figure 14 that the number of wavelet cones in the structure of $\tilde{U}(a, x_0)$ increases proportionally with N, but it is also clear that the cones become shorter as N increases. The width of the cones is larger at larger scales a, and thus the cones superimpose and cancel at these scales. However, at the smaller scales the cones do not cancel because the self-similar singularities remain isolated and sparse. *At these small scales, $\sigma_w(a)$ is independent of N.* The scales where $\sigma_w(a)$

is independent of N also become progressively smaller as N increases. In the case of the Morlet wavelet transformed spiral, the width of the cone $l(a) \propto a$. Indeed, in equation (III18), $\tilde{u}_r(a, x_0)$ is proportional to a Gaussian of width a. Thus, the thin ends of the cones will, on average, not cancel where $a < L/N$ (where L is the length of the data segments and N is the number of realisations).

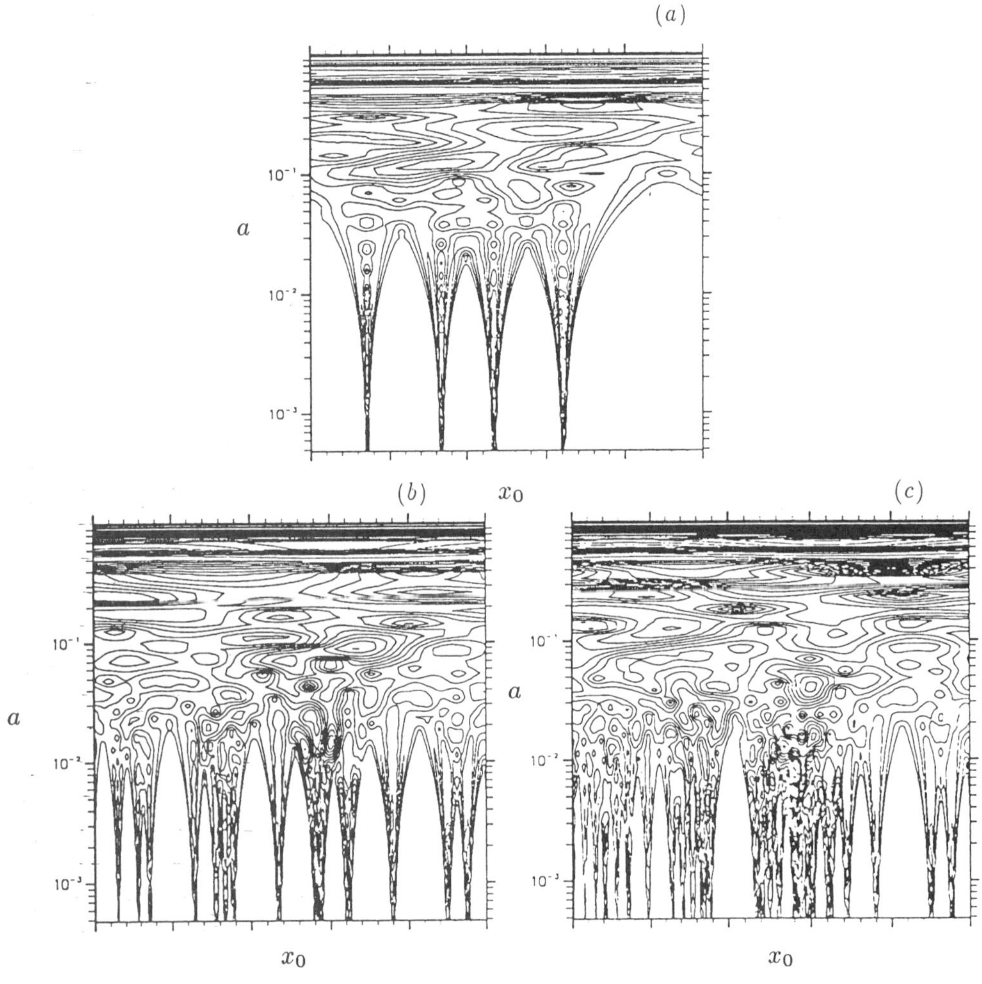

Fig. 14, modulus of the wavelet transform of (a) 4, (b) 16 and (c) 32 spiral functions (III16) showing the cone structure. Note how at the smaller scales the cones accumulate and do not cancel. (From Kevlahan & Vassilicos 1994).

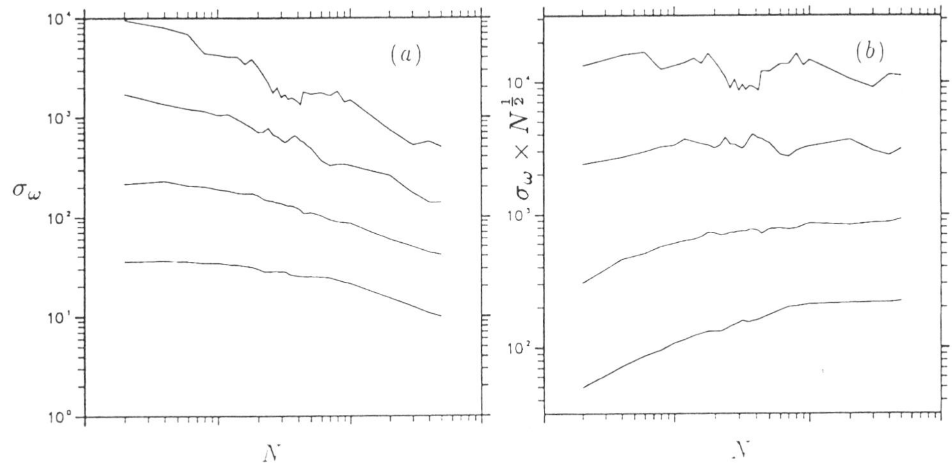

Fig. 15, (a) $\sigma_w(a)$ *at scales* $a = 1.94E - 2, 0.106, 0.575, 3.13$ *for the 'spiral' function* $sin(x^{-1/2})$*. (b) Same as (a), but* $\times N^{\frac{1}{2}}$*. Note how* $\sigma_w(a)$ *remains approximately constant at small scales and moderate* N*, and decreases like* $N^{-\frac{1}{2}}$ *at large scales and large* N*. (From Kevlahan & Vassilicos 1994).*

The independence of $\sigma_w(a)$ from N is based on equation (III18) which is valid with the restriction that $a \to 0$. Additionally, from the above argument, the scales where this effect is visible decrease with number of realisations N. For this method of distinguishing between local and global self-similar structure to work in practise we therefore require a very well resolved turbulence velocity signal with an extremely large range of length scales.

A question may be asked about the robustness of this method: how sensitive is it to the amount of noise in the signal? Figure 16 shows that the present method is relatively insensitive to phase noise. Even at 40% phase noise σ_w is still significantly different from the $N^{-\frac{1}{2}}$ random phase behaviour for $N < 10$. This is a useful result since most turbulent spirals will have some degree of noise and shows that this method should be useful in practise.

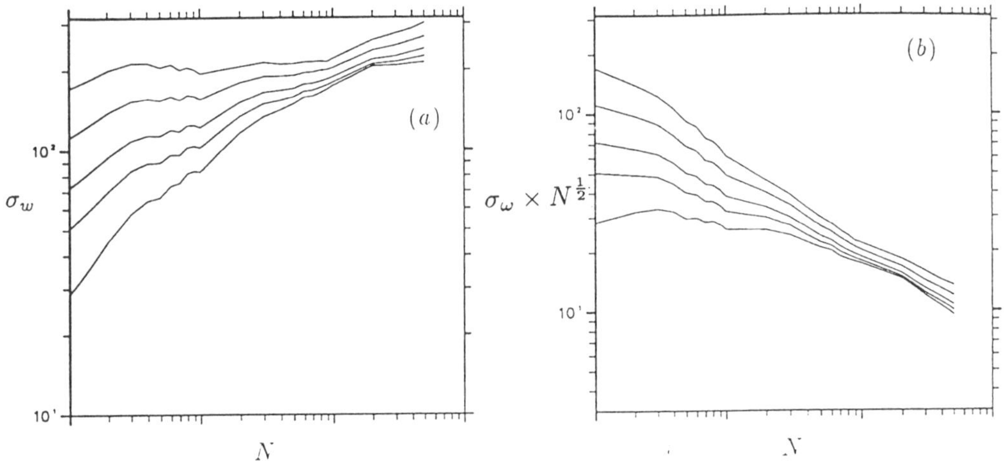

Fig. 16, (a) $\sigma_w(a)$ at a single scale for different levels of phase noise (see Figure 5); from bottom to top, 0%, 5%, 10%, 20% and 40%. (b) same as (a), but $\times N^{\frac{1}{2}}$. (From Kevlahan & Vassilicos 1994).

To summarise, the method is to wavelet transform successive stretches of at least one Taylor microscale in the turbulence signal. Then we collect the wavelet transforms of N 'realisations' of small-scale turbulence structure and average them together to obtain $\tilde{U}(a, x_0)$ and $\sigma_w(a)$. If $\sigma_w(a)$ decreases like $N^{-\frac{1}{2}}$ over all length scales then the turbulence singularities are *non-isolated*; if, on the contrary, $\sigma_w(a)$ is independent of N at small values of a, then the turbulence singularities are *isolated* and may be spiral.

Figure 10a shows the Morlet wavelet transform of the first 512 points of the experimental turbulence data shown and described in Figure 10b. There is cone-like branching at $x = 0.40m$ and $x = 1.60m$ which may be the result of locally self-similar structures (compare with the actual velocity trace in Figure 10b). In Figure 17a, σ_w indicates non-isolated singular behaviour ($\sigma_w \sim N^{-\frac{1}{2}}$) at all scales for large 17.0 m data segments (4096 points). For smaller 0.133 m segments (about 4 Taylor microscales), Figure 17b shows no $\sigma_w \sim N^{-\frac{1}{2}}$ behaviour for $N < 10$ and some faint evidence of isolated singular behaviour. As mentioned above, evidence of isolated singularities is expected to disappear eventually with increasing N because of the data's finite resolution.

Averaging segments much larger than λ may reduce existing traces of isolated singularities if the lateral size of the small-scale turbulence vortex tubes, assumed to have a spiral internal structure, is $O(\lambda)$. Segments much longer than λ would then include many spirals, thereby effectively increasing N. Note that the 17.0 m long segments are about $486a$ long. Unfortunately, the scales around the Taylor microscale λ and below are not resolved well-enough to provide strong evidence of isolated singularities, although some evidence does exist for averages of less than 10 realisations. Furthermore, the number of spiral events captured in the velocity data may be low because the probability of a single point probe intersecting a line vortex is small.

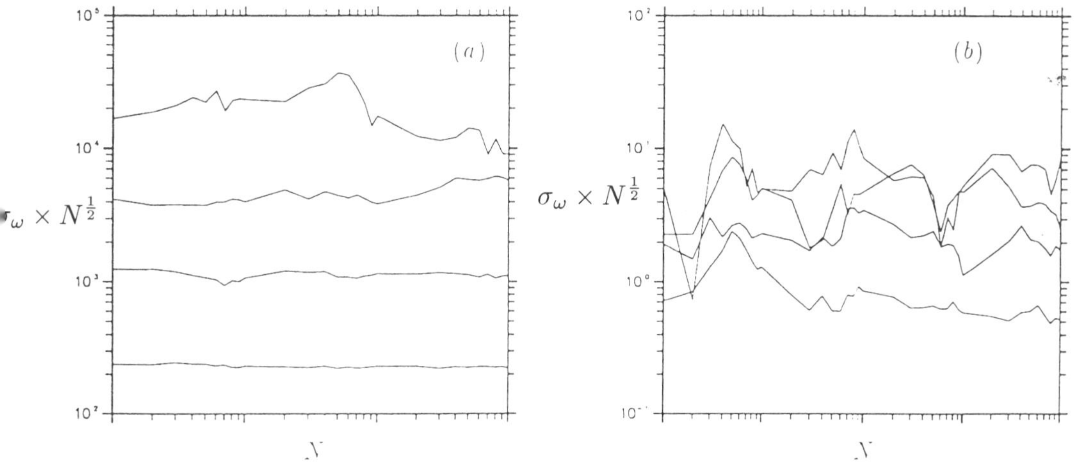

Fig. 17, (a) $\sigma_w(a) \times N^{\frac{1}{2}}$ *at scales* $a = 1.94E-2, 0.106, 0.575, 3.13m$ *for the 'spiral' function* $sin(x^{-1/2})$. *(b) Same as (a), but note the deviation from the H-fractal slope* $N^{-\frac{1}{2}}$ *for small N* $(N < 10)$. *(From Kevlahan & Vassilicos 1994).*

Conclusion

Wavelet transforms can detect singularities in both wavenumber and physical space. They provide alternative means of measuring spectral signatures and fractal dimensions. They can be used in conjunction with other eddy structure identification techniques to locate spectral signatures in the flow. Finally, they provide the basis for a singularity structure identification technique that can discriminate between isolated and non-isolated turbulence singularities.

REFERENCES

29. Aref, H. 1984 Stirring by chaotic advection. J. Fluid Mech. **143**, 1.

30. Arneodo, A., Grasseau, G. & Holschneider, M. 1988 Phys. Rev. Lett. **61**, 2281.

31. Delville, J., Chahine, Z. & Bonnet, J.P. 1987 Experimental study of an incompressible plane mixing layer by temporal and spectral analysis. In *Advances in turbulence*, ed. G. Compte-Bellot & J. Mathieu (Springer), 435.

32. Farge, M. 1992 The wavelet transform. Ann. Rev. Fluid Mech. **24**, 395.

33. Flandrin, P. 1993 Fractional Brownian motion and wavelets. In *Wavelets, fractals and Fourier transforms*, ed. M. Farge, J.C.R. Hunt & J.C. Vassilicos (Oxford University Press), 109.

34. Gagne, Y. 1987 Thèse de doctorat, Université de Grenoble, France.

35. Holschneider, M. 1988 J. Stat. Phys. **50**, 963.

36. Hunt, J.C.R., Kevlahan, N.-K.R., Vassilicos, J.C. & Farge, M. 1993 Wavelets, fractals and Fourier transforms: detection and analysis of structure. In *Wavelets, fractals and Fourier transforms*, ed. M. Farge, J.C.R. Hunt & J.C. Vassilicos (Oxford University Press), 1.

37. Kevlahan, N.K.-R. 1994 Ph.D. thesis. University of Cambridge.

38. Kevlahan, N.K.-R. & Vassilicos, J.C. 1994 The space and scale dependencies of the self-similar structure of turbulence. Proc. R. Soc. Lond. A **447**, 341.

39. Meneveau, C. 1991 Analysis of turbulence in the orthonormal wavelet representation. J. Fluid Mech. **232**, 469.

LOW DIMENSIONAL MODELS, WAVELET TRANSFORMS AND CONTROL

N. Aubry, G. Berkooz, B. Coller, J. Elezgaray, P. Holmes, J.L. Lumley and A. Poje

Cornell University, Ithaca, NY, USA

Notes of Lectures presented by J. L. Lumley for a Short Course *Eddy Structure Identification Techniques for Free Turbulent Flows* held under the auspices of the International Center for Mechanical Sciences, Udine, Italy, May 23-27, 1994. Parts of these Notes previously appeared in: Aubry, N., Holmes, P., Lumley, J.L. and Stone, E. 1990. The behavior of coherent structures in the wall region by dynamical systems theory. In *Near-Wall Turbulence* eds. S.J. Kline & N. H. Afgan, pp. 672-691. Washington, DC: Hemisphere; Berkooz, G., Elezgaray, J., Holmes, P., Lumley, J. & Poje, A. 1993 The proper orthogonal decomposition, wavelets and modal approaches to the dynamics of coherent structures. In *Eddy Structure Identification in Free Turbulent Shear Flows*, ed. J. P. Bonnet and M. N. Glauser, pp. 295-310. Dordrecht etc.: Kluwer; and Coller, B. D., Holmes, P. & Lumley, J. L. 1994. Interaction of adjacent bursts in the wall region. *Phys Fluids* 6 (2): 954-961; 137. Aubry, N., Lumley , J.L. & Holmes, P. 1988. The Effect of Drag Reduction on the Wall Region. *Theoretical and Computational Fluid Dynamics.* 1: 229-248.

---

[1]Supported in part by Contract No. F49620-92-J-0287 jointly funded by the U. S. Air Force Office of Scientific Research (Control and Aerospace Programs), and the U. S. Office of Naval Research, in part by Grant No. F49620-92-J-0038, funded by the U. S. Air Force Office of Scientific Research (Aerospace Program), and in part by the Physical Oceanography Programs of the U. S. National Science Foundation (Contract No. OCE-901 7882) and the U. S. Office of Naval Research. (Grant No. N00014-92-J-1547).

[2]Current address: Benjamin Levich Inst, Steinman 202, Convent Ave. at 140th St. City College of CUNY, NY,NY 10031

[3]Current address: CRPP, CNRS, Av. Schweitzer 33600 PESSAC, FRANCE

[4]Current address: IGPP MSC305, Los Alamos National Lab, Los Alamos, NM 87545

1 POD and Wavelet Representations

1.1 Coherent structures

If we examine pictures of various turbulent flows, we will discover that the proportion of organized and disorganized turbulence in each flow is different. For example, if we look at mixing layers from undisturbed initial conditions (with only thin laminar boundary layers on the splitter plate), we find that there is an energetically large organized component, which only relatively slowly becomes three-dimensional and disorganized, although the nearly two-dimensional organized structures have from the beginning a stochastic component, so that their occurrence is not precisely periodic, nor are their strengths equal. On the other hand, if we examine a mixing layer from quite disturbed initial conditions, with a thick, turbulent boundary layer on the splitter plate, we find that the proportion of organized component is considerably less - although the organized component is still visible, it is no longer dominant. Thus, in the same type of flow, we find that the initial conditions change the relative strength of the organized and disorganized components. We we examine a different flow, for example a jet, we find the same difference - that is, if the flow from the orifice is initially undisturbed - thin laminar boundary layers on the inner surface of the nozzle - then there is initially a laminar instability which gradually becomes three-dimensional and undergoes transition, leaving in the downstream development of the turbulence the remnants of the instability structure. On the other hand, if the boundary layers on the inside of the nozzle are initially thick and turbulent, there is no initial laminar instability, and there are no visible organized remnants in the turbulent motion. More than this, however, there is evidently a substantial difference between this flow and the mixing layer. Even in the undisturbed state, the organized component of this flow is a very great deal weaker than that in the mixing layer under the same circumstances. It is bearly discernible; when the initial conditions are disturbed, the organized structure becomes essentially undetectable. Thus, we can conclude that different flows, even under similar conditions, have different relative strengths of organized and disorganized components.

So far it seems that the organized structures, to the extent that they are present, are the remnants of initial laminar instability. Let us look at the wake of a flat plate normal to the stream. If the wake is visualized close to the plate, we see one kind of organized structure, which evidently is the remnant of the initial laminar instability. However, if the wake is visualized far from the body, we see a somewhat different, though similar organized structure. Evidently the organized structures initially present decayed, and new distinct, though similar, structures arose. If we examine the wake of a plate (still normal to the stream) which is sufficiently porous so as not to give rise to an initial instability, we find at first a turbulent wake without organized structures. However, after a time the organized structures present in the late part of the initially disturbed wake spontaneously appear in this initially undisturbed wake (we say undisturbed, but of course the wake is initially turbulent, and hence disturbed; however, organized structures are not initially present.) It seems reasonable to conclude that the organized structures that appear in the late wake are a type of instability of the developed turbulent flow, drawing on the mean velocity profile to obtain energy, and giving up energy to the turbulent transport. The precise form then, will be a function of the mean velocity profile, as well as of the distribution of the turbulent stresses. The energy budget for such existing organized structures is complex, because they have reached a non-linear energetic equilibrium, and their transport is modifying the mean velocity profile, as well as the turbulent stresses. Their initial growth is complicated also, since they must be imagined to grow from a mean velocity profile (and a profile of turbulent stresses) existing without the organized structures. What these profiles are is moot. We will return to these ideas later when we discuss the prediction of these organized structures.

Hence, in any given situation we may expect to find organized structures the relative strength of which are a function of the initial conditions of the flow, the type of flow, where we are in the flow (how far downstream), and which may be remnants of initial instabilities or may be a new instability of the turbulent flow.

If we restrict out attention to narrow two-dimensional shear flows (jets, wakes, mixing layers) we will find that the organized structures occur with more-or-less the same orientation and distance from the flow centerline each time; that, is, their orientation and position in the cross-stream, inhomogeneous direction is largely fixed by the boundaries of the flow; in the streamwise, or homogeneous direction, the location is more random; the existence of one structure seems to suppress the presence of another, but as soon as we are sufficiently distant from a structure, another one appears.

Organized structures are much more difficult to find in homogeneous flows, mostly because it is not clear where to look. There is nothing in the flow to pin them down to a particular location. Consider, for example, the homogeneous shear. In direct numerical simulations Moin and his coworkers have found hairpin vortices throughout the flow. The orientation of these hairpin vortices is determined by the direction of the mean velocity shear, but their location in three dimensions is random. They are thought to arise from a type of non-linear instability connected with the same instability that produces Langmuir cells in the ocean surface mixed layer, or that produces streamwise rolls in a turbulent boundary layer. Both of these flows are inhomogeneous in the direction normal to the surface, and hence the location of the organized structures is fixed by this inhomogeneity. The instability mechanism depends on transverse vorticity associated with the mean shear being deflected vertically (in the direction of the mean gradient), and then being transported in the streamwise direction at different rates at different heights, due to a gradient in the Stokes drift, resulting in a stretching and intensification of the streamwise component of vorticity.

We may probably conclude that *any* turbulent flow will have a more-or-less organized component, the strength of which will be a complex function of the type of flow, the age of the flow, and the initial conditions, and which may, depending on the situation, be random in orientation and location in up to three dimensions.

1.2 The role of coherent structures in turbulence dynamics

Whether it is necessary to take into account the presence of organized structures in a turbulent flow, when considering the dynamical behavior, will have a different answer in different situations. When we study a turbulent flow for practical purposes, we are seldom interested in more than the Reynolds stress. This is not a very sophisticated property of a turbulent flow; it is uninfluenced by subtle changes in the structure of the flow, and tells us little about the flow. Often, if the coherent structures in the flow scale in the same way as the disorganized motions, they can all be lumped together and the evident differences ignored. For example, in the turbulent mixing layer, if both the turbulence and the organized structures have the same origin, and have been growing together since the origin, then they will scale in the same way, and need not be considered separately. See Shih et al, [1], where a compressible turbulent mixing layer was successfully predicted using a second order turbulence model that completely ignored the presence of coherent structures. Many of the features of the mixing layer that are thought to be associated with the coherent structures (e.g.- the assymetry of the entrainment at the two sides of the mean velocity profile) are in fact mandated by the dynamics of the situation and must be produced by whatever physical mechanism is doing the transporting of momentum, whether organized or disorganized.

However, if a turbulence produced under one set of circumstances, and consequently having a given set of scales, is subjected to a different set of conditions (a new strain rate field, for example) it is quite likely that the new conditions will give rise to a new instability of the turbulent profiles, giving rise in its turn to a new organized structure. Until the scales have had a chance to equilibrate, we will have a situation consisting of background disorganized turbulence with one set of scales, on which are growing organized structures with a different set of scales. The transport produced by this combination will be quite difficult to predict unless explicit account is taken of the organized structure. We must also consider that the initial turbulence may already have an organized structure with which it is equilibrium.

1.3 The POD as a representation of coherent structures

The extraction of deterministic features from a random, fine grained turbulent flow has been a challenging problem. Zilberman et. al. [2] write: "there are no consistent methods for identification which are independent of the techniques and the observer" and "we cannot unambiguously define the signature of an eddy without a priori knowledge of its shape and its location relative to the observation station and cannot map such an eddy because we do not have a proper criterion for pattern recognition". In contrast, Lumley [3] proposed an unbiased technique for identifying such structures. The method consists of extracting the candidate which is the best correlated, in a statistical sense, with the background velocity field. The different structures are identified with the orthogonal eigenfunctions of the proper orthogonal or Karhunen-Loève decomposition theorem of probability theory (Loève [4]). This is thus a systematic way to find organized motions in a given set of realizations of a random field. The method applied here is optimal in the sense that the series of eigenmodes converges more rapidly (in quadratic mean) than any other representation.

* The use of these modes for a low dimension dynamical system study requires a very fast convergence of the series. The method we propose here is limited in application to certain types of flows in which large coherent structures contain a major fraction of the energy. It has been demonstrated that axisymmetric turbulent jet mixing layers (Glauser et al. [5]) and wall regions of turbulent boundary layers (Moin [6], Herzog [7]) belong to this group.*

Specifically, we will develop a model for the wall region of the boundary layer (from $x_2^+ = 0$ to $x_2^+ = 40$ in wall units (Tennekes and Lumley [8])), using the proper orthogonal decomposition of Lumley [3], [9], [10] in the direction normal to the wall, in which the flow is strongly non-homogeneous. In the streamwise and spanwise directions the flow is essentially homogeneous, and Fourier modes will suffice. Used in conjunction with Galerkin projection, the proper orthogonal decomposition yields an optimal set of basis functions in the sense that the resulting truncated system of ODEs captures the maximum amount of kinetic energy among all possible truncations of the same order. The method has obvious advantages over *a priori* decompositions, based on linear normal modes, but it does not appear to have been used before due to the difficulty of computing the proper orthogonal modes. For this one requires three dimensional autocorrelation tensors averaged over many realizations of the flow in question, data only obtainable from lengthy experiments and analyses or from detailed numerical simulations. In our case complete data is only available from experimental work in a glycerine tunnel (Herzog [7]), although Moin [6], has derived two-dimensional orthogonal modes from large eddy simulations. However, as we shall see, knowledge of the autocorrelation tensor, and use of the Navier Stokes equations, does allow one to uniquely determine the unsteady flow, in contrast to Cantwell's [11] expectation.

Lumley [3] proposed a method of identification of coherent structures in a random turbulent flow. An advantage of the method is its objectivity and lack of bias. Given a realization of an inhomogeneous, energy integrable velocity field, it consists of projecting the random field on a candidate structure, and selecting the structure which maximizes the projection in quadratic mean. In other words, we are interested in the structure which is the best correlated with the random, energy-integrable field. More precisely, given an ensemble of realizations of the field, the purpose is to find the structure which is the best correlated with all the elements of the ensemble. Thus we want to maximize a statistical measure of the magnitude of the projection, which can be given by the mean square of its absolute value. The calculus of variations reduces this problem of maximization to a Fredholm integral equation of the first kind whose symmetric kernel is the autocorrelation matrix. The properties of this integral equation are given by Hilbert Schmidt theory. There is a denumerable set of eigenfunctions (structures). The eigenfunctions form a complete orthogonal set, which means that the random field can be reconstructed. The coefficients are uncorrelated and their mean square values are the eigenvalues themselves. The Kernel can be expanded in a uniformly and absolutely convergent series of the eigenfunctions and the turbulent kinetic energy is the sum of the eigenvalues. Thus every structure makes an independent contribution to the kinetic energy and Reynolds stress.

The most significant point of the decomposition is perhaps the fact that the convergence of the representation is optimally fast since the coefficients of the expansion have been maximized in a mean square sense. The mean square of the first coefficient is as large as possible, the second is the largest in the remainder of the series once the first term has been subtracted, etc.

We have described here the simplest case, that of a completely inhomogeneous, square-integrable, field. If the random field is homogeneous in one or more directions, the spectrum of the eigenvalues becomes continuous, and the eigenfunctions become Fourier modes, so that the proper orthogonal decomposition reduces to the harmonic orthogonal decomposition in those directions. See Lumley [3], [9], [10] for more details.

The flow of interest here is three dimensional, approximately homogeneous in the streamwise direction (x_1) and spanwise direction (x_3), approximately stationary in time (t), inhomogeneous and of integrable energy in the normal direction (x_2).

We want a three dimensional decomposition which can be substituted in the Navier-Stokes equations in order to recover the phase information carried by the coefficients. We have to decide which variable we want to keep. Time is a good candidate since we are particularly interested in the temporal dynamics of the structures. Such a decomposition is possible and we do not need a separation of variables in the eigenfunctions of the type $\phi(\mathbf{x},\omega) = A(\omega)\,\psi(\mathbf{x})$ (as suggested by Glauser et al. [5]) if we do not use any decomposition in time and choose the appropriate autocorrelation tensor. The idea is to measure the two velocities at the same time and determine $< u_i(x_1,x_2,x_3,t)\, u_j(x'_1,x'_2,x'_3,t) > = R_{ij}$. Since the flow is quasistationary, R_{ij} does not depend on time, nor do the eigenvalues and eigenfunctions. The information in time is carried by the coefficients $a^{(n)}$ which are still "stochastic", but now evolve under the constraint of the equations of motion. Thus the decomposition becomes

$$u_i(x_1,x_2,x_3,t) = \sum_n \int a^{(n)}_{k_1k_3}(t) e^{2\pi i(k_1x_1+k_3x_3)} \phi^{(n)}_{ik_1k_3}(x_2) dk_1 dk_3 \tag{1}$$

$$\int \Phi_{ij}(x_2, x'_2)\phi^{(n)}(x'_2)dx'_2 = \lambda^{(n)}\phi_i^{(n)}(x_2), \tag{2}$$

and we have to solve equation (2) for each pair of wave numbers (k_1,k_3). Φ_{ij} now denotes the Fourier transform of the autocorrelation tensor in the x_1, x_3 directions.

Our second change to the decomposition is a transformation of the Fourier integral into a Fourier series, assuming that the flow is periodic in the x_1 and x_3 directions. The periods L_1, L_3 are determined by the first non zero wave numbers chosen. Finally, each component of the velocity field can be expanded as the triple sum

$$u_i(x_1,x_2,x_3,t) = \frac{1}{\sqrt{L_1L_3}} \sum_{k_1k_3n} e^{2\pi i(k_1x_1+k_3x_3)} a^{(n)}_{k_1k_3}(t)\phi^{(n)}_{ik_1k_3}(x_2), \tag{3}$$

In this case, a "structure" is defined by:

$$\frac{1}{\sqrt{L_1L_3}} \sum_{k_1k_3} e^{2\pi i(k_1x_1+k_3x_3)} a^{(n)}_{k_1k_3} \phi^{(n)}_{ik_1k_3}(x_2) \tag{4}$$

and the entire velocity field is recovered by the sum of all the structures (over n).

The candidate flow we are investigating is the wall region (which reaches $x_2^+ = 40$) of a pipe flow with almost pure glycerine (98%) as the working fluid (Herzog [7]). The Reynolds number based on the centerline mean velocity and the diameter of the pipe is 8750. The corresponding Reynolds number based on the shear velocity u_τ is 531. From this data the autocorrelation tensor at zero time lag (t-t' = 0 between the two velocities), $R_{ij}(x_1-x'_1, x_2,x'_2, x_3-x'_3)_{t-t'=0}$, was obtained and the spatial eigenfunctions were extracted by numerical solution of the eigenvalue problem. The results show that approximately 60% of the total kinetic energy and Reynolds stress is contained in the first eigenmode and that the first three eigenmodes capture essentially the entire flow field as far as these statistics are concerned. This very fast convergence of the decomposition in the near wall region is in good agreement with Moin's results [6]. From a large eddy simulation data base, Moin uses the proper orthogonal decomposition successively in one and two dimensions in the wall region (up to $x_2^+ = 65$). His first structure contains 60% of the total kinetic energy and 120% of the Reynolds stress (this apparent paradox occurs because the contribution of higher order structures to the Reynolds stress is negative). 90% of the kinetic energy is captured by the first three terms.

1.4 Low-dimensional models constructed using the POD

We decompose the velocity - or the pressure - into the mean (defined using a spatial average) and fluctuation in the usual way. We substitute this decomposition into the Navier-Stokes equations. Taking the spatial average of these equations we obtain, in the quasi stationary case, an approximate relation between the divergence of the Reynolds stress and the mean pressure and velocity.

$$\langle u_{i,j} u_j \rangle = - 1/\rho \, P_{,i} + \nu \, U_{,jj} \delta_{i1} . \tag{5}$$

This is substituted in the Navier-Stokes equations, giving an equation for the fluctuating velocity. Equation (5) may be solved to give the mean veloctity U in terms of the Reynolds stress $\langle u_1 u_2 \rangle$ in a channel flow in a manner which gives some feedback to the system of equations as the fluctuation varies. We will see that this feedback is necessarily stabilizing for the first structure (according to the experimental results) and increases as the Reynolds stress gets stronger. In other words, this term controls the intensity of the rolls, by reducing the mean velocity gradient as the rolls intensify, thus weakening the source of energy.

The expansion of the Fourier transform $\hat{u}_i$ of the fluctuating velocity u_i, defined by

$$u_i(x_1,x_2,x_3,t) = \frac{1}{\sqrt{L_1 L_3}} \sum_{k_1 k_3} e^{\, 2\pi i(k_1 x_3 + k_3 x_3)} \, \hat{u}_i(k_1,k_3, x_2, t)) \ , \tag{6}$$

is achieved by use of the complete set of eigenfunctions $\phi_{(n)}$'s in an infinite sum:

$$\hat{u}_i(k_1,k_3, x_2, t)) = \sqrt{L_1 L_3} \sum_{m=1}^{\infty} a_{k_1 k_3}^{(n)}(t) \phi_{i k_1 k_3}^{(m)}(x_2) \, . \tag{7}$$

Since we want to truncate this sum, we use a Galerkin projection which minimizes the error due to the truncation and yields a set of ordinary differential equations for the coefficients. After taking the Fourier transform of the Navier Stokes equations $N_i=0$ and introducing the truncated expansion, we apply Galerkin projection by taking the inner product

$$(\phi^{(n)}, N) = \int_0^{X_2} N_i(x_2)\phi_i^{(n)*}(x_2)dx_2 = 0. \tag{8}$$

Finally we obtain a set of ordinary differential equations of the form:

$$A\frac{da_k^{(n)}}{dt} = Ba_k^{(n)} + N.L. \tag{9}$$

where A and B are matrices. Here A is the identity matrix (since the complete set of eigenfunctions is orthogonal) and N.L. are non linear terms. The non linear terms are of two sorts: quadratic and cubic. The quadratic terms come from the non linear fluctuation-fluctuation interactions and represent energy transfer between the different eigenmodes and Fourier modes. Their signs vary. The role of the Reynolds stresses $<u_i u_j>$ on these terms should be mentioned. They vanish for all wave number pairs except for $(k_1,k_3) = (0,0)$ for which they exactly cancel the quadratic term. Therefore they prevent this mode from having any kind of quadratic interactions with other Fourier modes. Since the cubic terms are zero too, the (0,0) mode just decays by action of viscosity and does not participate in the dynamics of the system.

The cubic terms come from the mean velocity - fluctuation interaction corresponding to the Reynolds stress $<u_1 u_2>$ in the mean velocity equation (the other part of this equation leads to a linear term). Since the streamwise and normal components of the first eigenfunction have opposite signs, they make a positive contribution to the turbulence production and hence provide negative cubic terms which are thus stabilizing. We remark that this is not necessarily the case for higher-order eigenfunctions.

By use of the continuity equation and the boundary conditions

$$\phi_{2_{\mathbf{k}}}^{(n)} = 0 \text{ at } x_2 = 0 \text{ (at the wall)}, \tag{10}$$

and

$$x_2 \to \infty,\ \phi_{2_{\mathbf{k}}}^{(n)} \to 0 \text{ (decay at infinity)}, \tag{11}$$

it can be seen by integration by parts that the pressure term would disappear if the domain of integration covered the entire flow volume. Since this is not the case (rather the domain is limited to $X_2+ = 40$), there remains the value of the pressure term at the upper edge X_2 of the integration domain which represents an external perturbation coming from the outer flow.

The exact form of the equations obtained from the decomposition, truncated at some cut-off point (k_{1C}, k_{3C}, n_C), does not account for the energy transfer between the resolved (included) modes and the unresolved smaller scales. The influence of the missing scales will be parameterized by a simple generalization of the Heisenberg spectral model in homogeneous turbulence. Such a model is fairly crude, but we feel that its details will have little influence on the behavior of the energy-containing scales, just as the details of a sub-grid scale model have relatively little influence on the behavior of the resolved scales in a large eddy simulation. This is a sort of St. Venant's principle, admittedly unproved here, but amply demonstrated experimentally by the universal nature of the energy containing scales in turbulence in diverse media having different fine structures and dissipation mechanisms (see Tennekes & Lumley, [8] for a fuller discussion). The only important parameter is the amount of energy absorbed.

We begin by defining a moving spatial filter which removes from the total field the unresolved modes. The details of the definition are not important - it is sufficient to conceive of the possibility of such a filter. This filter is also an averaging operator. The velocity field may now be divided into the resolved and unresolved field by using this filter. The Reynolds stress of the unresolved field may now be defined as the average using our filter operator, of the product of the unresolved velocities; this acts on the resolved field. We suppose that the deviator of this Reynolds stress is proportional to the strain rate of the resolved field. We neglect the Leonard stresses, which essentially supposes that there is more of a spectral gap than really exists. This is what is done in the Heisenberg model, without ill effect. The way in which we are treating the effect of the unresolved modes on the resolved ones is very much like what is done in large eddy simulation, and is called sub-grid scale modeling; our model would probably be called a Smagorinsky model (there are minor differences in the definition of the equivalent transport coefficient). Let us agree to designate the resolved field as $u_{i<}$ and the unresolved field as $u_{i>}$, while an average of ϕ, say, over the unresolved modes (the filtering process) can be designated as $<\phi>_>$. Thus:

$$\tau_{ij>} = -2\alpha_1 \nu_T S_{ij<} \tag{12}$$

with $\tau_{ij>} = <u_{i>}u_{j>}>_> - <<u_{i>}u_{j>}>_>> - 1/3\,\delta_{ij}(<u_{k>}u_{k>}>_> - <<u_{k>}u_{k>}>_>>)$

and $S_{ij<} = 1/2\,(u_{i<,j} + u_{j<,i})$. Here

$<$ denotes the sum over all the modes (k_1,k_3,n) such that $k_1 \le k_{1C}$, $k_3 \le k_{3C}$, $n \le n_C$ and

$>$ denotes the sum over all the modes (k_1,k_3,n) such that $k_1 > k_{1C}$ or $k_3 > k_{3C}$ or $n > n_C$, where (k_{1C},k_{3C},n_C) is the cut-off mode. The characteristic scales of the parameter ν_T are those of the higher modes. We have introduced an explicit dimensionless parameter α_1, and will exclude adjustable constants from ν_T. By observation that the energy decreases rapidly with increasing n and **k**, we assume that these relevant scales are given by characteristic scales of the <u>first</u> neglected modes. This is probably a good approximation as far as the eigenmodes are concerned since they are separated by large gaps in the spectrum and it is a reasonable assumption for the Fourier modes since the steps of our Fourier series are also large.

Finally, the parameter ν_T is taken equal to

$$u_> l_> = \frac{\int_0^{X_2} \langle u_{i>} u_{i>} \rangle dx_2}{\left(X_2 \int_0^{X_2} \langle u_{i>,j} u_{i>,j} \rangle dx_2 \right)^{1/2}} \qquad (13)$$

(where $u_>$ and $l_>$ are characteristic scales of the neglected modes). This can be expressed in terms of the eigenvalues and eigenfunctions of the first neglected modes in the three directions (see Figure 1).

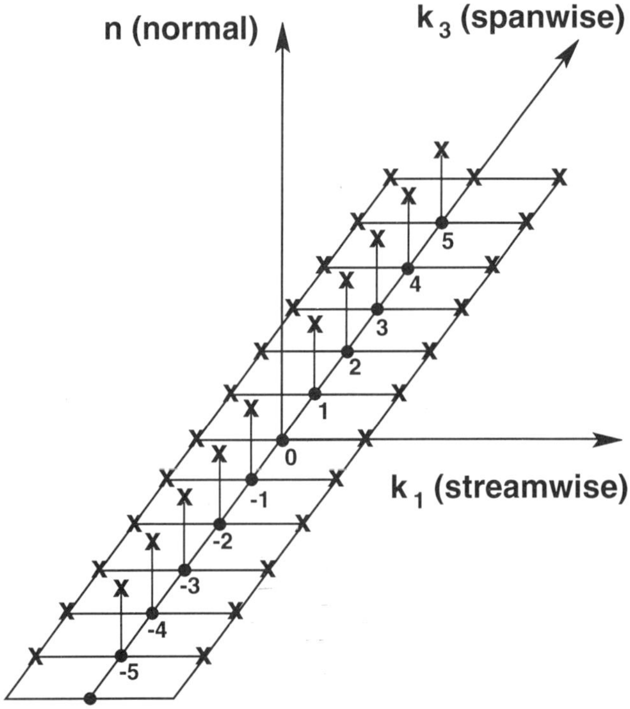

Figure 1: Inhomogeneous Heisenberg model applied to the specific truncation discussed in the text. Legend: • resolved modes x first neglected modes which are considered for the computation of the characteristic scales of the Heisenberg model.

We will refer to α_1 as a Heisenberg parameter. We will adjust α_1 upward and downward to simulate greater and smaller energy loss to the unresolved modes, corresponding to the presence of a greater or smaller intensity of smaller scale turbulence in the neighborhood of the wall. This might correspond, for example, to the environment just before or just after a bursting event, which produces a large burst of small scale turbulence, which is then diffused to the outer part of the layer.

A term $-1/3\delta_{ij}(<u_{k>}u_{k>}>_{>} - <<u_{k>}u_{k>}>_{>}>)$ appears in the equation for the resolved field. This term could be combined with the pressure term and would not have any dynamical effect if the integration domain covered the entire flow volume. In our case, it needs to be computed since, like the pressure term, it leads to a term evaluated at X_2. We assume that the deviation (on the resolved scale) in the kinetic energy of the unresolved scales is proportional to the rate of loss of energy by the resolved scales to the unresolved scales. This pseudo-pressure term gives some quadratic feed-back. The rate of loss of energy from the resolved scales to the unresolved scales is $2\alpha_1 \nu_T S_{ij<} S_{ij<}$, so that a free parameter appears also in this term. Because this approximation involves a further assumption, and to give ourselves greater flexibility, we call this parameter α_2, although in all work presented in this lecture, we have set $\alpha_1 = \alpha_2$.

Thus the Heisenberg model introduces two parameters in the system of equations, one, α_1, in the linear term, the other one, α_2, in the quadratic term. The equations therefore have the following form:

$$da^{(n)}_{k_1 k_3}/dt = L + (\nu + \alpha_1 \nu_T)L' = Q + \alpha_2 Q' + C \tag{14}$$

where L and L' represent the linear terms, Q the direct quadratic terms, Q' the quadratic pseudo-pressure term and C the cubic terms arising from the Reynolds stress.

1.5 Comparison with the wall region

This study has two conflicting requirements. On one hand we wish to keep as few modes as possible in order to obtain a low-dimensional system, permitting us to apply the techniques of dynamical system analysis. On the other hand, we would like to retain at least a qualitatively correct dynamical representation of the turbulence production phenomenon. A necessary condition for the second requirement consists of including as much of the energy and Reynolds stress as possible in our system. This is already satisfied in the inhomogeneous direction by the proper orthogonal decomposition itself, since the first structure is the most energetic. Given the energetic gap between the two first eigenfunctions (Figure 2), keeping only one structure seems quite reasonable. The choice of wave numbers is now of great importance, especially as far as the Reynolds stress is concerned (an important part of the Reynolds stress is contained in the higher modes. The best selection is probably the one for which the cut-off modes correspond to the experimental measurement cut-off. Since we retain only a few modes, the steps of our Fourier series are large. However, provided that the periodic length is larger than a few integral scales, the Fourier transform of the autocorrelation tensor is unchanged and the eigenfunctions are still the same.

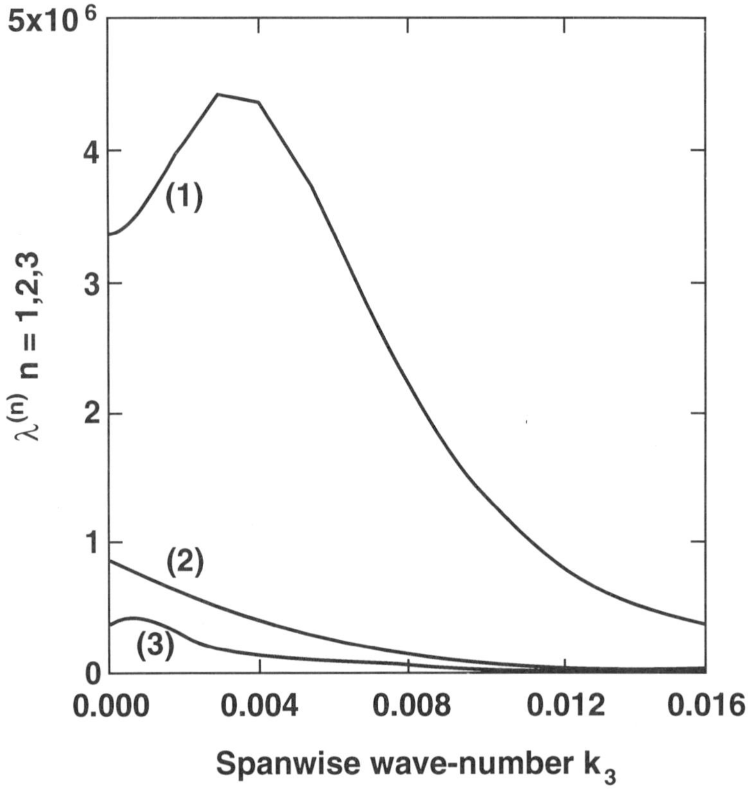

Figure 2: Convergence of the proper orthogonal decomposition in the near-wall region (x_2^+ + = 40) of a pipe flow according to experimental data. Turbulent kinetic energy in the first three eigenmodes, $\lambda^{(n)}$ (n = 1, 2, 3) function of: the spanwise wave number (from Herzog [1986])

We now seek a minimum truncation. The experimental results show that the ratio between the streamwise and spanwise characteristic length scales is of order ten (see Figure 2.). Our first approximation therefore neglects streamwise variations. We need at least three terms in the spanwise direction (see below). Thus the minimum truncation consists of one eigenmode, one streamwise wave number ($k_1 = 0$) and three (i.e. two active) spanwise wave numbers (0,k,2k). In this lecture, up to six spanwise wavenumbers (0,..., 5k) will be considered. In this case k has the value 3×10^{-3} and the lengths of the periodic box are $L_1 = L_3 = 333$. Even the model having six spanwise Fourier modes is still very crude, although the truncation seems to be the one which contains an optimally large amount of the total energy among those of the same dimension. The zero cut-off mode in the streamwise direction in particular is a very rough approximation. Such a truncation causes a drop of the spanwise and normal root mean square values of the velocity which is particularly significant in the upper half of the layer. For this reason, we do not expect our velocity field reconstructed without rescaling to have more than qualitative significance.

As a preliminary approach, we study the set of equations for a truncation limited to the first eigenfunction (n=p=q=r=1), the zero streamwise wavenumber $k_1 = 0$ and up to six spanwise wavenumbers $k_3 = 0, k, ...5k$, for a suitably chosen k.

When only the zero streamwise wavenumber is considered, the equations become much simpler. Indeed, because of the symmetries of the eigenfunctions (Herzog, [7]) in the (k_1,k_3) wavenumber plane, the first and second components are purely real and the third component is imaginary on the k_3 - axis (i.e. for $k_1 = 0$). Using these symmetries, the equations for the complex modal coefficents $a^{(1)}_{k_1} = 0, k_3$ can be readily derived. Letting k_3 take the values jk, $j = 0, 1, ..., 5$ and writing $a^{(1)}_{k_1} = 0, k_3 = jk = x_j + y_j$, a typical equation takes the form:

$$dx_1/dt = a_1x_1 \tag{15}$$

$$+ c_{-1,2}(x_1x_2 + y_1y_2) + c_{-2,3}(x_2x_3 + y_2y_3) + c_{-3,4}(x_3x_4 + y_3y_4) + c_{-4,5}(x_4x_5 + y_4y_5)$$

$$+ x_1(d_{11}(x_1^2 + y_1^2) + d_{12}(x_2^2 + y_2^2) + d_{13}(x_3^2 + y_3^2) + d_{14}(x_4^2 + y_4^2) + d_{15}(x_5^2 + y_5^2)$$

In these equations the zero Fourier mode $a^{(1)}_{00} = x_0+iy_0$ has been removed since its imaginary part is identically zero and its real part decays to zero under the influence of the viscosity ($dx_0/dt = a_0x_0$; $a_0 <0$). The coefficients of the quadratic and cubic terms are each the sums of two quantities: $c_{k',k-k'}$, $c_{k-k',k}$ and $d_{k,k'}$, $d_{k,-k'}$. The influence of the Heisenberg model appears in the linear and quadratic terms as follows:

$$a_k = a_k^1 + (1 + \alpha_1/Re_T)\, a_k^2$$

$$c_{k',k-k'} = c^1_{k',k-k'} + \alpha_2 c^2_{k',k-k'} \tag{16}$$

where α_1 and α_2 are the proportionality (Heisenberg) coefficients already mentioned. The computation of the nondimensionalized transport coefficient ν_T for this particular truncation gives the value $1/Re_T = 6.28$.

Numerical integrations of 3, 4, 5 and 6 mode models have been carried out, but we shall only report in detail on the 6 mode (5 active mode) simulations here.

We summarize here the behavior of the system. For $\alpha > 1.61$ a unique circle of globally attracting stable fixed points exists in the 2/4 subspace (with a window of instability between 2.0 and 2.3): the r_1,r_3,r_5 components, along with the trivial r_0 component, are zero. As α increases, the magnitude of the r_2 and r_4 components decreases until, at $\alpha \sim 2.409$, this non-trivial circle of fixed points coalesces with the origin, which for

$\alpha > 2.409$ is the unique and globally attracting fixed point for the problem. For $1.37 < \alpha < 1.61$, an S^1-symmetric family of globally attracting double homoclinic cycles Γ exists, connecting pairs of saddle points which are π out of phase with respect to their second (x_2,y_2) components. The points $\mathbf{r}+$, $\mathbf{r}-$ discussed above are typical members of this family. The existence of the cycles Γ implies that, after a relatively brief and possibly chaotic transient, almost all solutions enter a tubular neighborhood of Γ and thereafter follow it more and more closely. As they approach Γ, the duration of the "laminar" phase of behavior (in which r_2 and r_4 remain non-zero and almost constant and r_1,r_3,r_5 grow exponentially in an oscillatory fashion) increases while the bursts (in which r_1,r_3 and r_5 collapse) remain short. In an ideal, unperturbed system, the laminar duration would grow without bound, but small numerical perturbations, such as truncation errors, presumably prevent this occurring in our numerical simulations. More significantly, the pressure perturbation will limit the growth of the laminar periods. Thus there is an effective maximum duration of events, which is reduced as α is decreased from the critical value $\alpha_b \sim 1.61$.

For the present study the intermittent behavior exhibited by the six mode model for α between 1.35 and 1.61 appears to be of greatest interest, since it corresponds in a fairly clear way to the physical instability, sweep and ejection event observed in boundary layer experiments. We now describe this intermittent behavior, starting with some general remarks.

In the theory of dynamical systems three types of intermittency have been distinguished (Pomeau and Manneville [12]). They are associated with solutions repeatedly passing close to a weakly unstable fixed point or periodic orbit. The solution spends a long "laminar" phase near the point or orbit until it reaches a critical amplitude and a brief turbulent "burst" ensues, in which it travels far and fast in phase space before returning. (These terms were appropriated by the dynamical systems community and have been used in a predominantly metaphorical fashion thus far). The laminar phase is governed by the linearized dynamics near the fixed point or periodic orbit, but the burst and return are associated with a "global reinjection mechanism", usually a homoclinic orbit or heteroclinic cycle (Tresser *et al* [13], Silnikov [14], [15], [16], Tresser [17], Sparrow [18], Guckenheimer & Holmes [19, §6.5].)

While evidence of intermittency has been detected before in fluid systems and models (Pomeau & Manneville [12], Bergé *et al* [20], Maurer & Libchaber [21], Dubois *et al* [22]) there has been little evidence of type II, associated with a subcritical Hopf bifurcation, which is what we observe in the present model at $\alpha = 1.61$. We will therefore explore the connection between this dynamical phenomenon and its analogue in the turbulent boundary layer in some detail. Our model displays a "regular" intermittency, in contrast to the chaotic intermittency of Pomeau and Manneville [12], in which event durations are distributed randomly.

We describe here the reconstructions of the velocity field from the expansion, using the computed values of x_i and y_i. Probably the most interesting sets of solutions are those exhibiting intermittency, obtained for $1.3 \leq \alpha \leq 1.61$. In the flow field, the rapid event which follows the slowly growing oscillation and the repetition of the process reminds one of the bursting events experimentally observed (Kline et al. [23], Corino and Brodkey [24], etc.). For that reason we call it a "burst". We will analyse its effect on the streamwise vortices.

In Figure 4 we show an enlargement of the time histories of the modal coefficients during a burst for $\alpha = 1.4$ (cf. Figure 3). A description of the motion of the eddies during the burst is given in Figure 5 for $\alpha = 1.4$ by plotting u_2 and u_3 at the different times indicated on Figure 4. Before and after the event, two pairs of streamwise vortices are present in the periodic box, a structure very similar to that previously obtained for $1.61 \leq \alpha \leq 2.41$. Nonetheless pictures 1 and 14 are shifted in the spanwise direction by 333+/4, corresponding to the phase shift $\Delta\theta_2 = \pi$, $\Delta\theta_4 = 2\pi$. The bursting event leads to variations of positions and amplitudes of the basic streamwise rolls and formation

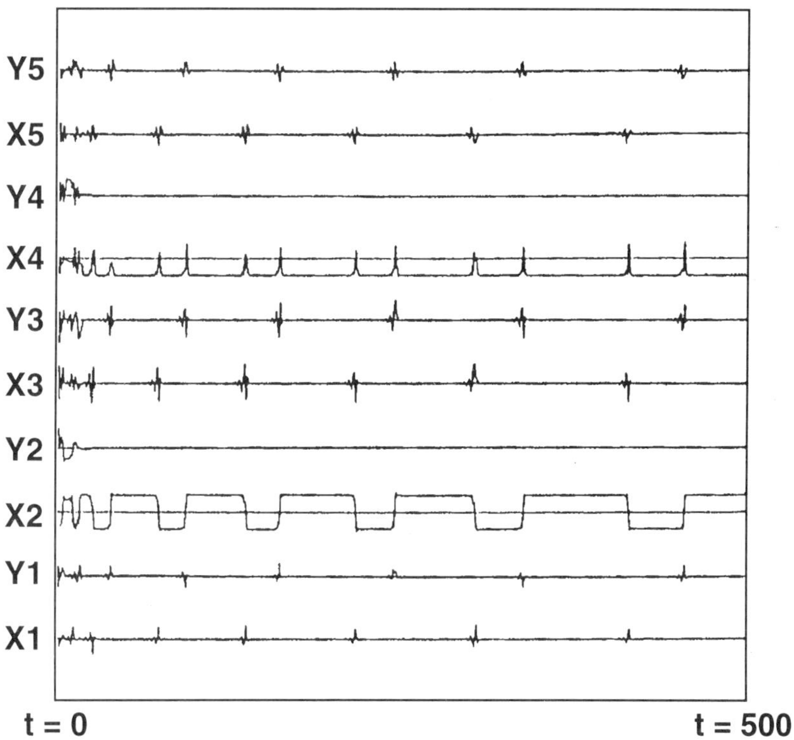

Figure 3: Time histories of the real (x_i) and imaginary (y_i) parts of the coefficients for a value of the Heisenberg parameter $\alpha = 1.4$.

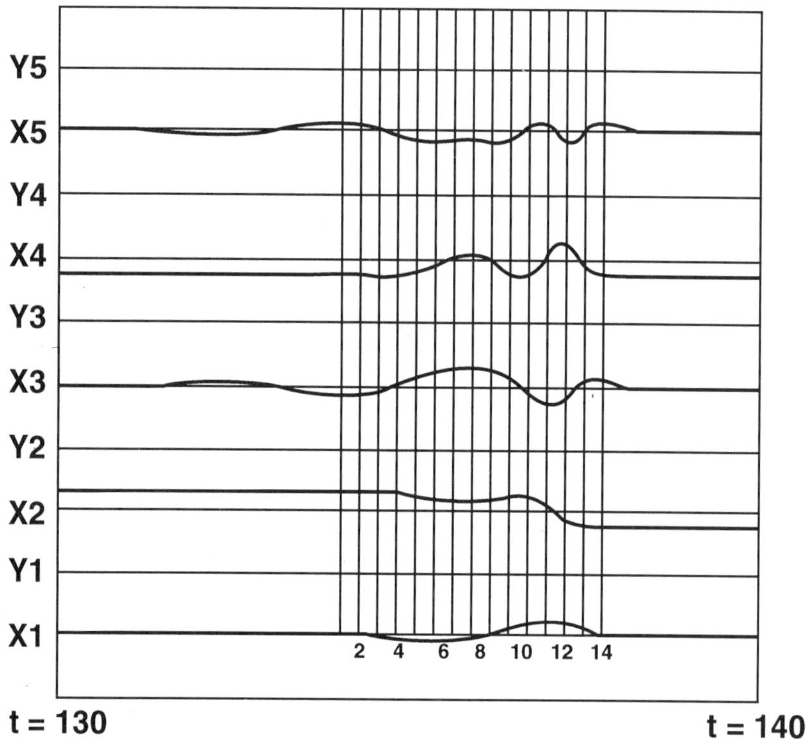

Figure 4: Intermittent solution corresponding to an Heisenberg parameter $\alpha = 1.4$. Time history of the real and imaginary parts of the model coefficients x_i, y_i ($i = 1, 2, ...5$)

of other vortices. The oscillation, death and rebirth of vortices make the streak spacing vary, in agreement with experimental results (Kline et al. [23]). However, since the intermittent solution is always very close to the real subspace or a rotation of it, the vortices remain symmetric and paired. Moreover it is possible to adjust the value of the viscosity parameter ($\alpha \sim 1.5$) so that the bursting period is 100 wall units as experimentally observed. It is found that, in this case, the "burst" lasts 10 wall units which is also the right order of magnitude.

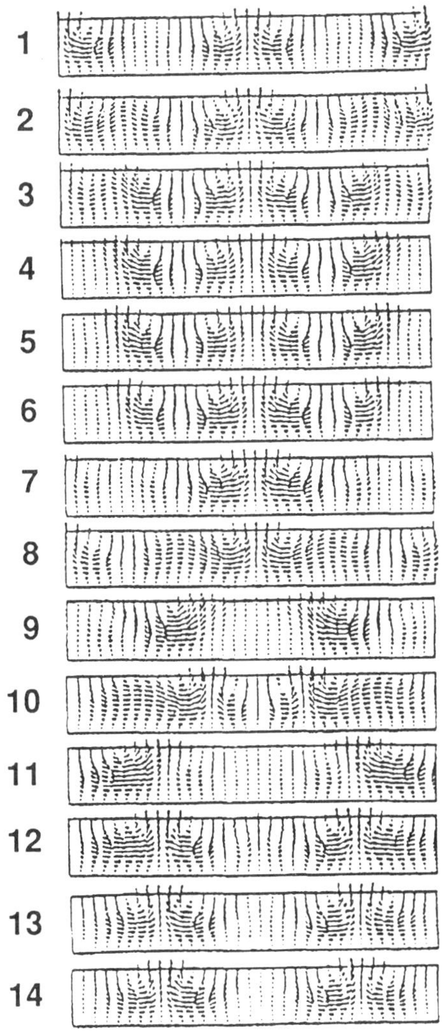

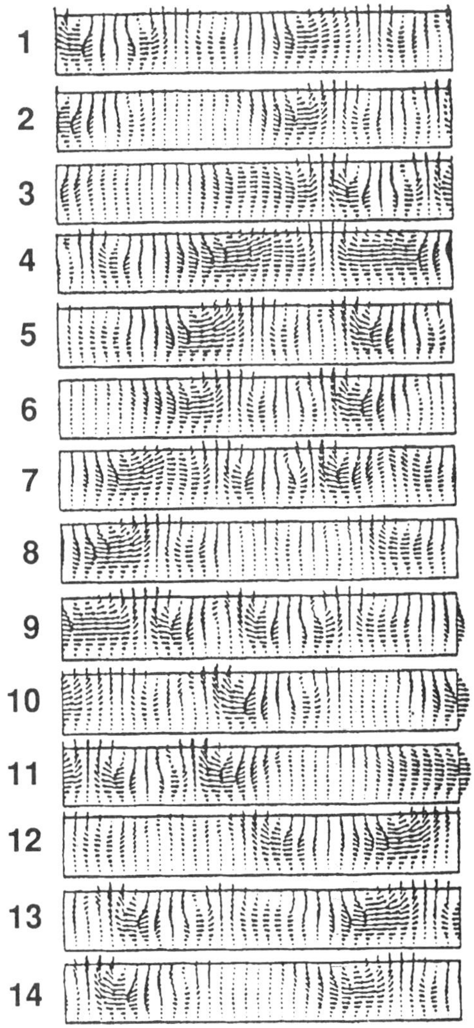

Figure 5: (left) Intermittent solution corresponding to an Heisenberg $\alpha = 1.4$. Vector representation of u_2, u_3 in an $x_2 - x_3$ plane at times indicated on Figure 4.

Figure 6: (right) Chaotic solution for Heisenberg parameter $\alpha = 1.2$. Vector representation of u_2, u_3 in an $x_2 - x_3$ plane at uniformly spaced times (close enough to resolve the motion).

The behavior of the eddies corresponding to a chaotic solution ($1.0 \leq \alpha \leq 1.3$) is shown in Figure 6, as before, by plotting the u_2, u_3 velocity components at specific times. The behavior is less regular and isolated vortices sometimes emerge. This is consistent with the flow visualization experiments of Smith and Schwarz [25] who observed a significant number of solitary vortices among the predominant vortex pairs. These patterns are also very rich in dynamics. We intend to study the regime further in future work.

Due to the truncation of our model to a single eigenmode and one streamwise Fourier component, the absolute and relative energy of the various velocity components is affected. First, the choice of streamwise length scale affects the absolute values of all the component energies, but not their relative values. We chose a streamwise length scale of 333. Keeping only the Fourier mode at zero streamwise wavenumber, this is equivalent to supposing that the distribution of the first eigenvalue with wavenumber is flat to 333, and drops to zero. Comparison with the true distribution of the first eigenvalue with streamwise wavenumber indicates that this should overestimate the energy in this eigenvalue. A more reasonable length scale would be between 400 and 500, a value determined by the requirement that the area between it and the origin is the same as the true area under the distribution of the first eigenvalue with streamwise wavenumber. However, examination of the equations indicates that changing the value of L_1 leaves the equations invariant. Hence all phenomena reported here would be unchanged, with simply a decrease in energy level of all the velocity components. This would decrease the overestimated streamwise energy contained in our system but would not necessarily improve our model since it would also decrease the energy, already deficient, in the normal and cross-stream components.

The truncation also affects the relative energy in the various velocity components. We have examined the distribution with x_2^+ of the energy in the three velocity components for truncations of 3, 6, and 17 cross-stream wavenumbers, 1, 2 and 7 streamwise wavenumbers, and 1 and 3 eigenmodes. It is clear that the energy in the u_2 and u_3 components is relatively low (compared to the u_1 component) when only one streamwise wavenumber is included. This is particularly true in the upper part of the layer. While inclusion of more than 6 cross-stream wavenumbers has no effect, addition of streamwise wave numbers does make a difference. In particular addition of another streamwise wavenumber helps considerably. We must conclude that there is energy in cross-stream and normal motions of higher streamwise wavenumbers, that makes little relative contribution to the streamwise velocity component. As a consequence, our predicted rolls are somewhat weak compared to reality, and we rescaled them. This rescaling will not be necessary in future work, when we will add streamwise wavenumbers.

One naturally asks if the addition of streamwise modes will drastically change the behavior of our model. In this respect, we observe that the same symmetries, inherited from invariance of the Navier-Stokes equations to spanwise translation and reflection, and streamwise translation, operate at any order of truncation. In particular, this implies that the subspace spanned by basis functions with zero streamwise wavenumber is invariant and that the coefficients of the equations restricted to this subspace remain unchanged as streamwise components are added. Thus, the model studied in this lecture, together with its dynamics, exists *unchanged*, as a *subsystem* in the larger set of differential equations, much as the 2-4 subsystem exists in the present model. This raises the interesting possibility that the 2-4 branch may be destabilized by streamwise modes as well as by the 1, 3, 5 "intersticial" spanwise Fourier modes.

We do not have present in our modal population a mechanism to represent the production of higher wavenumber energy when an intense updraft is formed, presumably as a result of a secondary instability. Thus, although our eddies are capable of exhibiting the basic bursting and ejection process, the labor is in vain - there is no sequel, no production of intense higher wavenumber turbulence. A contribution is made only to the low wavenumber part of the streamwise fluctuating velocity and the Reynolds stress. We could easily simulate the production of this high wavenumber turbulence, although we do not expect its inclusion to change the dynamics of our system qualitatively. However, we have held the value of our Heisenberg parameter constant , whereas its value should rise and fall with the level of this intense higher wavenumber turbulence in the vicinity of the wall. The transport effectiveness of the intense higher wavenumber turbulence would be expected to damp the system, suppressing the interesting dynamics until the higher levels are either blown downstream or lifted and diffused to the outer edge of the boundary layer. The production of higher values of the Heisenberg parameter could easily be parameterized by a single time constant first order equation tied to a measure of the amplitude of the coefficients. The major effect would simply be to cut off each burst more rapidly than at present. We plan to study the behavior of such a model in the future.

The occurrence of the bursts for values of the Heisenberg parameter between 1.4 and 1.6 appears to be pseudo-random. This is essentially due to round-off error in the calculations. The transitions from one solution to the other and back are extremely sensitive to the precise solution trajectory, and a minute change can make a considerable change in the time at which the next transition occurs. Initially we did not exercise the pressure term. Recall that the pressure term appeared due to the finite domain of integration. It represents the interaction of the part of the eddy that we have resolved with the part above the domain of integration, which is unresolved. The order of magnitude that we estimated for this term was small, and for that reason we at first neglected it. It has, however, an important effect, while not changing the qualitative nature of the solution. The term has the form of a random function of time, with a small amplitude. This slightly perturbs the solution trajectory constantly; away from the points **r**+ and **r**- this has little effect, but when the solution trajectory is very close to these points, the perturbation has the effect of throwing the solution away from the fixed point, so that it need not wait long to spiral outward. This results in a thorough randomization of the transition time from one solution to the other, while having little effect on the structure of the solution during a burst. While in the absence of the pressure term (and round-off error), the interburst time tends to lengthen as the solution trajectory is attracted closer and closer to the heteroclinic cycle (this effect is visible in Figure 3), with the pressure term, the mean time stabilizes (see Figure 7). The mean interburst times for various values of the pressure signal are shown in Figure 8.

Probably the most significant finding of this work is the identification of the etiology of the bursting phenomenon. That is, the bursts appear to be produced autonomously by the wall region, but to be triggered by pressure signals from the outer layer. Whether the bursting period scales with inner or outer variables has been a controversy in the turbulence literature for a number of years. The matter has been obscured by the fact that the experimental evidence has been measured in boundary layers with fairly low Reynolds numbers lying in a narrow range, so that it is not really possible to distinguish between the two types of scaling. The turbulent polymer drag reduction literature is particularly

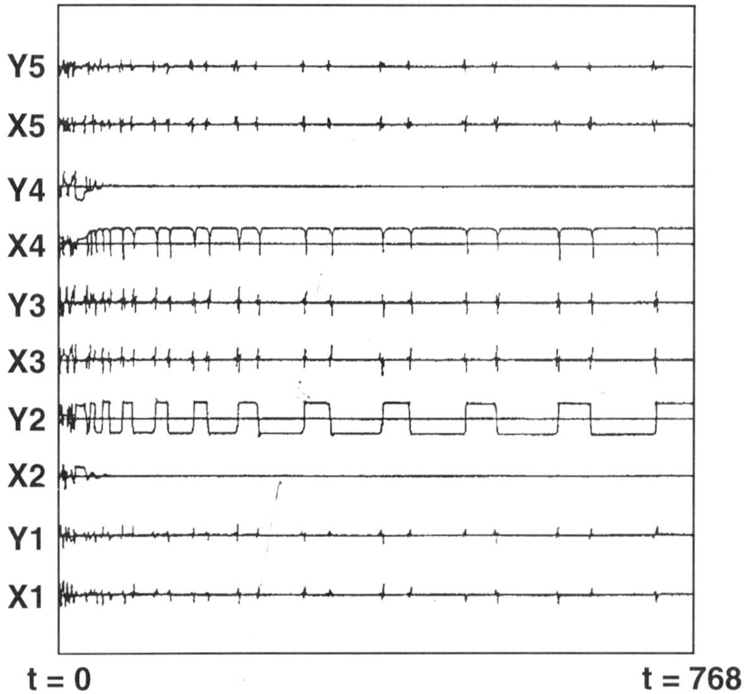

Figure 7: Pressure perturbations: solutions for $\alpha = 1.6$ with pressure term.

instructive, however, since the sizes of the large eddies, and the bursting period, all change scale with the introduction of the polymer (see, for example, Kubo and Lumley, [26], Lumley and Kubo [27]). The present work indicates clearly that the wall region is capable of producing bursts autonomously, but the timing is determined by trigger signals from the outer layer. This suggests that events during a burst should scale unambiguously with wall variables. Time between bursts will have a more complex scaling, since it is dependent on the first occurrence of a *large* enough pressure signal *long* enough after a previous burst; "long enough" is determined by wall variables, but the pressure signal should scale with outer variables.

2 Generation of eigenfunctions from stability arguments

2.1 Background

Ideally, one would like to apply the POD approach to a wide range of flows where coherent structures are known to play an important role in the dynamics. The POD procedure, however, requires the two-point velocity autocorrelation tensor as input, thus necessitating complete documentation of the flow before the analysis can proceed. For flows with very high Reynolds numbers or complicated geometries this can be prohibitively expensive given current computational and experimental capablities. In this

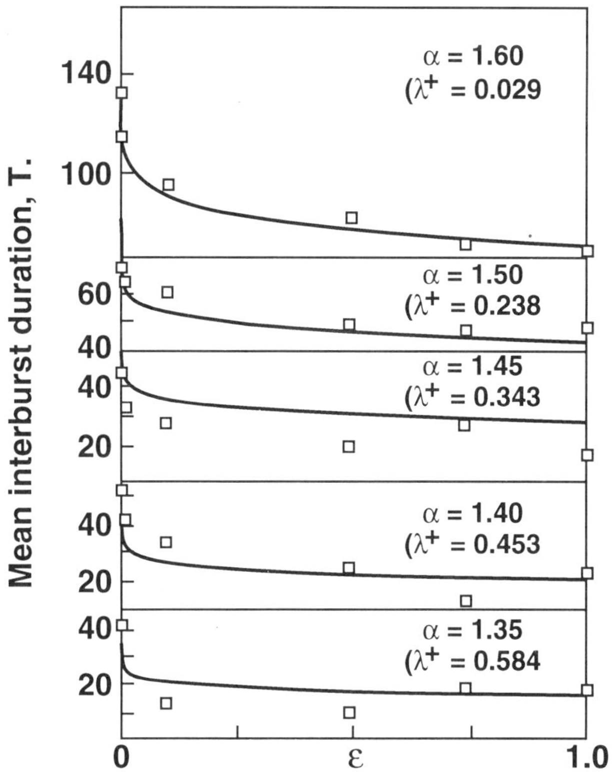

Figure 8: Mean interburst duration T as a function of eigenvalue λ_+ and pressure signal amplitude ε. Solid curves indicate prediction of Aubry *et al* (1989), open squares indcate results of numerical simuations.

section we will describe an analytic procedure for extracting basis functions (structures) which approximate those given by the POD but which requires much less *a priori* statistical information about the flow.

The method presented is based on energy stability considerations put forth by Lumley [28]. First, the instantaneous flow field is decomposed into three components in order to isolate the large scale structures. Evolution equations can then be written for the coherent velocity field and the coherent kinetic energy. A procedure can then be formalized to search for the structures which maximize the instantaneous growth rate of coherent energy, the rationale being that the structures which on average have the largest growth rates will compare well with the structures which contribute the most to the average turbulent kinetic energy (POD eigenfunctions).

As an example we consider turbulent channel flow assumed statistically homogeneous in both the downstream (x_1) and crosstream (x_3) directions. In order to extract spatial structures from the total velocity field, we avoid traditional Reynolds averaging and instead decompose the instantaneous velocity field into three components: the spatial mean (U), the coherent field (**v**) and the incoherent background turbulence (**u**').

$$u_i(\mathbf{x}, t) = U(x_2) + v_i(\mathbf{x}, t) + u_i'(\mathbf{x}, t) \qquad (17)$$

The spatial mean is an average over the x_1 - x_3 plane; we will indicate it by [...]. We introduce a second averaging procedure, denoted by <...>, which eliminates the small-scale turbulence while leaving the coherent field intact.

$$<u_i(\mathbf{x}, t)> = U(x_2) + v_i(\mathbf{x}, t) \qquad (18)$$

Practically this can be accomplished in several ways (Reynolds & Hussain, [29]; Liu, [30]; Brereton & Kodal, [31]; Berkooz, [32]). We will refer to this average as a phase average. For our purposes here it is sufficient that the phase average and space average commute, and that the cross correlations are negligible.

$$<u'_i v_j> = [u'_i v_j] = 0(\qquad (19)$$

Given these averaging procedures, we can manipulate the Navier-Stokes equations to arrive at evolution equations for the coherent velocity field.

$$Dv_i/Dt + v_j v_{i,j} = -\pi_{,i} + \nu v_{i,jj} + \tau_{ij,j} - S_{ij} v_j + [v_i v_j]_{,j} \qquad (20)$$

where D/Dt denotes the mean convective derivative, S_{ij} the mean rate of strain, and ν the kinematic viscosity. τ_{ij} represents the rectified effects of the small scale fluctuations on the coherent field and is defined by

$$\tau_{ij} = [u'_i u'_j] - <u'_i u'_j> \qquad (21)$$

This can be thought of as a perturbed Reynolds stress, which is unknown and will ultimately require modeling. In the limit of a completely random turbulence containing no structure (ie <...> = 0) this quantity is equal to the usual Reynolds stress. In the case when the turbulence is completely structured so that <...> = [...], τ_{ij} is identically zero.

We now follow classical energy method stability analysis for the coherent field. First, the growth rate of the volume averaged coherent energy E is defined as a functional of the coherent velocity field.

$$\lambda(\mathbf{v}, U, \nu\, \tau) = (1/2E)dE/dt \qquad (22)$$

Integration by parts and continuity are used to eliminate the nonlinear convective and pressure terms. We seek the solenoidal velocity field which maximizes λ. Application of the calculus of variations then gives the Euler equations for the maximizing **v** field in the form of an eigenvalue relation.

$$\lambda v_i + S_{ij} v_j = -\pi_{,i} + \nu v_{i,jj} + \tau_{ij,j}$$

$$v_{i,i} = 0 \tag{23}$$

We consider coherent fields which are periodic in the homogeneous directions. This allows a decomposition into poloidal and toroidal components which satisfy continuity exactly (Joseph, [33])

$$v_1 = \psi_{,12} - v_{,3}; \quad v_2 = -(\psi_{,11} + \psi_{,33}); \quad v_3 = \psi_{,23} + v_{,1} \tag{24}$$

The two scalar functions are then expanded in normal modes in the streamwise and spanwise directions.

$$\psi(\mathbf{x}) = \psi(x_2)\exp\{i(k_1x_1 + k_3x_3)\}$$

$$v(\mathbf{x}) = v(x_2)\exp\{i(k_1x_1 + k3x_3)\} \tag{25}$$

Substituting the above into equation (23) and eliminating the pressure π results in two coupled equations, forming a differential eigenvalue problem.

In order to procede we need to specify a mean velocity field and a model for the unknown stress terms (see Figure 9.).

We have investigated two different models for the unknown stress terms appearing in the eigenvalue relation. It should be noted that, modulo the modeled terms, equation (23) is linear in the coherent velocities, providing an inexpensive means of determining basis functions. This linearity is an essential advantage of the method and for this reason we will constrain any stress model to be both linear and homogeneous in the **v** field insuring that the governing equation remains a regular eigenvalue problem. Tensorially this requires

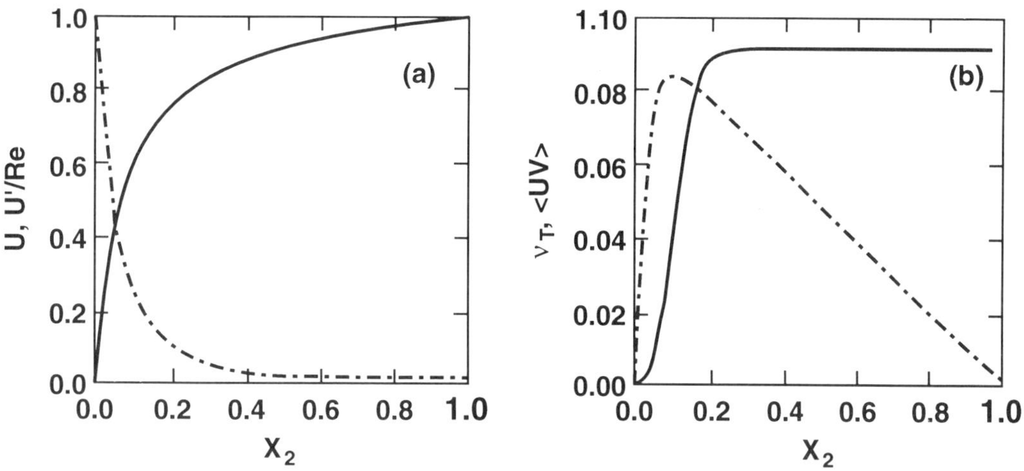

Figure 9: Model inputs; (a) mean velocity and mean gradient; (b) eddy viscosity and Reynolds stress.

$$\tau_{ij} - \tau_{kk}\delta_{ij}/3 = \Gamma_{ijkl}(v_{k,l} + v_{l,k}) \tag{26}$$

The nature of the averaging procedure implies that the scales of the coherent field and the background turbulence are different.

2.2 Isotropic eddy viscosity

Assuming that the background turbulence evolves on much shorter time and length scales than the structures, it seems plausible that a Newtonian stress-strain relationship like that for the molecular stresses will provide the basis for a model. We set

$$\tau_{ij} - \tau_{kk}\delta_{ij}/3 = \nu_t(v_{ij} + v_{j,i}) \tag{27}$$

Due to the inhomogenity of the turbulence in the wall normal direction, we specify ν_t as a function of x_2 corresponding to experimentally determined values of the traditional eddy viscosity. We will refer to this basic model as the isotropic eddy viscosity model.

Using the basic stress model and an analytic expression for the fully turbulent mean profile (Reynolds & Tiederman, [34]) we have solved the resulting equations numerically. Figure 10. shows comparisons between the calculated eigenvectors and the POD results of Moin & Moser [35] obtained from a numerical data base. Although there are qualitative similiarities in the shape of the structures, the modes predicted by the stability method fall off much more rapidly away from the wall than do the POD functions. The eigenvalue spectrum clearly shows that the stability analysis favors modes which have a much higher wavelength than the maximum energy modes of the POD.

2.3 Anisotropic eddy viscosity

Although there may be a number of reasons for this discrepancy, we choose to first examine more closely the closure model. Since both the POD analysis and the stability method favor modes which are infinitely long in the streamwise direction, we exmine our equations setting $k_1 = 0$. We find that the isotropic eddy viscosity model creates no coupling between the different components of the coherent velocity. When there is no streamwise variation of the coherent field the only coupling terms in the equations are those multiplying the mean gradient. For realistic mean profiles, regions of high shear are confined to thin regions near the wall, and the structures predicted may be expected to fall off as quickly as the shear.

We now seek to develop a stress model that allows for some anisotropy in the eddy viscosity, and thus couples the component equations through the stress terms. We begin with the evolution equation for the Reynolds stresses, where we are obliged to model a number of terms to obtain a closed system. We use standard second-order turbulence models. We model the pressure-strain correlation by a return-to-isotropy term and an isotropization-of-production term (Naot et al, [36]); we use an isotropic dissipation. We assume the stresses are in local equilibrium: $D[u_iu_j]/Dt = 0$. This reduces the evolution equation for the Reynolds stress to an algebraic expression.

Now we set up a perturbation expansion in terms of mean field quantities, taking the coherent field as an order ε perturbation to the spatial mean. On physical grounds, we argue that the perturbed stress field is due entirely to the presence of the structures and consequently we restrict the model to include only production due directly to coherent velocity gradients. This is in agreement with a cascade analogy for the complete flow: the coherent structures are fed energy directly by the mean gradients while the small scale turbulence is in turn fed by gradients of the coherent field. If we identify the 0th order stresses with an eddy viscosity tensor, then the closure model can be written as

$$\tau_{ij} - \tau_{kk}\delta_{ij}/3 = -\nu_{ik}s_{kj} \tag{28}$$

where the tensor viscosity has the following structure in this specific case: $\nu_{13} = \nu_{23} = 0$, $\nu_{33} = \nu_{22}$.

Despite the absence of mean production terms, this model is still a major improvement over the isotropic eddy viscosity formulation. In the simple model the effects of the mean field have been neglected entirely. Here we have allowed for modulation of the perturbation stresses by the mean field through the 0th order stresses appearing in the production terms. Also we have unconstrained the model in an important way since the tensorial form of the eddy viscosity allows the pricipal axes of the stress tensor to be unaligned with the axes of the rate of strain. This is more realistic considering the three-dimensionality of the coherent field. This model leads to the expected cross coupling of the equations through the stress terms.

Figure 12. shows eigensolutions for several values of k_3. The results compare well with the POD eigenvalues, especially for wavenumbers at or below the peak in the POD spectrum. The improvement with decreasing wavenumber is expected given the modeling considerations. The separation of scales between the background turbulence and the coherent structures increases as the wave number decreases adding to the expected accuracy of the stress model. The comparison of the two models indicates significant improvements in the results given by the anisotropic eddy viscosity form. The energy method procedure with the more refined closure model appears capable of extracting structures which closely approximate those given by the POD at least at the energy containing scales of motion.

Despite the general improvement, it is still clear that more needs to be done. From Figures 11 and 12, it is evident that the eigen spectrum produced by solution of Equation (17),while improved by the use of the anisotropic closure model, still predicts structures with maximum growth rate that are a factor of 2 smaller than those containing the most energy (as given by the POD). We next considered the effect on the spatial mean velocity field of the growing coherent perturbation (see Figure 11.).

2.4 The role of the mean velocity

At this point we consider the role of the mean velocity in the two methods. The POD structures are derived from solutions to the non-linear Navier-Stokes equations which allow for complicated interaction between the different scales of motion. The structures evolve in a mean velocity field that is changing due to the presence of the structures themselves. Conditionally averaged mean profiles clearly show the evolution of the local shear in the presence of coherent structures (see Figure 11, 12). We see that structures, in the relatively long period before bursting, act to erode the shear that they see. The POD eigenfunctions are given by averages of contributions from different mean profiles.

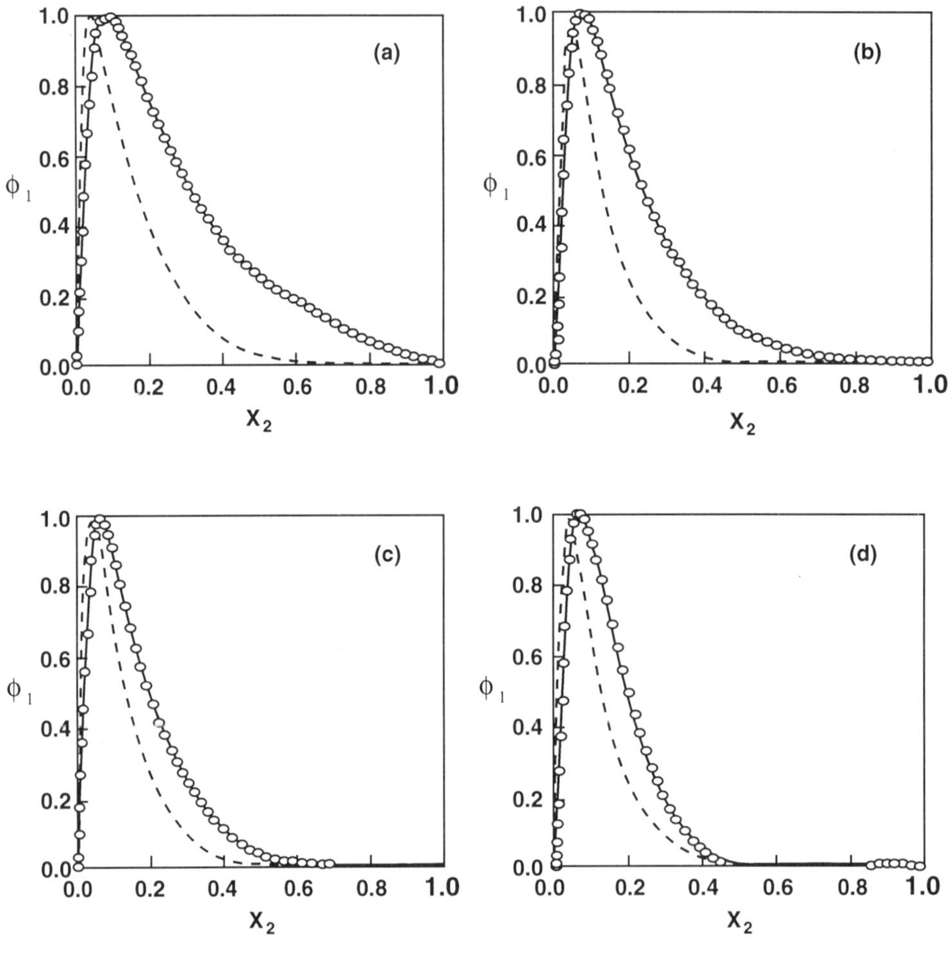

Figure 10: Isotropic eddy viscosity model: • POD; – – – – – , isotropic model; (a) k_3 = 6.00, (b) k_3 = 9.00, (c) k_3 = 12.00, (d) k_3 = 15.00.

The stability method on the other hand does not allow for any interaction between the mean and the coherent field. The mean flow is imposed and the resulting structures are calculated. The mean profiles we have used are time averages which mask any contribution from the coherent field. As such the stability analysis predicts that the highest growth modes are those which can best extract energy from the time averaged mean shear which is concentrated in the small near-wall region. Since the structures have an aspect ratio of about 1, the narrow region of high imposed shear leads to a peak in the eigen spectrum at a large wave number.

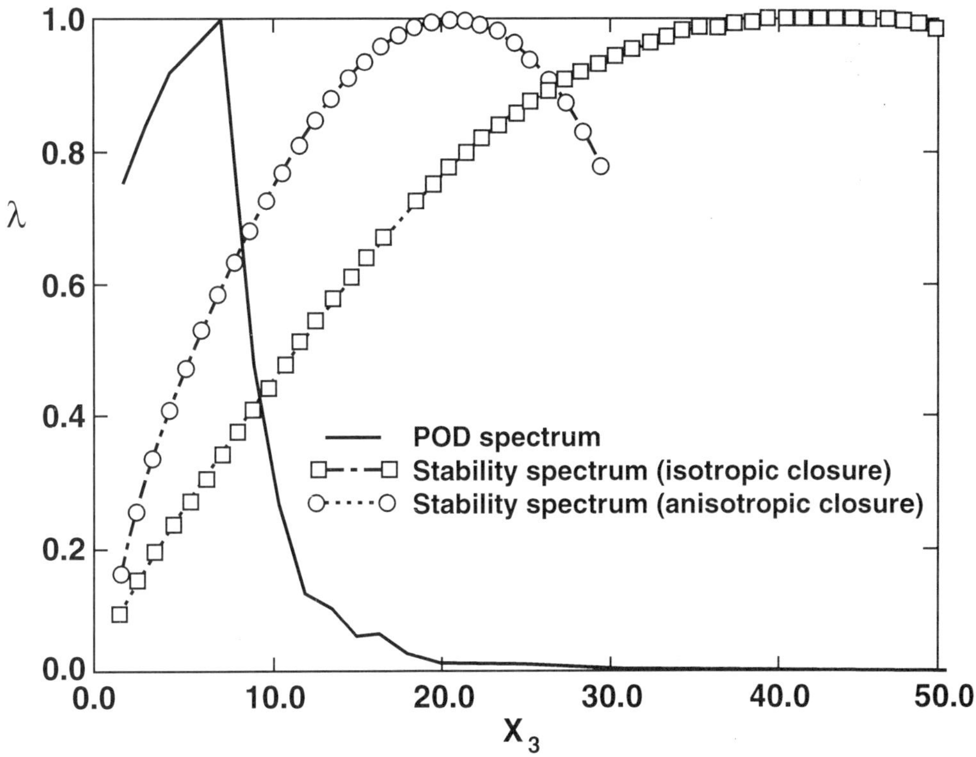

Figure 11: Comparison of the POD spectrum with the eigenspectrum of the stability problems.

To allow the mean field to evolve under the influence of the coherent field, we follow Liu [30] and write time evolution equations for the energy density of the coherent field. We allow the mean profile to depend on the coherent velocity as it does in reality. We expect that equilibrium solutions for the energy density as a function of cross stream wave number will approximate the average energy content as given by trhe POD spectrum.

We assume that the coherent field is given by the eigenvalues of the stability problem, but we now allow them to vary in time

$$v_i(\mathbf{x}, t) = A(t)\psi_i(\mathbf{x}) \tag{29}$$

$$\langle u_i u_j \rangle(\mathbf{x}, t) = E(t)B_{ij}(\mathbf{x}) \tag{30}$$

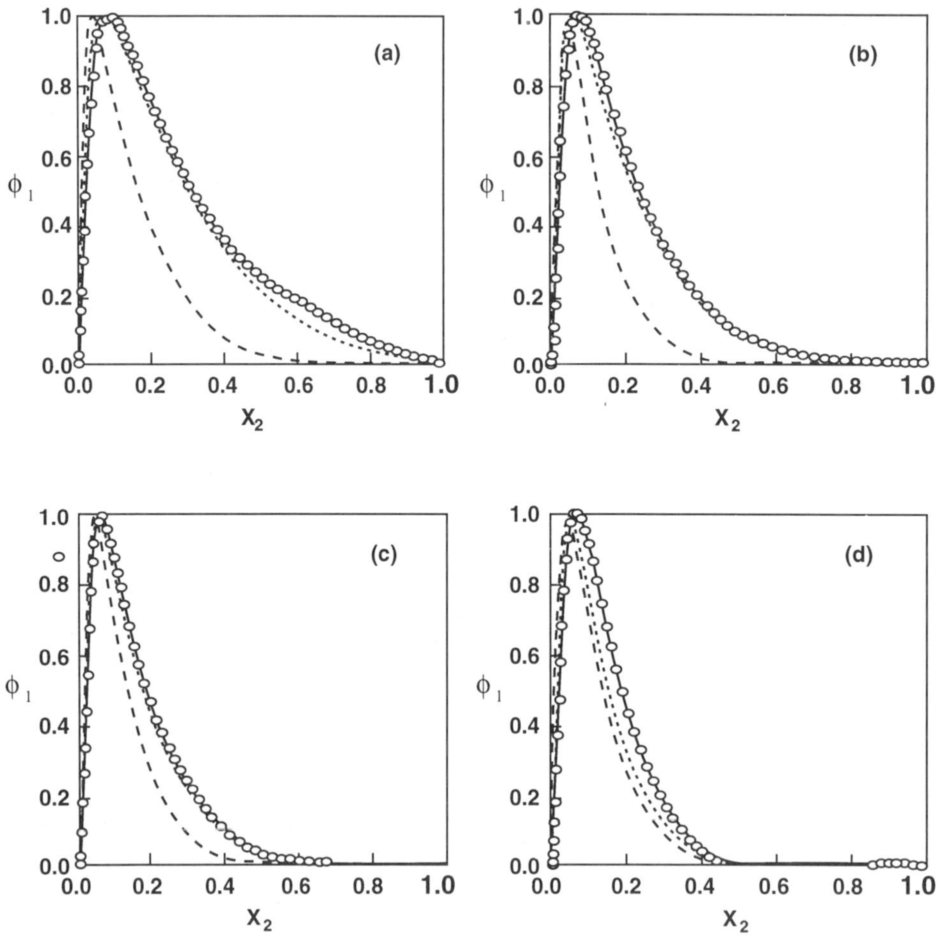

Figure 12: Anisotropic eddy viscosity model: •, POD; – – – – – – – –, isotropic model; - - - - - - - anisotropic model: (a) $k_3 = 6.00$. (b) $k_3 = 9.00$. (c) $k_3 = 12.00$. (d) $k_3 = 15.00$.

where $\psi_1 = v \exp\{ikx_3\}$, $\psi_2 = ik\psi \exp\{ikx_3\}$, $\psi_3 = -D\psi \exp\{ikx_3\}$. By examining the evolution equation for $r_{ij} = \langle u_i u_j \rangle - [u_i u_j]$, the forcing terms are of the form $\langle u_i u_j \rangle v_{i,j}$. Consequently, we assume the perturbation stresses also to be a product:

$$r_{ij}(\mathbf{x}, t) = A(t)E(t)R_{ij}(\mathbf{x}) \tag{31}$$

Since we have used an eddy viscosity in obtaining the coherent forms we further assume that:

$$R_{ij}(\mathbf{x}) = -\nu_{ik}(\psi_{j,k} + \psi_{k,j}) \tag{32}$$

where $\nu_{11} = \nu_{22} = \nu_{33} = \nu_\tau$, $\nu_{12} = \nu_{21} = \beta_\tau$, $\nu_{13} = \nu_{23} = 0$. All that remains is to model the mean profile. For this we adopt the quasi-steady model used in Aubry *et al* [37]. This allows the mean to respond to growing structures providing the necessary feed-back to the evolving modes. Using the friction velocity, u_τ, and the channel half height, a, the scaled equation for the mean gradient is:

$$\partial U(x_2, t)/\partial x_2 = Re_\tau\{<u_1u_2> + <v_1v_2> - x_2\} \tag{33}$$

The rate of dissipation of turbulent energy is given by a simple model adopted from second order closures schemes.

$$D\varepsilon/Dt = c_1(\varepsilon/k)<u_iu_k>U_{i,k} - c_2\varepsilon^2/k + \text{Transport} \tag{34}$$

Substituting these various models into the energy equations results in a set of three, coupled ODEs for the temporal evolution of the energies and dissipation.

In order to evaluate the integrals appearing in these equations, we need to assume the spatial form of the averaged turbulence quantities B_{ij} and the dissipation $D(x_2)$. For this simple model we have assumed that while the intensity of the turbulence varies its spatial dependence remains unchanged. We use experimental data for fully developed turbulent channel flow to determine both B and D. The coherent structures are found by energy stability analysis, as described above.

We show in Figure 13. the temporal evolutionof the coherent energy density for parameter values corresponding to different wavenumbers.

In order to quantify comparisons of the single mode evolution model and the POD eigenspectrum, we define an average of the lifetime of an individual structure. Taking $A^2(t_i) = A^2(t_f)$, where t_i is the initial time and t_f ois the final time, we designate

$$\{...\} == T^{-1}\int...dt \tag{35}$$

where the integral is over t_i to t_f, and $T = t_f - t_i$.

Comparison between $\{A^2\}(k_3)$ and the spectrum of the POD is shown in Figure 14. The single mode model, while not capturing the shape of the POD spectrum exactly, gives a good indication of which wavenumbers are the most energetic. The discrepancy at high wavenumbers is not surprising when we consider the simplifications involved in the model. Our equations describe only the interactions of a single mode with the local mean and the background turbulence. The effects of interactions between modes may be negligible for the large scale structures, but must become important for the smaller scales. We will consider such interactions below.

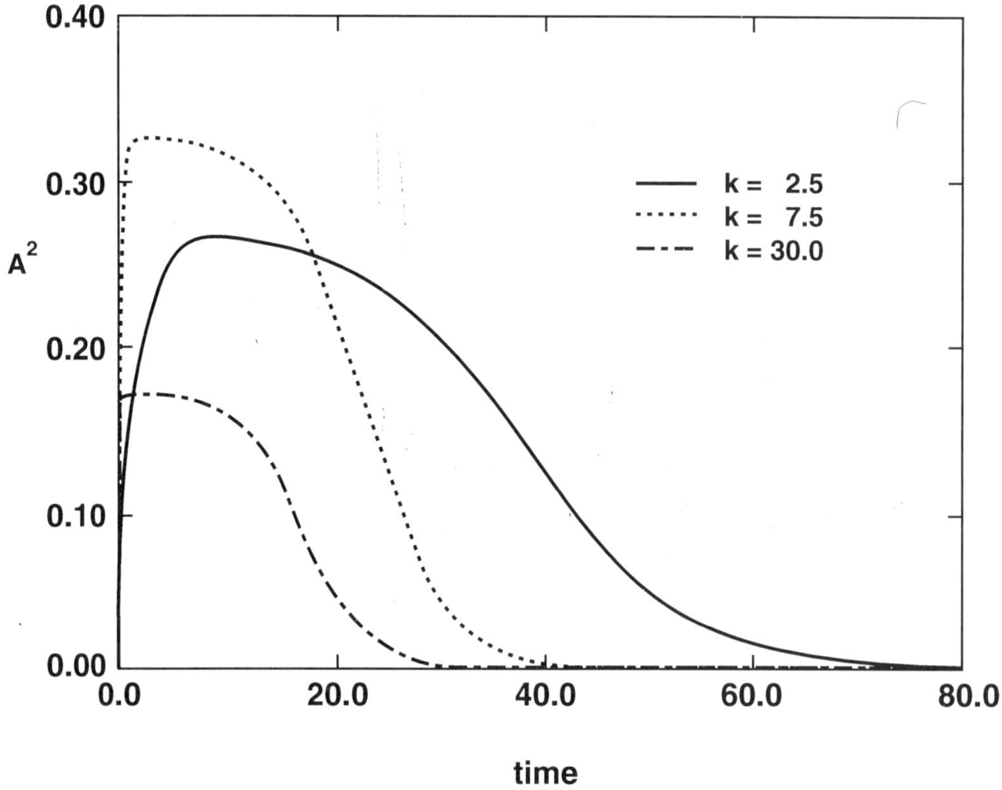

Figure 13: Temporal evolution of the coherent energy density for parameter values corresponding to different wavenumbers.

2.5 Mode/mode interaction

In order to estimate the effects of larger scale motions on the evolution of the smaller scales, we examine the interaction between a fundamental wave disturbance (small scale) and its subharmonic (large scale). The coherent field is then made up of two components, $v_i = v_i' + v_i''$, where the fundamental v_i'' is periodic in x_3 with a wavelength $\lambda/2$ and the subharmonic v_i' has a wavelength λ.

Since the periods of the disturbances have been artificially prescribed, two phase averages can be defined to decompose the velocity field analytically:

$$\langle f \rangle = N^{-1}\Sigma_n f(x_3 + n\lambda) \tag{36}$$

$$\langle\langle f \rangle\rangle = N^{-1}\Sigma_n f(x_3 + n\lambda/2) \tag{37}$$

In this way the coherent contribution from the instantaneous velocity is given by

$$\langle u_i \rangle = U_i + v_i' + v_i'' \tag{38}$$

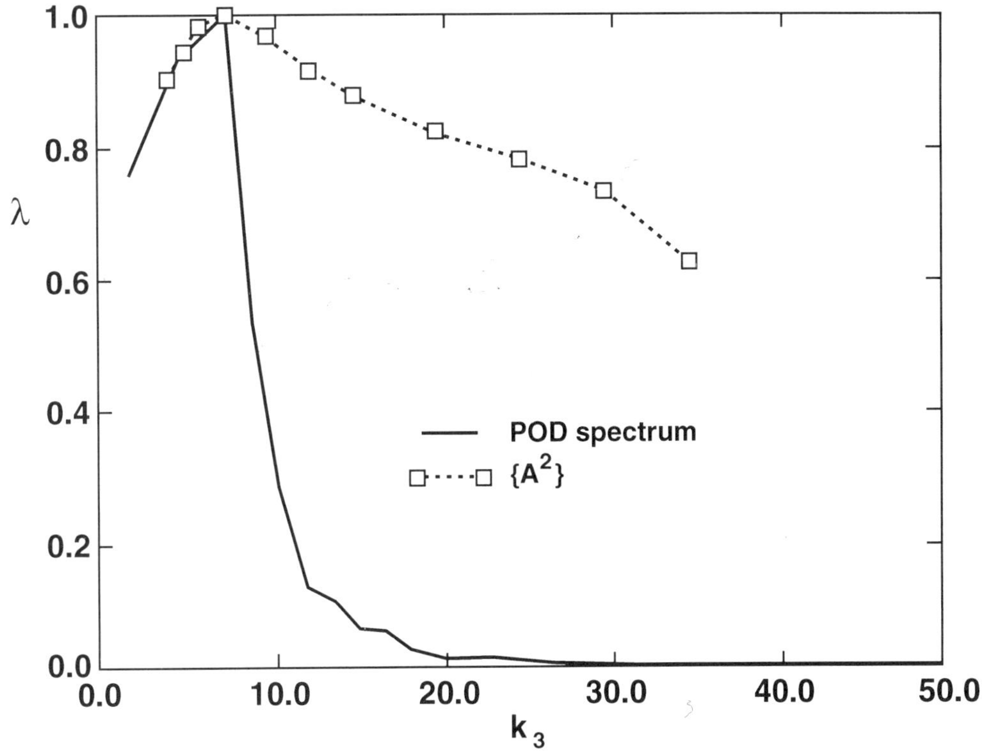

Figure 14: Comparison of ensemble averaged energy content with POD spectrum.

and the fundamental can be separated from the subharmonic using the <<.>> average:

$$\langle\langle v_i' + v_i'' \rangle\rangle = v_i'' \tag{39}$$

The volume averaged kinetic energy equations for the two components of the coherent field are:

$$\partial_t \int (v_i' v_i')\, dV = - \int (v_i' v_j' U_{i,j} + v_i' v_j' v''_{i,j})\, dV + \ldots \tag{40}$$

$$\partial_t \int (v''_i v''_i)\, dV = - \int (v''_i v''_j U_{i,j} - v'_i v'_j v''_{i,j})\, dV + \ldots \tag{41}$$

As before, the coherent velocities are taken as the eigenfunctions of the stability problem:

$$v''_i = A_1(t)\phi^1{}_i(x) \tag{42}$$

$$v'_i = A_1(t)\phi^2{}_i(x) \tag{43}$$

and the mean velocity model now contains contributions from both the fundamental and the subharmonic:

$$U_{1,2} = <u_1u_2> + (v'_1 + v''_1)(v'_2 + v''_2) + \ldots \tag{44}$$

New coupling terms appear in the evolution equations for the coherent energy densities due to the interaction between the different size modes:

$$\partial_t A_1{}^2 = A_1{}^2(I_1{}^1 - I_2{}^{11}A_1{}^2 - I_2{}^{12}A_2{}^2 - EI_3{}^1) - A_1A_2{}^2I_4$$

$$\partial_t A_2{}^2 = A_2{}^2(I_1{}^2 - I_1{}^{21}A_2{}^2 - I_2{}^{22}A_2{}^2 - EI_3{}^2) + A_1A_2{}^2I_4$$

$$\partial_t E = E(A_1{}^2I_3{}^1 + A_2{}^2I_3{}^2) \tag{45}$$

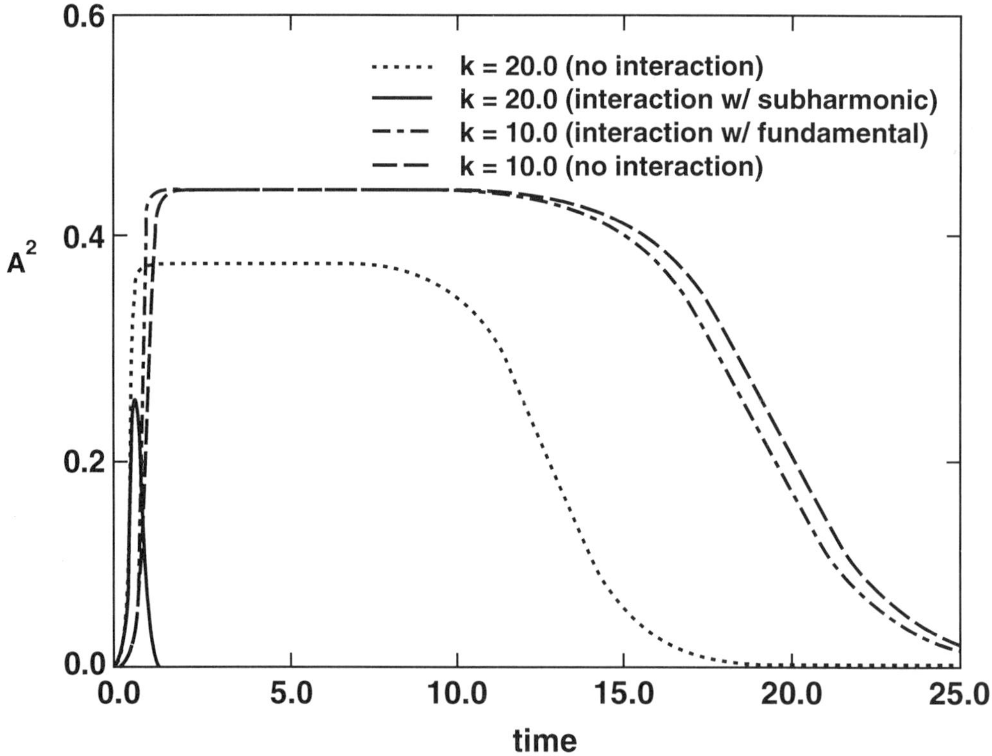

Figure 15: Temporal evolution of coherent energy densities showing the effect of mode-mode interaction.

$I_2{}^{12}$ and $I_2{}^{21}$ represent the effect on the mean shear production of fundamental (subharmonic) coherent energy due to the presence of the subharmonic (fundamental). I_4 is a measure of the direct energy transfer between the two modes due to the working of the sunharmonic stresses against the fundamental rate of strain. This quantity appears with opposite sign in each equation.

As an example, we have calculated the interaction integrals using the mode $k_3 = 20$ as the fundamental. The results of integrating equations (45) for these coefficient values are shown in Figure 15. The mode-mode dynamics are dominated by terms due to the mean velocity feed-back model. The direct interaction term I_4 is much smaller. The effect of the large scale subharmonic on the behavior of the small scale fundamental is dramatic. Although I_2^{21} is twice as large as I_2^{12}, the slowly growing subharmonic mode is practically unaffected by the presence of the fundamental. The fundamental, however, is quickly damped by the larger scale motion. This agrees with our intuitive picture of the physics. The small scale motions see not only the mean shear, but the strain rate due to all larger scales.

While the situation analyzed here is admittedly artificial, since a real turbulent flow will contain structures of all sizes at any given instant, there is no reason not to expect a qualitatively similar effect on the smaller scale motions in the real flow. Since the larger eddies are relatively unaffected by the smaller scales, the arguments presented earlier and the small wavenumber results shown in the figures will carry over to the case of many interacting eddies. The spurious slow fall-off in the spectrum of $\{A^2\}$, however, would be eliminated.

3 *Wavelet representations*

3.1 *Convergence*

All the flows of interest to us have one or more homogeneous directions. We are accustomed to use in these directions the Fourier transform, which is the homogeneous equivalent of the POD. However, the Fourier transform is not nearly so appropriate in the homogeneous case as is the POD in the inhomogeneous case. This is because the Fourier modes are not confined to a neighborhood, but extend to infinity without attenuation. All disturbances in a fluid, and coherent structures in particular, are localized. There is therefore considerable motivation to find another representation that is more appropriate.

In Tennekes & Lumley [8] it was suggested that a more appropriate quantity would be the energy surrounding a wavenumber κ, say from κ/a to $a\kappa$, where $a = 1.62$. In physical space, this packet with appropriate phase relations is confined to a region, essentially dropping to zero in about $2\pi/\kappa$ from the origin. Tennekes & Lumley called these "eddies", but they are an example of what are now called wavelets.

While wavelets appear to make more physical sense, we might worry because we would be discarding the optimality of the Fourier representation; would convergence be much slower, so that we would need many more terms, or would we lose considerable energy if we used the same number of terms? A main result of a recent lecture (Berkooz et al, [38]) is that very little energy is lost when using a wavelet basis instead of a Fourier basis. Although wavelets are physically appealing, it would also be nice to have reassurance from calculations that physical behavior would be preserved in a wavelet representation. To set our minds at rest on this point, Berkooz et al [38]) also display a *relatively* low-dimensional wavelet model of the Kuramoto-Sivashinsky equation that shows dynamical behavior similar to the full equation.

Without getting involved with mathematical details, an orthonormal wavelet basis is constructed by starting with a function, say $\psi(x)$, similar to the eddy suggested by Tennekes & Lumley [8]. From this, construct a set $\psi_{j,0}(x) = \psi(x2^j)$, j = 1, 2, ... Each of these is shrunk affinely, but is geometrically similar to the original function. Now consider the translates of $\psi_{j,0}$: $\psi_{j,k}(x) = \psi_{j,0}(x\text{-}k2^j)$, k = 1, 2 ....

Berkooz *et al* [38] consider a periodic, homogeneous stochastic process. It is then obvious that the POD decomposition becomes identical to the Fourier decomposition. Now if, for a given ε, we need $N(\varepsilon)$ POD modes in order to satisfy

$$\left| \sum_{i\geq 1} < |a_i(t)|^2 > - \sum_{i=1}^{N(\varepsilon)} < |a_i(t)|^2 > \right| \leq \varepsilon \tag{46}$$

then, if the $\{b_i\}$ are the coefficients in a wavelet basis, we get for some constant C, depending only on the process (and not on ε):

$$\left| \sum_{i\geq 1} < |a_i(t)|^2 > - \sum_{i=1}^{\mathcal{N}(\varepsilon)} < |b_i(t)|^2 > \right| \leq C\varepsilon \tag{47}$$

for some $\mathcal{N}(\varepsilon)$ slightly bigger than $N(\varepsilon)$ (the precise statement is given in Berkooz *et al* [38].

3.2 *Dynamics with the wavelet representation in a simple equation*

Berkooz *et al* [38]wished to apply these ideas to a simple situation. The three-dimensional, three component Navier Stokes equations are too complicated for a first effort.

The one-dimensional, scalar Kuramoto-Sivashinsky (K-S) equation appears in a variety of contexts, such as quasi-planar fronts, chemical turbulence, etc . It shares some properties with Burgers' equation and the Navier Stokes equations, but is much easier to deal with. Up to some rescaling, this equation can be written as:

$$u_t + u_{xx} + u_{xxxx} + u_x u = 0, \qquad 0 \leq x \leq L \tag{48}$$

u periodic on [0, L], where L, the length of the spatial domain, is the only free parameter in the problem.

Although the dynamical behavior for small values of L is fairly well understood (see Hyman & Nicolaenko, [39] for an overview), many open questions remain concerning the limit $L \to \infty$ (Zaleski, [40]; Pomeau *et al*, [41]. As can be seen from the numerical simulations, for $L > 30$, a chaotic regime involving both space and time disorder occurs (see Figure 20, where we plot a space-time representation of a typical solution, $L=400$, $0 \leq t \leq 100$).

In order to check the estimate (47), we compare the energy resolved by a given number of modes using either a Fourier (POD) or wavelet basis. Note that, to compare the Fourier and wavelet bases, all the translates in the wavelet basis must be considered. Here are some results (see Figure 17):

No. of modes	wavelets (m = 6)	wavelets (m = 8)	Fourier
64 (j = 6)	70.84%	71.5%	72.2%
96 (j = 6,5)	79.1%	79.43%	83.3%
127(0≤j≤6)	84.1%	84.9%	89.7%
255(0≤j≤7)	99.9%	99.9%	99.9%

The scale j=6 which captures most of the energy on the average, corresponds to a characteristic length $2^{-6}L \approx q_m^{-1}$, which is also the length scale associated with the most unstable wavelength q_m. In agreement with the general shape of the energy spectrum, the scales $0 \leq j \leq 5$ are shown to capture more energy than the scales in the dissipative range ($j \geq 7$). The above figures show that (for sufficiently smooth splines) the wavelet projection captures almost the same amount of energy as the Fourier (= POD) decomposition (within 5%).

These results prompted us to conclude that from an average energy point of view wavelets are a reasonable candidate for a modal decomposition of the K-S equation. The localized nature of the wavelets may give us a unique view of the spatial attributes of the coherent structures. We outline our approach. We conjecture the existence of a dynamically relevant length scale L_C such that interactions between physical regions of distance greater than L_C are dynamically insignificant (a dynamical St. Venant principle). Determining the validity of this conjecture is part of our study. We use this conjecture to remove terms in a wavelet-Galerkin projection that correspond to interactions between regions of distance greater than L_C. To study whether the dynamics of coherent structures are indeed locally determined we construct truncations corresponding to a small box size L_B in the larger box of size L.

We need to address the role of unresolved physical space (i.e.- modes located outside the box of size L_B). It is obvious that the Dirichlet type of boundary condition imposed will create a boundary layer which will affect the dynamics, especially in small boxes which are of interest to us. There are two plausible approaches to remove this effect. One approach uses a stochastic boundary condition (which is hard to implement numerically and treat analytrically). The other approach appeals to the conjecture on the existence of L_C. One takes a box of size L_B greater that $2L_C$ so that one can periodize the small model using resolved relatively distant modes instead of unresolved ones. We opted for the second approach.

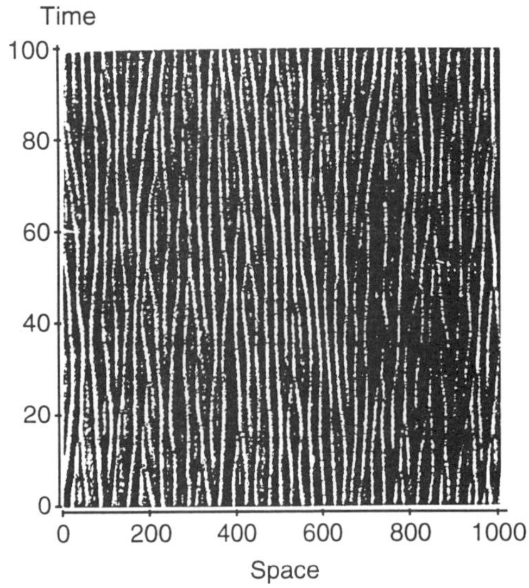

Figure 16. A space-time representation of a typical solution of the one-dimensional, scalar Kuramoto-Sivashinsky equation, L = 400, 0 t 100.

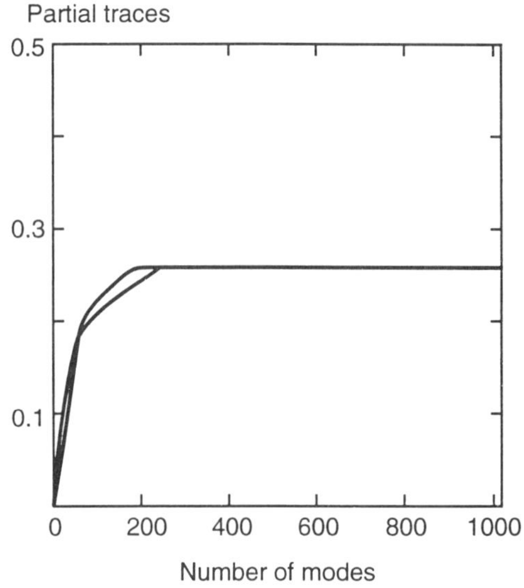

Figure 17. The energy resolved by a given number of modes using either Fourier or a wavelet basis. Note that all translates of the wavelet basis are included.

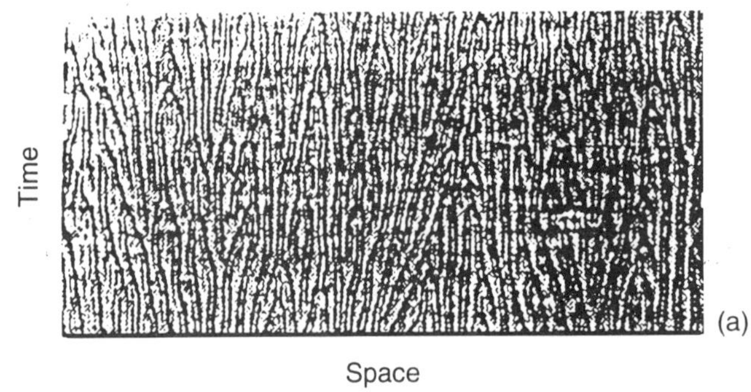

Figure 18. The spatio-temporal evolution of the full system for the Kuramoto-Sivashinsky equation, with a Fourier basis.

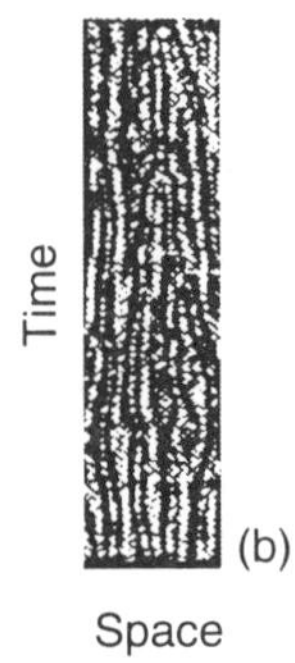

Figure 19. The spatio-temporal evolution of a (rescaled) model for the Kuramoto-Sivashinsky equation, with L_C = 50x3/8 (see text), which is in excellent qualitative agreement with the dynamics of the full system.

We present some preliminary results of the integration of one such model. We resolved a box of size $L_B = 50$ (this is 1/8 of the original box). Figure 18 shows the spatio-temporal evolution of the full system, with a Fourier basis. Figure 19 shows the spatio-temporal evolution of a (rescaled) model with $L_C = 50 \times 3/8$, which is in excellent qualitative agreement with the dynamics of the full system. If L_C is too small, after a long initial transient, the system eventually settles down to a periodic oscillatory state, in which no interaction between the localized structures is observed. This can be avoided by an increase in L_C, which adds significant non-linear interactions between relatively distant physical space locations. It might also be avoided by a stochastic boundary condition.

4 *Drag reduction*

4.1 Turbulent drag reduction

Since Toms [42] observed that the addition of a few ppm of certain polymers could reduce significantly the skin friction, there has been a strong interest in this technique for reducing the turbulent skin friction drag. Submerged vehicles are the main concern since for these the skin friction is the majority of the drag, and in any case polymers are only effective in liquids; increasing the speed and decreasing the effective cost have been the primary motivations. There are many other commercial applications: for instance, it is possible to increase significantly the surface irrigated by a pump.

Many other techniques have been examined in addition to polymers, such as compliant surfaces, heated bodies, riblets, microbubbles, etc. It is not our intention to recall all the possibilities which have been considered. A review is given in Lumley & Kubo [27]. We will focus our attention on a few techniques which are currently under investigation, namely the addition of polymers and the injection of micro air bubbles. Lumley and Kubo [27] believe that these two techniques are related to the same physical mechanism.

Ten years ago, the effect of the presence of polymers on the mean velocity and fluctuations was not clear even though experiments had been carried out since the mid-sixties. The main reason was the great variety of conflicting data, much of it obtained with tools the response of which was influenced by the polymers, such as hot wire and hot film anemometers and Pitot tubes. In 1979, excluding such techniques, Kubo and Lumley accomplished a general survey of available experimental data in the literature between 1970 and 1979. They were able to draw the general conclusions concerning both the mean velocity and the fluctuation intensities that we summarize here.

The mean velocity field is modified in the buffer layer only. In particular, its asymptotic behavior at the wall is not affected. There is no significant change in the viscous sublayer, which is slightly thicker and reaches $x_2^+ = 8 - 10$ ($^+$ indicating renormalization with friction velocity and kinematic viscosity). All the changes are observed in the buffer layer, which is significantly thicker. When the buffer layer reaches the center of the pipe or the edge of the boundary layer, the drag reduction phenomenon reaches its limit. Figure 1 of Kubo and Lumley shows that the mean flow velocity is higher in the drag reduced flow than in the Newtonian flow. All the data agree that the peak in the intensity of the streamwise components is shifted to higher x_2^+. Although there is a lot of controversy concerning the peak intensity, Kubo and Lumley conclude it is close to or slightly higher than that in Newtonian fluids. However, most data indicate a decrease in the intensity in the region below $x_2^+=20$. Transverse intensity is lower than in the Newtonian case in the buffer layer.

All data agree that non-dimensional streak spacing increases gradually as drag decreases, clearly indicating that the coherent eddies (streamwise rolls) in the viscous sublayer and buffer layer grow in size. The effect of polymers on the bursting period is much less obvious. Flow visualization indicates that the bursting period in polymer fluids is unchanged from that in Newtonian fluids at the same shear velocity, whereas the autocorrelation method indicates that it increases. Finally, observation of the frequency spectra of the fluctuating signals shows relative attenuation of the high frequency components. This interesting finding is common to wall regions of boundary layers, water jets and grid-generated turbulence.

Lumley proposed a mechanism consistent with these observations. He postulates that the only difference in the turbulent boundary layer structure between Newtonian and drag-reduced flows is the scaling in the different regions. His theory is based on the polymer molecular expansion outside of the viscous sublayer due to the fluctuating strain rate. Such an expansion increases the effective viscosity, which damps the small eddies and reduces the momentum transfer. The consequence is a decrease of the slope of the mean velocity profile and a thickening of the wall region. The large eddies adapt themselves to such a change in the mean velocity profile and expand as well.

In this lecture, we use some of these experimental results in a very rudimentary way to model drag reduced flows. Since we have already derived a simple model to study the dynamics of the large structures of the wall layer (Aubry et al.[37) and in particular to explain the birth, nature and characteristics of the intermittent bursting events, we attempt here to modify our flow in order to reduce the drag. Using dynamical systems theory tools, we investigate the effect of these modifications on the dynamics of the wall region.

4.2 Model for a drag reduced flow

In this study, we do not concern ourselves with the question of how the presence of a few ppm of polymers directly affects the structure of the flow. Whatever the physical mechanism is, we assume that the presence of polymers changes the velocity. Given a few simple modifications -easily controllable- concerning the fluctuating velocity field based on experimental observations, we intend to investigate the consequences of these changes on the dynamics of the coherent structures of the sub and buffer layer. In order to implement these ideas, we use the model outlined in section 2.

Our first attempt consists in simulating the thickening of the buffer layer (which seems to be one of the most relevant phenomena for drag reduction) by stretching the velocity field in the direction normal to the wall. Note that it is not obvious from existing experimental data that the viscous layer (up to $y^+ = 5$) is being stretched as well as the buffer layer. However, for simplicity, we will expand our whole layer ($0 < y^+ < 40$). A simple way of changing the scale is to apply a linear stretching transformation to the fluctuations in the following manner:

$$u'_2(x_1, x_2, x_3, t) = u_i(x_1, \frac{x_2}{\beta}, x_3, t) \tag{49}$$

where the prime refers to the modified field. β is a real parameter called the stretching factor.

Our new fluctuation u' can be expanded in the same way as u, i.e. in proper orthogonal modes in the normal direction and in discrete Fourier modes in the streamwise and cross-stream directions (assuming periodic conditions in the two last directions):

$$u'_i(x_1, x_2, x_3, t) = \frac{1}{\sqrt{L_1 L_3}} \sum_n \sum_k e^{2\pi i k \bullet x}\, a_k^{(n)}(t)\, \phi_{ik}^{(n)}\left(\frac{x_2}{\beta}\right). \tag{50}$$

This is equivalent to stretching our eigenfunctions in the normal direction in the following way:

$$\phi_{ik}^{(n)'}(x_2) = \phi^{(n)}{}_{ik}\left(\frac{x_2}{\beta}\right). \tag{51}$$

The new set of eigenfunctions still form a complete set and the velocity field can be expanded in them. The orthogonality of the eigenfunctions is preserved under the stretching transformation with the expansion of the integration domain from $[0, X_2]$ to $[0, \beta X_2]$. Indeed, by a change of variable, the equation:

$$\int_0^{X_2} \phi_{ik}^{(n)}(x_2)\, \phi_{ik}^{(m)*}(x_2)\, dx_2 = \delta_{nm}. \tag{52}$$

implies:

$$\int_0^{\beta X_2} \phi_{ik}^{(n)'}(x_2)\, \phi_{ik}^{(m)*'}(x_2)\, dx_2 = \delta_{nm}\, \beta. \tag{53}$$

Our stretched eigenfunctions are thus orthogonal but not normalized. In order to obtain orthonormal eigenfunctions, we are led to the following definition:

$$\phi_{ik}^{(n)'}(x_2) = \frac{\phi_{ik}^{(n)}\left(\frac{x_2}{\beta}\right)}{\sqrt{\beta}}. \tag{54}$$

The original eigenfunctions satisfy the continuity equation. This can be shown by considering the continuity equation for the fluctuating velocity, multiplying this equation by $a^{(m)}$, averaging and using the fact that the coefficients $a^{(n)}$ are uncorrelated with each other. However, our new eigenfunctions, defined by (54), do not yet satisfy the continuity equation, which, of course, they must. That is, we suppose that the new eigenfunctions we are seeking are proper orthogonal eigenfunctions of the modified flow. Their properties are thus the same as those of the original eigenfunctions. In other words, the first eigenfunction $\phi^{(1)}$ is such that the projection on the fluctuating, random and energy integrable velocity field is maximum in the Hilbert space of the energy integrable functions with the inner product defined by:

$$(\phi, \psi) = \int_0^{\beta X_2} \phi_i^* \psi_1 \, dx_2 . \tag{55}$$

(The upper limit of this integral is βX_2 instead of X_2.) The second eigenfunction is defined in the same way once the first eigenfunction has been subtracted from the velocity field. One important property is the lack of correlation of the coefficients of the expansion which implies that our newly constructed eigenfunctions must satisfy the continuity equation. It is thus necessary to modify the definition (54) and the expansion (49).

This matter is more subtle than it at first appears. There are two ways of forcing the new eigenfunctions to satisfy the continuity equation. We can either divide the first and third components by the stretching coefficient β (or equivalently multiply the second component by β) or divide the streamwise and cross-stream wavenumbers by β. Physically, for a stretch ($\beta > 1$), either the flow is relatively faster in the normal direction in the modified flow than in the unmodified flow and the corresponding length scales remain the same, or the cross-stream and streamwise length scales are larger and the velocities are approximately unchanged.

It has been experimentally observed that in the unmodified as well as the modified flow, the rolls are indeed approximately round. In other words, not only is the wall region vertically thickened but also the average streak spacing is larger. In our model (before stretching), the vertical scale of the rolls is fixed since there is only one proper orthogonal eigenmode. Even though the eigenfunctions used in Galerkin projection encode the basic velocity field structure, the relative importance of the different cross-stream Fourier modes (which determines the spanwise scaling) is autonomously determined by the dynamics (through the coefficients $a^{(n)}$). In other words, provided that we input enough Fourier modes, the dynamics will determine the shape of the rolls: circular, flat or tall. For high values of the Heisenberg parameter, there is a unique attractor, a globally attracting fixed point -or rather an S^1-family of such fixed points corresponding to a shift of the solution in the cross-stream direction, due to the homogeneity of the flow in this direction. Thus, whatever the initial condition is, flat or tall rolls, the solution becomes steady very fast, recovering the circular shape of its rolls with a diameter of about 70 wall units. When the wall region is vertically thickened, we expect to see the same phenomenon happen. In order to give enough freedom for our dynamical system to exhibit this, it therefore seems reasonable to stretch the spanwise length of the box by the same factor as the vertical stretching. In this way, we give the rolls the extra lateral room they need in order to expand. Note, however, that the rolls could eventually rearrange themselves in a box with the previous lateral dimension (although this would be more difficult since they have to appear in pairs in their steady state due to the statistical reflection symmetry). Since at the moment we have two pairs, for a stretching factor between 1 and 2, our box would probably be too big for one pair and not big enough for two. For a stretching factor bigger than 2, our present box would be definitely too small.

On physical grounds, therefore, it seems best to adjust all the length scales. However, before doing this, we consider for a moment modification of the relative importance of the different components of the velocity. The resulting eigenfunctions are not orthogonal in the sense of the inner product (55). One way to avoid this inconvenience is to consider the following inner product:

$$(\phi, \psi) = \int_0^{\beta X_2} \beta^2 \phi_i^* \psi_1 + \phi_2^* \psi_2 + \beta^2 \phi_3^* \psi_3 \, dx_2 . \tag{56}$$

The new eigenfunctions defined as:

$$\phi_{i_k}^{(n)'}(x_2) = \frac{1}{\beta\sqrt{\beta}} \phi_{i_k}^{(n)}\left(\frac{x_2}{\beta}\right) \text{ if } i = 1 \text{ or } 3 .$$

$$\phi_{2_k}^{(n)'}(x_2) = \frac{1}{\sqrt{\beta}} \phi_{2k}\left(\frac{x_2}{\beta}\right) \tag{57}$$

are orthonormal in the sense of the inner product (56) and satisfy the continuity equation. Via Galerkin projection in the new Hilbert space, we can get sets of ordinary differential equations in a straightforward manner. The only problem concerns the fluctuating pressure term. In the unmodified layer, we showed by Galerkin projection, integration by parts and use of the incompressibility equation that this term disappears except at the upper boundary (at the lower boundary, the term is also zero because of the no-slip condition) and thus acts as a small forcing term. In the modified flow, when the eigenfunctions (57) are introduced, the pressure term remains in the interior of the layer since the Galerkin projection defined with the new inner product does not introduce the continuity equation which would cancel the remaining integral of the integration by parts. In this case, the pressure would be an internal agent acting at each level of our layer. This would be a major inconvenience and for this reason and the physical observations mentioned above, we will not pursue further the possibility of simultaneously stretching the layer in the vertical direction and keeping the same periodic length in the cross-stream direction.

We thus take the second option and expand our box in all directions by the same factor. The resulting differential equations will determine the time evolution of stretched structures. In particular, we would like to answer questions such as: Do extra flat or extra tall rolls dynamically recover their circular shape? On the basis of experimental evidence, we expect that the answer will be yes, that the new eddies will be bigger but essentially similar in shape to those of the unmodified flow. Do larger rolls (if indeed this is the case) have the same dynamical characteristics? Would this rescaling preserve the turbulent interactions and the intermittent dynamics as it is observed in boundary layer flows with polymers? How does the mean velocity adjust itself? (Note that we could have proceeded the other way around, namely modifying the mean velocity and questioning the corresponding shape of the big eddies.) Does this adjustment imply reduction of the skin friction drag as speculated? Finally, we are interested in the effect of slightly perturbing the coefficients of our ordinary differential equations. The method introduced here is a systematic and particular way to do so, and, in addition to answering the questions above, this study will also give us some insights into the robustness of our model.

To summarize, we have devised a consistent scheme for stretching the eigenfunctions both normal to the wall and in the cross-stream direction, which matches observation in drag reducing flows. The stretched eigenfunctions satisfy all appropriate requirements. The mean velocity consistent with these stretched eigenfunctions, derived through an average in planes parallel to the wall, is (in time mean) larger at the outer edge of the layer for the same wall shear i.e. - drag reducing.

4.3 Consequences for the set of ODE's and the nature of the solutions.

Provided with our modified flow and expansion, we take the Fourier transform of the equation for the fluctuating velocity where the mean flow has been replaced by its expression as a function of the Reynolds shear stress and apply a Galerkin projection using the eigenfunctions (14) as basis vectors. Recall that the coefficients of the expansion are normalized with the square root of the eigenvalues which are given by equation (16). Since the eigenfunctions are orthogonal and satisfy the continuity equation, the structure of our new system of ODEs is identical to that of the unmodified flow. The coefficients of the various terms change as follows:

- The coefficients of the viscous terms are divided by the stretching factor β.
- The (linear) interactions between the fluctuations and mean velocity (through the mean pressure gradient) are more complex because of the explicit dependency on x_2. The part of the coefficient which is linear in x_2 in the integral is multiplied by β^2 although the other part is multiplied by β.
- The coefficients of the quadratic (inertia) terms (fluctuation/fluctuation interactions) are divided by β.
- The coefficients of the pseudo-pressure term are divided by β.
- The coefficients of the cubic terms (fluctuation/Reynolds shear stress interaction) are divided by β.
- The effective transport coefficient is unchanged. The characteristic length scale of the non-resolved scales is multiplied by β whereas the characteristic velocity is divided by β.
- The linear and cubic terms involving streamwise gradients are modified differently but they are ignored in this model.

The truncation studied is that outlined in section 1 which consists in the first eigenmode (n=1), the zero streamwise wavenumber (k_1=0) and six spanwise wavenumbers k_3 = 0,k,..5k for a chosen k.

As the layer is stretched, there is apparently less viscous dissipation and more energy supply among the linear terms corresponding to the resolved modes. (In this discussion, we exclude the influence of the smaller scales through possible variations of the Heisenberg parameter α which are considered in the next section.) However, in the interaction between fluctuations and mean velocity, the part coming from the mean pressure gradient ($\beta^2 b_k{}^{12}$) is negative. This term is very small in the wall region compared to the interaction with the wall shear stress (typically 10%; this is due to the fact that 1 >> x_2/H). The sum of both terms, which represents the global fluctuation/mean velocity interaction, is generally positive so that the mean velocity supplies energy to the fluctuating flow. When the layer is stretched in the manner discussed in section 4.1, this picture may change. The stretching transformation ($\beta > 1$) accentuates the mean pressure gradient/fluctuation interaction so that the source of energy is relatively decreased. Eventually the layer could be stretched enough to cancel all possible energy supply on all the modes considered here. In that case the flow would be of course stable. In the present case this takes place for a stretching factor of 12.95, which would bring the top of our layer to $x_2^+ = 518$. However, the experimental flow that we are investigating is a pipe flow. The diameter of the pipe is 0.285m and the viscous length is 0.538mm. The centerline of the pipe is thus located at $x_2^+ = 264.9$. If we limit the stretching process when the top of the buffer layer ($x_2^+ = 100$ in an unmodified flow) reaches the centerline, then we cannot allow the stretching factor to exceed 2.65. This limitation imposed by the flow itself eliminates the possibilities of large stretching factors with no energy supply from the mean flow. In the following, we will consider stretching factors between 1.00 and 2.65. We will see in the next section that stretching stabilizes the linear terms. It also makes the quadratic and cubic interactions weaker. Recall that the cubic terms are necessarily stabilizing since their coefficients are all negative.

4.4 Discussion and conclusions.

We have found that our stretched boundary layer corresponds to higher values of α, the Heisenberg parameter. As we have parameterized the loss to the unresolved motions, there is no real distinction made between this loss and the direct loss to viscosity. In the discussion above, we have been thinking of the Heisenberg parameter as representing primarily the turbulent loss to the unresolved modes. Let us, however, consider that it represents here the direct viscous loss. With that interpretation, when we observe that the solution bifurcates from the fixed point at a higher value of the Heisenberg parameter, this is equivalent to finding that the stretched boundary layer is more unstable than the unstretched, that it requires a higher value of viscous loss to restore it to marginal stability. Considering for a moment experimentally observed boundary layers, the classical interpretation of the drag reduction phenomenon in polymers (Lumley, [42]) is that the instability leading to the bursting phenomenon is stabilized by the increased (extensional) viscosity due to the uncoiled molecules in the turbulent part of the flow, while the viscous sublayer near the wall is unaffected because the molecules are coiled up there. Hence, an increased viscosity in the turbulent part of the flow leads to the instability occurring only when the scales are larger. This is completely consistent with our model: a larger scale and an increased viscosity at the first bifurcation. We have sidestepped the problem of having a different viscosity in the turbulent part of the flow and in the sublayer. This is necessary in nature to avoid having the increased viscosity slow down the mean flow. Here, however, we have arbitrarily adjusted the disturbances relative to the mean flow, in effect allowing a different viscosity for the disturbances and for the mean flow. In the case of microbubbles, it is thought (Lumley & Kubo, [27]) that these also selectively change the viscosity in the turbulent part of the flow (relative to the sublayer) so that the mechanism is the same. Our model for a drag reduced flow is rudimentary. It is based on the experimental observation that the wall region is generally thicker when polymers or microbubbles are added. Although it is consistent with the main changes experimentally observed in the mean and fluctuating velocity profiles and the increase of the size of the large structures, it still ignores many aspects of the problem. If drag reduction in polymers and microbubbles flows is due to changes in the distribution of the effective viscosity with distance from the wall, a more complete model should include this effect.

We have stretched the layer by a factor β and we have held this value constant. This is a rough approximation. In our model, the value of β that we impose corresponds to how much the layer is stretched and thus how much the drag is reduced. However, if our model were sophisticated enough to include the equations concerning the polymer and its interaction with the solvant (perhaps similar to the model of Sarkar and Lumley [44]), β would be determined by the physics and would be selected by the dynamical system itself.

In particular, there is no reason why it should be a constant for a given flow. For example, as the eddies grow and shrink, they should stay approximately round. The vertical scale of the eigenmodes should thus adjust to the cross-stream wavelength selected by the dynamical system. However, the amplitude of this adjustment is probably much smaller than the global thickening of the wall region due to the presence of polymers. For this reason and for simplicity, we make β vary only as a bifurcation parameter.

Stretching the eigenfunctions results in modification of the coefficients of the ODEs. The qualitative behavior of the dynamical system is similar to that in the unmodified layer. For large values of the Heisenberg parameter, the solution is a stable fixed point representing two pairs of streamwise rolls. Although the periodic box is larger and the layer is thicker, the eddies recover their round shape, as expected. As the Heisenberg parameter decreases, an intermittent solution appears due to the presence of an attracting heteroclinic cycle. The scenario by which intermittency appears through a subcritical Hopf bifurcation and disappears through a bifurcation to travelling waves is identical for the different values of the stretching factor considered here ($1 \leq \beta \leq 2.65$), although the Heisenberg parameter values at which bifurcations occur increase as the layer is stretched. The morphology of the bifurcation diagram is not identical, however. In particular, 1) although the value of α increases as β increases, it decreases compared to the largest bifurcation value from the trivial solution, and 2) as the layer is more stretched, the window of intermittency becomes shorter compared to the total size of the bifurcation diagram.

To summarize, all dynamical phenomena observed in the unmodified flow are present in the stretched layer. However, everything changes scale with the introduction of stretching. This is in agreement with experimental results for flows whose drag is reduced but still display intermittency.

These studies demonstrate that our model equations retain much of their physically significant behavior under transformations resulting from stretching the flow. The mathematical reason for this robustness lies in the structural stability of the O(2)-equivariant heteroclinic cycles which the equations possess in certain parameter ranges (Armbruster et al.[45]). Structural stability implies that the qualitative features of the system do not change under (sufficiently low) perturbation within the appropriate class of systems (Guckenheimer and Holmes [19]).

In this lecture, we did not include the fluctuating pressure term which acts as a forcing term of small amplitude. In the unmodified flow, the role of the term is to trigger the burst. Since the nature of the intermittent solution is the same (due to the presence of a heteroclinic cycle) when the layer is stretched, the effect of the pressure will be identical. The scaling of the bursting period should thus depend on both the characteristics of the heteroclinic cycle (inner variables) and the amplitude of the noise (possibly outer variables). The complexity of this scaling could explain the controversy in the polymer literature concerning the effect of polymers on the bursting period.

5 *The Possibility of Control*

5.1 *Background*

We are interested in the possibility of controlling the turbulent boundary layer. That is, we wish to control the bursting process in the wall region, in order to reduce the drag, or increase the mixing. We envision an array of sensors, to detect the current state of the layer, and an array of actuators to generate perturbing velocity fields to interact with the existing velocity field. The sensors would provide input to software, which would drive the actuators.

In order to make this concrete, we must answer several questions: dynamically, what is happening during the bursting process? What sort of velocity field can influence this process? What control strategy can we devise to bring this about?

During the last decade, we have constructed a heirarchy of low-dimensional models which mimick the behavior of the bursting process in the wall region. We believe that these models answer the first question, as to what is happening dynamically during the bursting process.

Here we suggest answers for the second and third questions; we suggest what sort of velocity field can influence the coherent structures in the wall region, and we suggest a control strategy to bring this about. Our interpretation of the influence of the velocity field is based on an analysis of our low-dimensional model, and our proposed control strategy is evaluated on the same low dimensional model. Briefly, we find that the perturbing effect of an artificially created pair of streamwise vortices manifests itself most strongly in a new parameter that can be used to stabilize the saddles of the unperturbed dynamical system and thus reduce the bursting rate.

More specifically relevant to the subject of this lecture, we have obtained a number of important results indirectly and directly related to control and drag reduction. Polymer drag reduction is known to increase the time between bursts, and to increase the scale of the wall region, leaving it geometrically the same [27]; this has been presumed to result from stabilization of the bursting process. In our model, stretching of the wall region to increase its scale reduced the drag, and required stabilization of the bursts, to increase the time between bursts [46. It is thus completely consistent with the observations, and with the presumed mechanism. Polymer drag reduction is, of course, a passive means of control. However, we could envision using our sort of relatively simple model as a "black box" in an active feed-back system to control the boundary layer. Bloch & Marsden [47] considered such a possibility, and showed that feeding back eigenfunctions with the proper phase (in the presence of noise) could delay the bursting, (the heteroclinic jump to the other fixed point), thereby decreasing the average drag. It is also possible to speed up the bursting, thus increasing mixing, to control separation Making an interesting connection to passive control, Bloch & Marsden [47] also showed that polymer drag reduction is formally equivalent to delaying the bursts by active feedback. Hence, we may expect an actively controlled layer to resemble a passively polymer-drag-reduced layer.

Our active control strategy consists of detecting the location of the system by an array of sensors. When the system comes into the neighborhood of a saddle, it is held near the saddle by means of small feedback terms for as long as possible, delaying the jump to the opposite saddle. Once the system escapes from the vicinity of the saddle, control is abandoned until the system again enters the neighborhood of a saddle (as it must, due to the dominant global dynamics).

Active control can be accomplished in many ways. We are envisioning here the possibility of producing pairs of streamwise vortices by raising a bump on the surface, using a piezo-electric device. Measurements taken at the Pennsylvania State University by A. Fontaine and S. Deutsch, of the velocity field produced in the wall region by a Gaussian bump, indicate that a pair of streamwise vortices is produced. The decision to raise a bump in a particular location would be taken by software using a low-dimensional model of the boundary layer as described above, working from input from the sensors. Such a pair of induced vortices would have a circulation the reverse of the naturally occuring pair. There are many possible ways in which such an induced pair could interact with a naturally occuring pair. The induced pair could be created immediately underneath the naturally occuring pair; this is the configuration that comes first to mind. One would then expect the downdraft between the induced pair to weaken the updraft between the naturally occuring pair.

We are going to examine here a second type of interaction. We will imagine that the induced pair is adjacent to the naturally occuring pair, at the same distance from the surface, on one side or the other. We will consider at the same time, the interaction of two adjacent naturally occuring pairs. In this preliminary study, we consider only the simplest in a heirarchy of models of increasing complexity.

We refer in the above to naturally occurring and induced vortex pairs. The induced vortices may, in fact, be expected to occur in more-or-less equal pairs. The observed naturally occuring vortices, however, are nearly always of unequal strengths and may occur singly. We can consider a generic natural occurrence to be two vortices of arbitrarily unequal strength, taking the isolated vortex as a special case, in which the strength of one is zero. Our low-dimensional model, on the other hand, deals with periodic arrays of vortices (our calculations encompass four vortices, with periodic boundary conditions). Our model and the observed natural occurrences are clearly different. We are taking steps to develop a model that more nearly represents reality, using wavelet transforms to replace the Fourier transforms, so that we can represent isolated events, either isolated unequal pairs or singles, instead of periodic arrays. In the meantime, what reason do we have to believe that our results, based on periodic arrays, have any relevance for isolated unequal pairs or singles?

In fact, it is the purpose of this lecture to demonstrate that the presence of adjacent vortices can affect the dynamical behavior of otherwise isolated vortices or unequal pairs. If we accept these results, then it seems unlikely that the dynamical behavior of a periodic array and of an isolated unequal pair will be the same, since vortices in the periodic array would be influenced by their neighbors. However, we have seen that the predicted dynamical behavior of periodic arrays *is* qualitatively and even quantitatively similar to the experimentally observed natural behavior of isolated unequal pairs. We may thus anticipate that the predicted behavior of isolated unequal pairs will remain qualitatively similar to that of periodic arrays, with probably only small changes in parameter values. We may perhaps speculate that this is due to the fact that, in a periodic array, the influence (of the type we will discuss below) of vortices on one side of a pair is approximately cancelled by the vortices on the other side; in the symmetric case, the cancellation is exact.

Finally, an isolated vortex will produce a velocity field similar to that of an isolated unequal pair (see below), and hence will have a similar influence on a periodic array or an isolated unequal pair.

5.2 Interaction of adjacent bursts

We consider the velocity field due to an isolated vortex pair near the wall in a boundary layer. This pair may be induced by an actuator, or might occur naturally. Our being transformed to coordinates moving with the gentle lateral motion) until the lateral motion brought the second pair into the immediate vicinity of the first, the time required of course depending on the initial spacing. The effect of the shear is more interesting. Below, we will estimate the effect of the shear from the changes in the equations. Reasoning physically, however, the rolls will be distorted, tipped laterally, their aspect ratio will change and the intensity of the internal shear will change. We will find that these effects change the bursting rate. Note that "open loop" flow modification which simply introduces rapid cross-stream oscillations has been shown by numerical simulation, to significantly reduce drag, apparently by reducing the bursting rate[48].

From a physical point of view, the rolls are influenced primarily by the part of the mean velocity profile next to the wall, where the gradient is approximately linear. There the strain rate is a maximum, and it is from this part of the mean velocity profile that the greatest energy is extracted. If the lateral boundary layer is thick enough so that its linear region is comparable with this region in the streamwise profile, then our physical situation is similar to rolls at a small angle to the streamwise direction. There is certainly evidence from non-linear stability theory that the angle to the streamwise direction can have a considerable influence on the stability and growth rate. Rolls in the streamwise direction are known to grow faster than rolls normal to the stream, although there is no reason to believe that the growth rate is monotone with angle. Note that, if the extent of a linear profile is much different in the streamwise and the cross-stream directions, then the direction of the mean velocity will change with distance from the surface, and the stability problem is more nearly like the stability of the Ekman layer. Again, however, this may be expected to be different from the stability of the streamwise vortices. Hence, in any event, we may expect that the presence of a pair will have an influence on the bursting behavior of an adjacent pair.

We assume no streamwise variation, so that the streamwise and cross-stream velocities decouple. We may thus consider the evolution of the velocities in the cross-stream plane in isolation. We suppose initially that the motion in the cross-stream plane is irrotational, with the exception of two isolated vortices; we will then consider this the outer solution, and construct a boundary layer on the wall. We will assume that all variables have been non-dimensionalized by wall parameters, i.e.- u_τ and ν. Such variables would ordinarily be indicated by a $^+$, but we will suppress this for simplicity of notation. We take as our coordinate system $x_1 = x$ in the direction of the undisturbed free stream, $x_2 = y$ parallel to the surface at right angles to the free stream, and $x_3 = z$ normal to the surface. Suppose, in an inviscid fluid, we have two vortices a distance b from the wall, and 2a apart, with circulations of $+\Gamma$ on the left and $-\Gamma$ on the right (we take + to be counterclockwise looking upstream). This corresponds to a naturally occuring pair; an induced pair will have the opposite signs. Just above the wall, for y large, the velocity parallel to the plate is

$$v \approx -4ab\Gamma/\pi y^3 \tag{58}$$

This may be understood by considering that each vortex has an image under the surface. The velocity in the vortex falls off as $1/y$, and when cancellation due to the image is taken into account, we get $1/y^2$. If both vortices have equal and opposite strengths, again the two pairs cancel to first order, and we are left with $1/y^3$. If the vortices were not of equal (but opposite) strengths, they would not cancel, and we would have behavior of $1/y^2$ at a distance. Unequal strengths are more likely in a naturally occuring pair.

idea is, that another vortex pair at some distance from the first one must exist in this velocity field. Thus, to a first approximation, a knowledge of this velocity field will indicate the strength and nature of the interaction we may expect. See Figure 20.

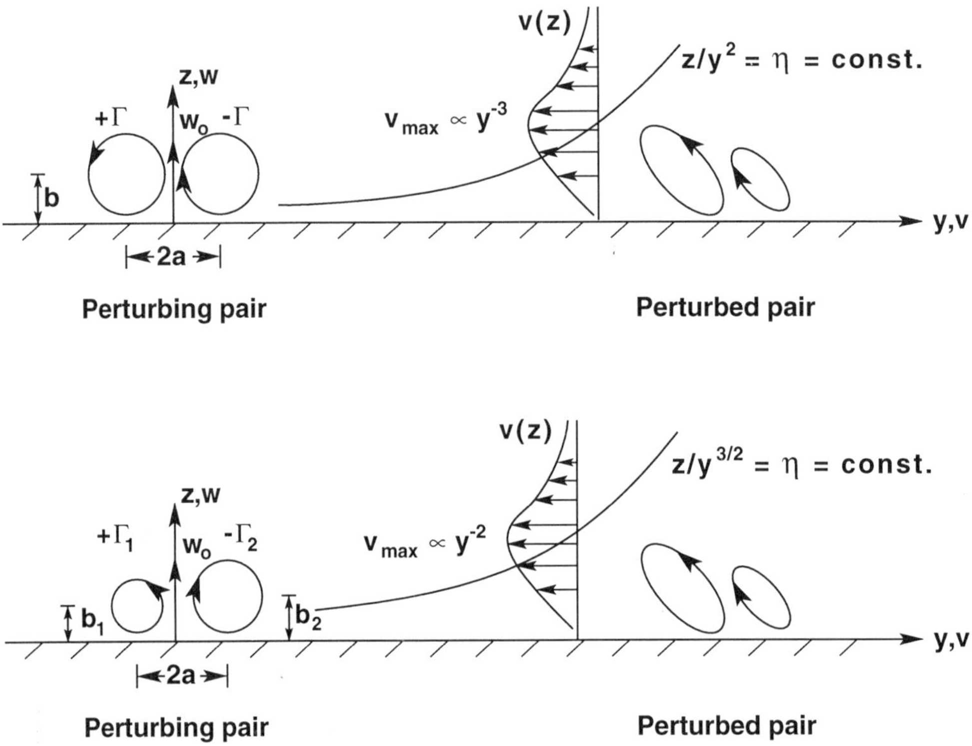

Figure 20: Cartoon of an isolated naturally occurring vortex pair inducing a boundary layer that shears an adjacent naturally occurring pair. The upper figure shows relations when the perturbing pair are of equal strength. If the velocity directions were reversed, the perturbing pair might be an induced pair. The lower figure indicates relations when the perturbing pair are of unequal strength, corresponding to the more usual situation for a naturally occurring pair.

We may anticipate the results of the rudimentary calculations below: the velocity field generated by an isolated vortex pair near the wall, in the vicinity of the wall at some distance from the pair, is essentially a flow toward the pair parallel to the wall (if the pair is naturally occuring; the induced pair rotates in the opposite direction, so that the flow is away from the pair), and is relatively weak. The thickness of the boundary layer on the wall increases rapidly with distance from the pair. Hence, another pair at some distance from the first will see a weak lateral velocity field with a weak shear. To first order, the lateral velocity will simply sweep the second pair slowly toward or away from the first. This would have no influence on the dynamical evolution of the second pair (it simply

If we take this to be the outer solution, and construct an inner solution (also a boundary layer, but internal to the primary boundary layer, and at right angles to it), we obtain the equation (in the standard boundary layer approximation)

$$vv_y + wv_z = -3\beta^2/y^7 + v_{zz} \tag{59}$$

$$v(0) = 0,\ v(\infty) = -\beta/y^3 \text{ where } \beta = 4\Gamma ab/ \tag{60}$$

This problem can easily be shown to have a similarity solution. If we introduce a stream function,

$$v = \psi_z,\ w = -\psi_y \tag{61}$$

then the solution can be written

$$\psi = y^{-1}3^{1/4}\beta^{1/2}g(\eta),\ \eta = 3^{1/4}\beta^{1/2}z/y^2 \tag{62}$$

where g satisfies the equation

$$gg'' - 3g'^2 = -1 + g''' \tag{63}$$

$$g'(\infty) = -1/3^{1/2},\ g'(0) = 0. \tag{64}$$

In the case of unequal vortices, proceeding in exactly the same way, we obtain a boundary layer equation with a term $-2\beta^2/y^5$, with a boundary condition $v(\infty) = -\beta/y^2$. This also has a similarity solution of the form

$$\psi = y^{-1/2}2^{-1/2}\beta^{1/2}f(\eta),\ \eta = 2^{1/2}\beta^{1/2}z/y^{3/2} \tag{65}$$

where f satisfies the equation

$$-f'^2 = -1 + f''' \tag{66}$$

$$f'(0) = 0,\ f'(\infty) = -1 \tag{67}$$

Of course, the equation for $g(\eta)$ must be solved numerically (although $f(\eta)$ can be obtained analytically). However, we may obtain a simple solution, which will be adequate for our purposes, if we neglect the subtantial derivative in the x_2 direction. This is equivalent to reducing the equation for the unperturbed x_2 velocity to $g''' = 1$. Here, the wall shear is not known *a priori*, and it is necessary to make an assumption. We have assumed that the v velocity reaches $-1/3^{1/2}$ at a finite distance from the wall, and that it is a maximum at that point. The distance from the wall at which this happens is $z = 2^{1/2}/3^{1/4}$, which is about 1. This assumption is based on estimates of the velocity field due to the full solution which are beyond the scope of this lecture. The approximate solution becomes:

$$v(y,z) = [3^{1/2}\beta/y^3][\eta^2/2 - 2^{1/2}\eta/3^{1/4}] \tag{68}$$

If we express the vortex strength Γ in terms of the updraft velocity between the vortices at a height b, say $v^+{}_0$, and if we take a = b, which is approximately true, we obtain

$$\Gamma = 5\pi\nu b_+ v^+{}_0/4 \tag{69}$$

Now we will consider the interaction of two vortex pairs. We take a first pair symmetric about the origin, which we will refer to as the perturbing pair, and another pair at some distance in the positive y direction, which we will refer to as the perturbed pair. We wish to determine the effect of the perturbing pair on the bursting behavior of the perturbed pair, due to the v velocity shear generated by the perturbing pair in which the perturbed pair must evolve. We will proceed very much as we did in Aubry *et al* [49], but note that our coordinate axes are not the same as those used by Aubry *et al* [49]. We are using z normal to the plate, whereas they used y.

The spanwise velocity generated by the perturbing pair is given by

$$U_2 = \int \langle u_2 u_3 \rangle \, dx_3 + (3^{1/2}\beta/y^3)(\eta^2/2 - 2^{1/2}\eta/3^{1/4}) \tag{70}$$

where the limits of integration are from the wall, z = 0 to an arbitrary value of z. In dimensionless terms, β is given by

$$\beta = 5\pi w_0 b^3 \tag{71}$$

where b is half the distance between centers of the perturbing pair (equal to the distance from the wall of the perturbing pair), say nominally 25 in wall units, and w_0 is the updraft velocity between the perturbing pair, at the height of their centers. The u_j denote the velocity components of the perturbed vortices.

The first term in the expression for U_2 arises from the modification to the mean shear due to transport by the perturbed vortices themselves. In order to connect the real perturbing pair with the simplified model of the perturbing pair (which consists of a pair of concentrated vortices), we may take the updraft velocity produced between the idealized perturbing pair at the height of their centers and identify this with that produced by the real perturbing pair.

For future reference, we note here the expression for the z derivative of the y velocity:

$$U_{2,3} = \langle u_2 u_3 \rangle + (\beta^{3/2} 3^{3/4}/x_2{}^5)(\eta - 2^{1/2}/3^{1/4}) \tag{72}$$

The Navier Stokes equations in wall variables become

$$u_{i,t} + u_{i,j}U_j + U_{i,j}u_j + u_{i,j}u_j - \langle u_{i,j}u_j \rangle = -p_{,i} + u_{i,jj} \tag{73}$$

We now need an expression for u_i in the perturbed pair, which we represent as in Aubry *et al* [49]

$$u_i = \Sigma a_\kappa \exp[i\kappa x_2]\phi_{i\kappa}(x_3) \tag{74}$$

where the sum is over various values of cross-stream wavenumber κ, which are taken to be positive and negative multiples of L_2. The $a_\kappa(t)$ are time dependent complex variables and the $\phi_{i\kappa}(z)$ are components of empirical eigenfunctions. We need only concern ourselves with the second and third terms in the Navier Stokes equations. We will multiply by exp[-$i\kappa' y]\phi_{i\kappa'}(z)$ and integrate in y from 0 to L_2 and in z from 0 to X_3, the upper edge of the resolved region. All other terms will be the same as in Aubry *et al* [1]. We will need an expression for the Reynolds stress:

$$<u_2u_3> = L_2\Sigma a_\kappa a^*_\kappa \phi_{2\kappa}\phi_{3\kappa}^* \tag{75}$$

where, of course, the a_κ are functions of t, and the $\phi_{i\kappa}$ are functions of z. After projection only values of κ' will remain, so we will drop the prime for simplicity. The projection of the first term becomes

$$L_2 a_\kappa i\kappa \int U_2 \phi_{i\kappa}\phi_{i\kappa}\, dz \tag{76}$$

where the integration is over [0, X_3]. This leads to

$$L_2 a_\kappa i\kappa\{L_2\Sigma a_\lambda a_\lambda^* \int\int \phi_{2\lambda}\phi_{3\lambda}^*\phi_{i\kappa}\phi_{i\kappa}^* dz'\, dz + (3\beta^2/\,2y^7)\int\phi_{i\kappa}\phi_{i\kappa}^* z^2\, dz$$

$$- (6^{1/2}\beta^{3/2}/y^5)\int\phi_{i\kappa}\phi_{i\kappa}^* z\, dz\} \tag{77}$$

Note that, in the terms arising from the cross flow due to the perturbing pair, we are assuming that the perturbed pair is far from the perturbing pair, and that the spatial extent of the perturbed pair is small relative to that distance. This allows us, to a first approximation, to remove the terms in y^n from the integral; the value of y, properly speaking, is that at the centroid of the perturbed pair. The inner integration in the first term is from 0 to z; all others are from 0 to X_3. The first term, of course, is the reduction in the mean velocity shear due to the vertical momentum transport by the eddy in the perturbed pair. This is a familiar mechanism and the form is also familiar, although now we have an integral of the 2-3 eigenfunctions, instead of the 1-3, as in Aubry *et al* [49]. The usual 1-3 term will, of course, appear, arising from the third term in the Navier Stokes equations. This has the form

$$L_2 a_\kappa \int U_{1,3}\phi_{1\kappa}^*\phi_{3\kappa}\, dz + L_2\Sigma a_\kappa \int U_{2,3}\phi_{2\kappa}^*\phi_{3\kappa}\, dz \tag{78}$$

The first term is the usual coefficient, leading to the cubic term representing the reduction in the mean shear in the 1-direction due to the 3-momentum transport by the eddy. The second term is a new term, which consists of three parts:

$$L_2 a_\kappa\{\, L_2\Sigma a_\lambda a_\lambda^* \int\phi_{2\lambda}\phi_{3\lambda}^*\phi_{2\kappa}^*\phi_{3\kappa}\, dz + (\beta^2 3/y^7)\int x_3\phi_{2\kappa}^*\phi_{3\kappa}\, dz$$

$$- (6^{1/2}\beta^{3/2}/y^5)\int\phi_{2\kappa}^*\phi_{3\kappa}\, dz\}. \tag{79}$$

Again, the first of these is, in principle, a reduction of the 2-3 shear by the 3-momentum transport of the perturbed eddies. This would produce a term which would be of the same form as the customary cubic term, and of the same form as that obtained above. They would all untimately be lumped together, serving only to change the values of the coefficients. In fact, in the model with no streamwise variation, we find that $<u_2u_3> \equiv 0$, so that there is no contribution from this term. The second and third terms will make a contribution to the term linear in a_κ, changing the value of the coefficient. Hence, no terms of new form have been introduced in the equations for the a_κ, so that the dynamical equations continue to have the same structure, but the coefficients of the linear terms have been changed.

We can attempt to determine at least the sign of these various coefficients. Very roughly speaking, we will get the right phase relations if we take as an approximation for the components of the leading eigenfunction:

$$\phi_{1\kappa} = -0.5\kappa f \tag{80}$$

$$\phi_{3\kappa} = \kappa f \tag{81}$$

$$\phi_{2\kappa} = if' \tag{82}$$

The forms of $\phi_{2\kappa}$ and $\phi_{3\kappa}$ are dictated by continuity; the form of $\phi_{1\kappa}$ is determined to give the right sign and magnitude to the Reynolds stress. The function $f = f(x_3, \kappa)$ is quadratic near the wall, proportional to $\exp[-\kappa z]$ for large z, and is unimodal. Substituting these crude forms in the integrals allows us to estimate at least the sign and phase of each of the terms. Note first that the terms appear on the right hand side of the equation, and hence with a minus sign. The terms will be of the form (where we indicate an imaginary term as (i) and a real one by (r))

$$a_{\kappa,t} = a_\kappa\kappa\{(r_1) - (i_2)/y^7 + (i_3)/y^5\} + a_\kappa\{-(r_4) + (i_5)/y^7 - (i_6)/y^5\} + \ldots \tag{83}$$

We may write $\mathbf{a} = \mathbf{x} + i\mathbf{y}$, and recall from Aubry *et al*(1988) that the equations for the **x** and **y** components have the forms

$$y_{1,t} = b_1y_1 + c\ldots + y_1\,[d_{11}(x_1^2 + y_1^2) + \ldots + d_{15}(x_5^2 + y_5^2)] \tag{84}$$

(where a subscript 1 designates the first Fourier mode). We need not concern ourselves with the c terms, since our considerations here will not change them. The first and fourth terms in our equation for $a_{\kappa,t}$ (the real ones) are **d** terms, and the others contribute to **b**. The contributions to the **d** terms are identically zero, as noted above. The imaginary terms, however, that contribute to **b**, result in cross-coupling between the real and imaginary components; that is, a term in bx_1 appears in the equation for y_1, while a term in by_1 appears in the equation for x_1. The additional coefficients are of the form

$$x_{1,t} = y_1\{[i(i_5) - \kappa i(i_2)]/y^7 - [i(i_6) - \kappa i(i_3)]/y^5\} + \ldots \tag{85}$$

Hence, at a given distance we may expect the coefficients to change sign as wavenumber increases, and at a given wavenumber we may expect them to change sign as distance increases. The latter effect is an artifact of our simplified solution for g, and can be ignored. The real part of the **b** term results in radial motion in phase space; the addition of the imaginary part to the **b** term results in spiral motion.

We may summarize: raising a bump produces a vortex pair; the vortex pair produces a cross-stream shear; the cross-stream shear produces cross-coupling of the coefficients. Since the original terms in the model are all purely real, the only qualitative change is in the linear coefficients, which now read:

$$a_{\kappa,t} = (\mu_\kappa + i\nu_\kappa)a_\kappa + \dots \tag{86}$$

If the perturbing vortex pair may be actuated at will, then the coefficients ν_κ can in principle be varied at will (although their relative magnitudes are fixed by the form of the eigenfunctions). Qualitatively, since the real coefficient μ_κ is present in the absence of cross-stream shear, weak variations in it due to perturbations will have little influence on the dynamics. In contrast, we can expect the wholly new imaginary term to have a significant effect, and it is terms of this type that we will take as our control variables.

Since we are primarily interested, at this stage in the investigation, in the qualitative effects of the control strategies, we have not attempted to compute numerical values for the additional coefficients such as ν_κ, nor have we considered the effect on the (complex) coefficients of models possessing streamwise variations, which we leave to future studies. The main conclusion of this analysis is that additional linear terms of a qualitatively new type result in our models from the presence of the perturbing vortices. The remainder of this lecture is devoted to a study of how such terms can be used to delay heteroclinic bursting.

5.3. Feedback Control

To illustrate our control strategy, we recall a specific example, the smallest in the hierarchy of models for the turbulent wall region described by Holmes [50], cf. Aubry *et al.* [49]. It has been analyzed in some detail by Armbruster *et.al.* [45], who subsequently showed[23] that it resides on a center-unstable manifold for the O(2)-symmetric Kuramoto-Sivashinsky equation. It is, in fact, the cubic order normal form for an O(2)-symmetric dynamical system near a codimension 2 degenerate bifurcation point, and since such system can be transformed into an equation of this type, the model has relevance considerably beyond the turbulent boundary layer motivation. It takes the form

$$a_{1,t} = \mu_1 a_1 + a_1{}^* a_2 + a_1(e_{11}|a_1|^2 + e_{12}|a_2|^2) \tag{87}$$

$$a_{2,t} = \mu_2 a_2 - a_1{}^2 + a_2(e_{21}|a_1|^2 + e_{22}|a_2|^2) \tag{88}$$

where $a_1(t)$, $a_2(t)$ are the complex time dependent coefficients of two (spanwise) Fourier modes with spatial wave numbers in the ratio 1:2, * denotes complex conjugate and the coefficients e_{ij} and μ_j are real, with the latter thought of as bifurcation parameters. As shown by Armbruster *et.al.* [45], this equation set possesses attracting heteroclinic cycles connecting diametrically opposite points on the circle of saddle points

$$|a_2| = [-\mu_2/e_{22}]^{1/2}, \; a_1 = 0, \qquad (89)$$

provided $\mu_1, \mu_2 > 0 > e_{ij}$ and $\mu_1 - \mu_2 e_{12}/e_{22} < 0$ (these conditions are sufficient, but not necessary). A typical solution spends relatively long periods near the saddle points, followed by rapid heteroclinic transitions to the opposing saddle.

In the absence of external perturbations, the durations between such transitions grow without bound (although finite numerical precision limits them in simulations [49]), but as we have noted, small external noise or other forcing terms should be added to the system to account for the effect of the outer flow in the boundary layer application. When this is done, the duration between transitions equilibrates and a well defined mean emerges.

The analysis of the previous section suggests that we add the parametric control terms $i\nu_j a_j$ to our system. Including random excitation, this leads to the system

$$a_{1,t} = (\mu_1 + i\nu_1)a_1 + a_1 a_2 + a_1(e_{11}|a_1|^2 + e_{12}|a_2|^2) + \varepsilon\xi_1(t) \qquad (90)$$

$$a_{2,t} = (\mu_2 + i\nu_2)a_2 - a_1^2 + a_2(e_{21}|a_1|^2 + e_{22}|a_2|^2 + \varepsilon\xi_2(t) \qquad (91)$$

where the $\xi_j(t)$ are independent, complex-valued Wiener processes with independent and identically distributed real and imaginary parts. For now we set $\varepsilon=0$. Once we have developed a control strategy for the completely deterministic system, we will include the noise terms again.

To select suitable feedback laws $\nu_j = \nu_j(a_1, a_2)$, we transform the deterministic part of our equation set into polar coordinates, thus considerably simplifying the problem. Letting $a_j = r_j \exp[i\theta_j]$, this yields

$$r_{1,t} = r_1 r_2 \cos\phi + r_1(\mu_1 + e_{11}r_1^2 + e_{12}r_2^2) \qquad (92)$$

$$r_{2,t} = -r_2^2 \sin\phi + r_2(\mu_2 + e_{21}r_1^2 + e_{22}r_2^2) \qquad (93)$$

$$\phi_{,t} = 2\nu_1 - \nu_2 - (2r_2 - r_1^2/r_2)\sin\phi \qquad (94)$$

where only the phase difference $\phi = 2\theta_1 - \theta_2$ appears, thus explicitly demonstrating the reduction of dimension possible due to the symmetry. In the absence of control, $\nu_j = 0$, the circle of equilibria reduces to the fixed points $(r_1, r_2, \phi) = (0, \rho, 0), (0, \rho, \pi)$ (where $\rho = [-\mu_2/e_{22}]^{1/2}$) and we see that only the linear combination $2\nu_1 - \nu_2 = \gamma$, say, of the controls appears as a term which, *if it were constant,* would move the equilibria to $(0, \rho, \phi_\gamma)$, where

$$\phi_\gamma = \sin^{-1}[\gamma/2\rho] \tag{95}$$

While a feedback control will not be constant, this "static" observation suggests our strategy, which is similar to that of Ott *et al* [51]. We fix a neighborhood $U = \{r_1 \leq \beta;\ |r_2 - \rho| < \beta_1;\ |\phi - \pi| < \beta_2\}$ of the saddle point $(0, \rho, \pi)$ and monitor recurrent solutions attracted to the heteroclinic cycle. Once in U, the contracting eigendirection transverse to the circle (in the r_2 direction) keeps solutions close to a plane spanned by the ϕ and r_1 directions and, since solutions approach $r_1 = 0$, we need only attempt to control their motion in the ϕ direction, in which they tend to escape. This is done by moving the fixed point according to the rule

$$\gamma(\phi) = 2\rho\sin(\phi_d) \tag{96}$$

where the desired position

$$\phi_d(t) = \pi + k(\phi(t) - \pi) \tag{97}$$

is computed in terms of the current location $\phi(t)$ and $k > 1$ is an adjustable parameter. This translates the fixed point and its stable manifold to the other side of the current position, thus shepherding the solution towards the saddle point: see Figure 21. The mechanism is similar to that of lateral motions of the hand in balancing a vertical stick. Other feedback laws which produce the same effect can, of course, be chosen.

In the absence of noise, a solution entering the feedback sector will remain within the sector and asymptotically approach the saddle point. Heteroclinic bursting in this case would be completely eliminated. If we include the noise terms in our system, however, almost all solutions eventually get "kicked" out of U, primarily near the saddle point where, because of the bilinear structure of the controller, the noise has greater effect than the bounded control input.. Hence, with noise present, the best one can do is delay heteroclinic bursting by holding the state in the feedback sector for as long as possible. This system does not seem amenable to much analysis beyond the geometric interpretations offered above. Consequently, to predict the amount by which bursting can be delayed, it is necessary to examine a simpler model which reproduces the key aspects of the feedback scheme described above and which does permit approximate analysis. This is done in Coller *et al* [52], who predict passage time distributions in good agreement with simulations such as those we now report.

5.4 Numerical simulations

In this section we numerically integrate our equations , computing the control input $v_1 = \gamma(\phi)/2$ (with $v_2 = 0$) at each time step, where $\gamma(\phi)$ is given by the expression above. In the simulations we set the system parameters to $\mu_1 = 0.04$, $\mu_2 = 0.10$, $e_{11} = -4.0$,, $e_{12} = -1.0$, $e_{21} = e_{22} = -2.0$, $\varepsilon = 6.7 \times 10^{-4}$. This implies that $\rho = 0.2236$, and we somewhat arbitrarily set the control gain to $k = 1.5$.

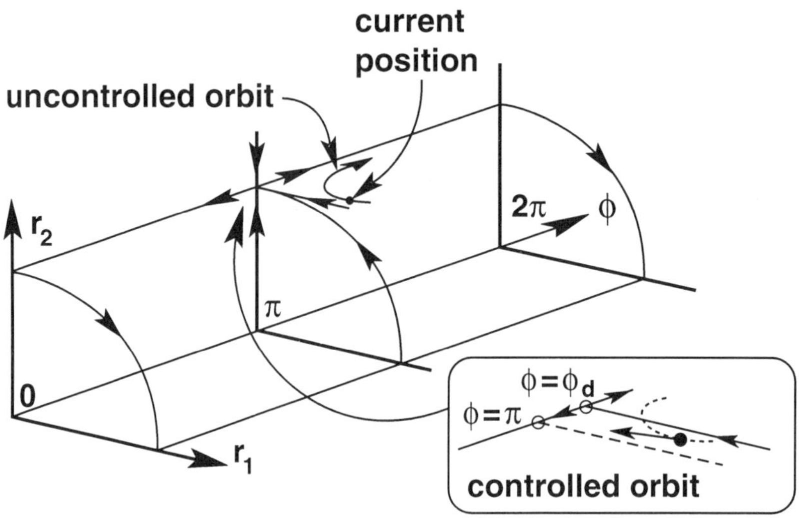

Figure 21: Phase space for our two-(complex) dimensional system, showing the effect of control.

In Figure 22 we show the distribution of passage times through the neighborhood U (in the same figure these data are compared with analytical estimates obtained in Coller *et al* [52], under variation of a parameter not discussed here). Also shown in Figure 22 is the distribution of passage times for the uncontrolled system $\nu_1 = \nu_2 = 0$. The use of control, in this case, dramatically increases the mean time spent in the fixed neighborhood U of the saddle.

Outside U, the solution hugs the heteroclinic connection closely. Via numerical integration, we find the transition time for the particular system described above to be 28.85 time units. When noise is added, we find the standard deviation of these outer transition times to be less than 0.2 percent of the total outer transition time. Hence, with the use of control as described in Section 2, we increase the mean global round-trip time (i.e. the time it takes for a solution, after entering the fixed neighborhood, U, of an equilibrium point, to escape the neighborhood and enter a similar neighborhood of another equilibrium) from 48.4 to 81.9 units of time (when using the maximum amount of control possible without turning the saddle points into sinks), an increase of nearly 70 percent. Since control is only applied locally, the heteroclinic excursions and hence burst intervals themselves, are not altered. Thus, the ratio of interburst to burst times is increased by the same amount. The magnitude of the effect produced is of the same order as that produced by polymer drag reduction.

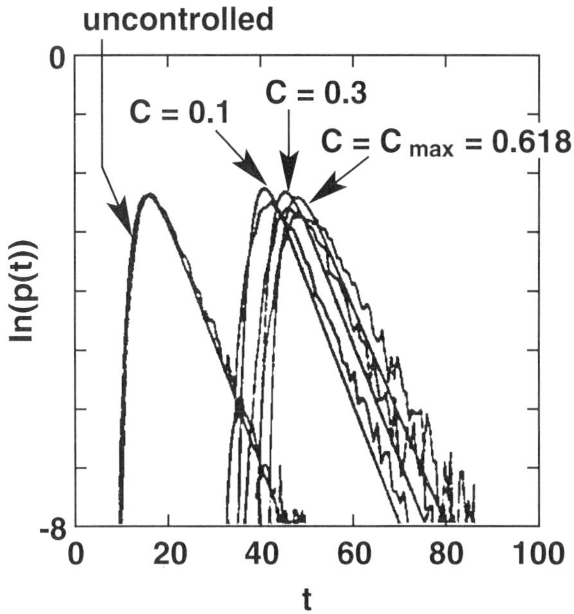

Figure 22: Numerical estimates for passage time distribution through the neighborhood $\mathcal{U}$ in the uncontrolled and controlled cases (jagged graphs). Also indicated on the figure are analytical estimates from Coller *et al* (1993) (smooth curves).

5.5 Conclusions and Implications for Boundary Layer Control.

In this lecture we have described and analyzed a nonlinear feedback control strategy which reduces the bursting rate of a randomly perturbed system possessing heteroclinic cycles connecting saddle points. A rotational vector field in phase space is used to manipulate one of the stable eigenvectors in a manner which retains solutions in the neighborhood of the saddle for substantially longer periods than in the uncontrolled system. This has much in common with the methods for stabilizing periodic orbits in chaotic attractors due to Ott *et al.* [51]. Implementation requires that one sense the current position in phase space and apply control only when solutions enter the relevant neighborhoods. The perturbations imply that almost all solutions eventually escape, at which point control is abandoned and one relies on the global dynamics to return solutions to the neighborhood of a saddle once more.

Motivated by our interest in control of the turbulent boundary layer, we have sketched how actuation of streamwise vortices on either side of naturally occurring vortices leads to additional terms of rotational type in low dimensional dynamical models. Our control analysis focuses on the simplest (two mode) model which captures the key features of the bursting process.

We have not tried to compute the quantitative magnitudes of control vector fields that specific actuators might afford; nor have we considered the costs of our control strategy.

REFERENCES

1. Shih, T.-H., Lumley, J. L. & Janicka, J. 1987. *Second order modeling of a variable density mixing layer. J. Fluid Mech.* 180: 93-116.

2. Zilberman, M., Wygnanski, I. , and Kaplan, R.E. 1977. Transitional boundary layer spot in a fully turbulent environment. *Phys. Fluids Supp.* 20(10): S258-S271.

3. Lumley, J.L. 1967. The structure of inhomogeneous turbulent flows. In *Atmospheric Turbulence and Radio Wave Propagation*, A.M. Yaglom and V.I. Tatarski:, eds.: 166-178. Moscow: Nauka.

4. Loeve, M. 1955. *Probability Theory.* Van Nostrand, New York.

5. Glauser, M.N., Leib, S.J. and George W.K. 1985. Coherent structure in the axisymmetric jet mixing layer. *Proc. of the 5th Symp. of the Turbulent Shear Flow Conf. Cornell University.* (Springer selected lectures from TSF).

6. Moin, P. 1984. Probing turbulence via large eddy simulation. AIAA 22nd Aerospace Sciences Meeting.

7. Herzog, S. 1986. *The large scale structure in the near-wall region of turbulent pipe flow.* Ph.D. thesis, Cornell University.

8. Tennekes, H. and Lumley, J.L. 1972. *A first course in turbulence.* Cambridge, MA: M.I.T. Press.

9. Lumley, J.L. 1970. *Stochastic tools in turbulence.* Academic Press, New York.

10. Lumley, J.L. 1981. Coherent structures in turbulence. *Transition and turbulence*, edited by R.E. Meyer, Academic Press, New York: 215-242.

11. Cantwell, B.J. 1981. Organized motion in turbulent flow. *Ann. Rev. Fluid Mech.* 13: 457-515.

12. Pomeau, Y. and Manneville, P. 1980. Intermittent transition to turbulence. *Comm. Math. Phys.* 74, 189-197.

13. Tresser, C. Coullet, P. and Arneodo, A. 1980. On the existence of hysteresis in a transition to chaos after a single bifurcation. *J. Physics (Paris)* Lett. 41, L243-246.

14. Silnikov, L.P. 1965. A case of the existence of a denumerable set of periodic motions. *Soviet Math. Dokl* 6: 163-166.

15. Silnikov, L.P. 1968. On the generation of a periodic motion from trajectories Doubly asymptotic to an equilibrium State of Saddle type. *Math. U.S.S.R. Sbornik* 6, 427-438.

16. Silnikov, L.P. 1970. A contribution to the problem of the structure of an extended neighborhood of a rough equilibrium state of saddle-focus type. *Math U.S.S.R Sbornik* 10 (1), 91-102.

17. Tresser, C. 1984. About some theorems by L.P. Silnikov. *Ann de l'Inst. H. Poincaré,* 40: 441-461.

18. Sparrow, C.T. 1982. *The Lorenz Equations: Bifurcations, Chaos and Strange Attractors*. Springer-Verlag, New York, Heidelberg, Berlin.

19. Guckenheimer, J. and Holmes P. J. 1983. *Nonlinear Oscillations, Dynamical Systems and Bifurcations of Vector Fields*. Springer-Verlag, New York. Corrected second printing. 1986.

20. Bergé, P., Dubois, M. Manneville, P. and Pomeau, Y. 1980. Intermittency in Rayleigh-Bénard convection. *J. Phys. Lett* 41 L341.

21. Maurer, J. and Libchaber, A. 1980. Effects of the Prandtl number on the onset of turbulence in liquid helium. *J. Phys. Paris Lett.* 41, L-515-.

22. Dubois, M., Rubio, M.A. and Bergé, 1983. Experimental evidence of intermittencies associated with a subharmonic bifurcation. *Phys. Rev. Lett.* 51, 1446-1449.

23. Kline, S.J., Reynolds, W.C., Schraub, F.A. and Rundstadler, P.W. 1967. The structure of turbulent boundary layers. *J. Fluid Mech.* 30(4): 741-773.

24. Corino, E. R., and Brodkey, R.S. 1969. A visual investigation of the wall region in turbulent flow. *J. Fluid Mech.* 37(1):1-30.

25. Smith, C.R. and Schwarz, S.P. 1983. Observation of streamwise rotation in the near-wall region of a turbulent boundary layer. *Phys. Fluids* 26(3) 641-652.

26. Kubo, I. and Lumley, J. L. 1980. *A study to assess the potential for using long chain polymers dissolved in water to study turbulence.* Annual Report, NASA-Ames Grant No. NSG-2382. Ithaca, NY: Cornell.

27. Lumley, J. L. and Kubo, I. 1984. Turbulent drag reduction by polymer additives: a survey. In *The Influence of Polymer Additives on Velocity and Temperature Fields.* IUTAM Symposium Essen 1984. Ed. B. Gampert. pp. 3-21. Springer Berlin/Heidelberg:

28. Lumley, J. L. 1971. Some comments on the energy method. In *Developments in Mechanics 6*, L. H. N. Lee and A. H. Szewczyk, eds. Notre Dame, IN: N. D. Press.

29. Reynolds, W. C. & Hussain, A. K. M. F. 1972. The mechanics of an organized wave in turbulent shear flow. Part 3. Theoretical models and comparisons with experiment. *J. Fluid Mech.* 54: 263-287.

30. Liu, J. T. C. 1988. Contributions to the understanding of large scale coherent structures in developing free turbulent shear flows. *Advances in Applied Mechanics* 26: 183-309.

31. Brereton, G. J. & Kodal, A. 1992. A frequency domain filtering technique for triple decomposition of unsteady turbulent flow. *J. Fluids Engineering.* To appear.

32. Berkooz, G. 1991b. *Turbulence, coherent structures and low dimensional models.* Ph.D. Thesis, Cornell University.

33. Joseph, D. D. 1973. *Stability of Fluid Flow.* Berlin etc.: Springer.

34. Reynolds, W. C. and Tiederman, W. G. 1967. Stability of turbulent channel flow with application to Malkus' theory. *J. Fluid Mech.*, 27: 253-272.

35. Naot, D. Shivit, A. and Wolfshtein. Interactions between components of the turbulent velocity correlation tensor. *Isreal J. Tech.,* 8:259, 1970.

36. Moin, P. and Moser, R.D. 1989. Characteristic-eddy decomposition of turbulence in a channel. *Journal of Fluid Mechanics*,200:471-509.

37. Aubry, N., Holmes, P., Lumley, J.L., & Stone, E. 1988 The Dynamics of Coherent Structures in the Wall Region of a Turbulent Boundary Layer. *J. Fluid Mech.* 192, 115-173.

38. Berkooz, G., Elezgaray, J. & Holmes, P. 1992. Coherent structures in random media and wavelets. *Physica D.* in press.

39. Hyman, J. M. & Nicolaenko, B. and Zaleski, S. 1986. Order and complexity in the Kuramoto Sivashinsky model of weakly turbulent interfaces. *Physica D* 18: 113.

40. Zaleski, S. 1989. A stochastic model for the large scale dynamics for some fluctuating interfaces. *Physica D* 34: 427-438.

41. Pomeau, Y., Pumir, A. & Pelce, P. 1984. Intrinsic stochasticity with many degrees of freedom. *J. Stat. Phys* 37: 39-49.

42. Toms, B.A. *Proc 1st Inter. Congr. Rheol.,* 2, 135-41 North Holland, Amsterdam, 1948.

43 Lumley, J. L. 1973. Drag reduction in turbulent flow by polymer additives. *J. Polymer Sci..: Macromolecular Reviews.* 7: 263-290.

44. Lumley, J.L. & Sarkar, S. 1987 Angular dispersion of material lines in isotropic turbulence. *Phys. Fluids*, 30 (5), pp.1269-71.

45. D. Armbruster, J. Guckenheimer & P. J. Holmes, "Heteroclinic cycles and modulated travelling waves in systems with {O}(2) symmetry", *Physica D*, 29, 257(1988).

46. N. Aubry, J. L. Lumley & P. J. Holmes, "The effect of modeled drag reduction on the wall region" *Theoret. Comput. Fluid Dynamics,* 1, 229 (1990).

47. A. M. Bloch & J. E. Marsden, "Controlling homoclinic orbits", *Theoret. Comput. Fluid Dynamics*, 1, 179 (1989).

48. J. W. Jung, N. Mangiavacchi & R. Akhavan, "Suppression of turbulence in wall-bounded flows by high-frequency spanwise oscillations", *Phys. Fluids A*, 4, 1605 (1992).

49. N. Aubry, P. Holmes, J. L. Lumley & E. Stone, "The dynamics of coherent structures in the wall region of a turbulent boundary layer," *J. Fluid Mech.* 192, 115 (1988).

50. P. J. Holmes, 1989. "Can dynamical systems approach turbulence?", *Whither Turbulence? Turbulence at the crossroads.* Lecture Notes in Physics Vol. 357 edited by J. L. Lumley. (Springer, Berlin, 1989).

51. E. Ott, C. Grebogi & J. A. Yorke, "Controlling chaos", *Phys. Rev. Lett.*, 64, 1196 (1990).

52. B. D. Coller, P. J. Holmes & J. L. Lumley, "Control of noisy heteroclinic cycles", *Physica D* 72 135-160 (1994).

DIRECT AND LARGE-EDDY SIMULATIONS OF COHERENT VORTICES IN THREE-DIMENSIONAL TURBULENCE:

GEOPHYSICAL AND INDUSTRIAL APPLICATIONS

O. Métais

Polytechnical Institut of Grenoble, France

and

University J. Fourier, Grenoble, France

Abstract:

We first discuss the concept of coherent vortex. Some elementary vortex interactions which appear in transitional or turbulent shear flow are described such as the formation of spiral vortices, and the pairing of two vortices of same sign. The three-dimensional stretching and turning mechanisms leading to longitudinal hairpin vortices (aligned or staggered mode) are presented. Then the formalism of large-eddy simulations (LES) is recalled in details and classical as well as more recent subgrid-scale models are considered: Smagorinsky's model, Kraichnan's spectral eddy-viscosity model, structure-function models, and the dynamic model. Various examples of numerically-simulated three-dimensional flows and of coherent vortices imbedded in them are displayed such as three-dimensional isotropic turbulence, free-shear flows, separated flows and boundary-layers. Comparisons with experimental data are presented and geophysical and industrial applications are discussed. We finally show how a solid-body rotation may drastically modify the free-shear layer topology. The shear/Coriolis linear instability yields a purely longitudinal mode for anticyclonic moderate rotation. Numerical simulations show how this mode evolves non-linearly into concentrated longitudinal hairpin vortices of absolute vorticity.

1 Introduction

1.1 Coherent vortices identification

It has become more and more clear that developed turbulent flows contain a lot of coherent vortices of various sizes, which play a major role in all heat and mass transfer processes. These vortices, which correspond to local concentrations of vorticity $\vec{\omega} = \vec{\nabla} \times \vec{u}$ in space, are present in isotropic two- and three-dimensional turbulence, and in free-shear flows like mixing layers, wakes, jets, separated flows. They can also be found in flows close to boundaries, such as boundary layers, channel flows or pipe flows. To define a coherent vortex, it is important to include the notion of time evolution (see e.g. Lesieur & Métais, 1994; Lesieur, Comte & Métais, 1994)

Lets consider at some time a region of space with a local concentration of vorticity ω, and of recognizable shape. This region is supposed to move at a velocity U_c. It will be said to be a coherent vortex if it has a life time T_c much larger than the local turnover time within the eddy, ω^{-1} (let's say, $\approx 5\ \omega^{-1}$). We follow the isolated vortex in a Galilean frame moving with a constant velocity of the order of U_c. Consideration of the order of magnitude of the different terms of the equation of motion (in the constant-density case, and neglecting viscosity) shows that the vorticity $\vec{\omega}$, the velocity $\vec{u}$ (induced by the vortex) and the generalized dynamic pressure P satisfy, in a coherent vortex, the cyclostrophic balance

$$\vec{\omega} \times \vec{u} = -\frac{1}{\rho} \vec{\nabla} P \quad . \tag{1}$$

This shows that the pressure gradient is oriented towards the exterior of the vortex: the centre of the coherent structure is a pressure trough.

The same phenomenology can be explained by considering the Poisson equation for the pressure in a constant density flow. Bradshaw & Koh (1981) have written this equation under the following form:

$$2\frac{\Delta P}{\rho} = \omega^2 - S^2 \tag{2}$$

where ω^2 is the square of the vorticity vector magnitude and S^2 is twice the square of the strain-rate tensor:

$$S^2 = \frac{1}{2}\left(\frac{\partial \bar{u}_i}{\partial x_j} + \frac{\partial \bar{u}_j}{\partial x_i}\right)\left(\frac{\partial \bar{u}_i}{\partial x_j} + \frac{\partial \bar{u}_j}{\partial x_i}\right) \ , \tag{3}$$

The physical consequences of eq. (2) is that the pressure minima will coincide with space regions of high vorticity and low strain (vortex cores) when the pressure maxima will be situated at the saddle points inbetween the vortices (see Brachet, 1991).

Low pressure regions will then be extensively used in the present notes as an efficient way of coherent vortices identification. Vortices will also be characterized with vortex line concentrations on the basis of Kelvin's theorem, which states that, in the limit of perfect barotropic fluid, vortex tubes and filaments are material and move with the particules they contain.

In section 5, the effect of a solid-body rotation of angular velocity $\vec{\Omega}$ on turbulence will be considered. In that case, the coherent vortex satisfies

$$(\vec{\omega} + 2\vec{\Omega}) \times \vec{u} = -\frac{1}{\rho}\vec{\nabla}P \quad . \tag{4}$$

If rotation $\vec{\Omega}$ is fast, which can be measured in terms of the local Rossby number $R_o << 1$, with

$$R_o = |\frac{\omega}{2\Omega}| \tag{5}$$

eq. (4) is approximated by the geostrophic balance

$$2\vec{\Omega} \times \vec{u} = -\frac{1}{\rho}\vec{\nabla}P \quad , \tag{6}$$

which replaces the previous cyclostrophic balance. In this case, cyclonic vortices (of vorticity parallel and of same sign as $\vec{\Omega}$) will correspond to pressure lows, while anticyclonic vortices will be pressure highs. In fact, eq. (4) indicates that, for anticyclonic vortices (still parallel to $\vec{\Omega}$), the bifurcation from low to high pressures when the rotation rate is increased occurs at a local critical Rossby number of one. Cyclonic vortices are depressions whatever the value of the Rossby number.

1.2 Direct numerical simulations (D.N.S.) and large-eddy simulations (L.E.S.)

Direct-numerical simulations of turbulence (DNS) consist in solving explicitly all the scales of motion, from the largest l_I to the Kolmogorov dissipative scale l_D. It is well known from the statistical theory of turbulence that l_I/l_D scales like $R_l^{3/4}$, where R_l is the large-scale Reynolds number $u'l_I/\nu$ based upon the rms velocity fluctuation u'. Therefore, the total number of degrees of freedom necessary to represent the whole span of scales of a three-dimensional turbulent flow is of the order of $R_l^{9/4}$ in three dimensions. With the presently available computers, the DNS are then limited to Reynolds numbers which are several orders of magnitude smaller than those encountered in the atmosphere, the ocean or most of the industrial facilities. For a weakly viscous fluid, it is, therefore, not possible in the near future to simulate explicitly all the scales of motion from the smallest to the largest. Generally, we are interested in the description of the largest scales of the flow only since those are often the most energetic. This is no longer a direct-numerical simulation of the turbulence, but a large-eddy simulation (LES). These LES need the representation of the energy exchanges with the small scales which are not explicitly simulated, called the subgrid-scales. Thus, subgrid-scale models have to be developed. The first of these models was proposed by Smagorinsky (1963) for numerical studies related to atmospheric flows. Since then, modern methods of large-eddy simulations have been developed, and will be presented in these notes.

2 Vortex dynamics

2.1 Vortex interactions

Lets first consider the two-dimensional vortices resulting from the well-known Kelvin-Helmholtz instability. For the temporal mixing layer the velocity profile is of hyperbolic-tangent form, i.e.,

$$\bar{u}(y) = U_0 \tanh \frac{2y}{\delta_i} \ , \tag{7}$$

where $2U_0$ is the velocity difference accross the layer, and δ_i the vorticity thickness. Since $d\bar{u}/dy$ presents an extremum within the flow, the mixing layer is unstable from an inviscid point of view. This is a consequence of the Rayleigh inflection-point inviscid-instability criterion (see e.g. Drazin & Reid, 1981). In the viscous case, the Orr-Sommerfeld equation has to be solved. For a given Reynolds number, there exists a range of unstable longitudinal modes α. It may be shown that their amplification rate is maximum for a wave number α_a, called the most-amplified wave number. At low-Reynolds number R_e, its amplification rate is affected by viscosity. This is, however, no longer the case, when R_e exceeds values of the order of $30 \sim 40$. The most amplified wave number, calculated by Michalke (1964), is then given by

$$\alpha_a = 0.88 \ \delta_i^{-1} \quad , \tag{8}$$

corresponding to a spatial wave-length

$$\lambda_a = 7 \ \delta_i \quad . \tag{9}$$

The first stage of the instability yielding spiralling Kelvin-Helmoltz vortices has been described by Lesieur (1990). The subsequent two-dimensional evolution of the mixing-ayer can be investigated by means of direct-numerical simulations. The figure 1 taken from Comte (1989) present a two-dimensional computation. The temporal assumption is performed, that is to say that flow periodicity is assumed in the mean flow direction (longitudinal direction). The length of the computational domain is equal to 8 fundamental wave lengths given by eq. 9. The vorticity iso-contours clearly show that the vortices correspond to high-vorticity regions. The calculations also demonstrate that these are pressure lows. Two-pairs of the originally formed Kelvin-Helmoltz vortices undergo a pairing. The first pairing occurring in the row of primary vortices may be understood in terms of a subharmonic instability of wave length $2\lambda_a$ (see e.g. Lesieur, 1990). This pairing, where the vortices rotate about each other and eventually pair, is due to very simple velocity induction mechanisms: each vortex creates irrotational motion in its environment, and tends to entrain the other around it, so that the vortices rotate about each other. Since the irrotational motion induced a distant r apart from each vortex has an angular velocity $\propto r^{-2}$, the outer part of each vortex turns slower than the inner part, and both vortices develop a tail. During the pairing, these tails mix into spirals.

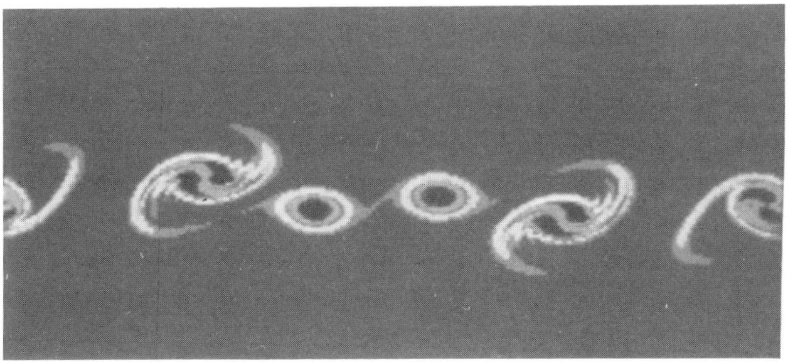

FIG. 1 - Direct-numerical simulation of the vorticity in an incompressible mixing-layer (from Comte, 1989).

We must point out that in two-dimensional numerical simulations of spatially-growing mixing-layer (Normand *et al.*, 1988) as well as in the laboratory (see e.g. Winant & Browand, 1974), the pairing between vortices is also observed if we follow them with the average motion.

Another interesting two-dimensional vortex interaction is the dipole: when two vortices of opposite vorticity come close to each other, they will tend to entrain mutually under their induced velocity, and will travel together. These "Batchelor's couples" (see Batchelor, 1967) form for instance in two-dimensional isotropic turbulence, and also in wakes, as was shown experimentally by Couder (1984). Evidence of dipole formation has also been found in various numerical simulations (see e.g. Fouillet, 1991, for the weakly-compressible plane wake; Couder & Basdevant, 1986, for the incompressible plane wake).

2.2 Vorticity stretching and turning

In two-dimensional incompressible flows, vorticity may not increase. In three dimensions, however, it may be amplified through vortex stretching mechanisms. In particular, in shear flows, three-dimensional instabilities of the vortex filaments, when submitted to a weak vertical oscillation in the spanwise direction, are stretched longitudinally by the basic shear. This may occur in any type of basic velocity (inflexional, constant shear, non-inflexional) and yields longitudinal hairpin vortices of opposite sign. In regions where the vorticity is small in front of the deformation, such as the stagnation regions between spiral vortices, one can assume that the vorticity stretching has a negligible effect on the deformation. It can then be shown that the vortex filaments are stretched exponentially along the principal axis of deformation of the basic shear, that is, 45^0 with respect to the horizontal plane defined by the streamwise and spanwise directions (see Lesieur & Métais, 1994). It explains why longitudinal hairpin vortices are stretched between the basic vortices (see e.g. Lin & Corcos, 1984). This is observed in numerous laboratory experiments of mixing layers (see e.g. Konrad, 1976, Breidenthal, 1981, Bernal & Roshko, 1986), jets (Lasheras & Meiburg, 1990) or cavitating wakes (Franc *et al.*, 1982). The velocity induced by the hairpin will pump up fluid between the legs of the "peak", and down in the "valley" (this terminology is due to Klebanoff *et al.*, 1962). Therefore, and since the lower fluid is slower than its upper counterpart (due to the shear), the peak will correspond (speaking of the longitudinal velocity) to a "low-speed streak", and the valley to a "high-speed streak".

The way the initial or upstream perturbed vortex filaments are organized is essential in determining the further vortex topology. Basically, two distinct organizations can be distinguished. On the one hand, if the vortex filaments are in phase in the streamwise direction, they give rise to hairpins will have a global in-phase sinuous oscillation: we will call it the aligned mode following the terminology due to Klebanoff *et al.* (1962), who observed an aligned pattern of hairpin-vortices (also called Λ-vortices) in the forced transitional boundary layer. In this situation where the hairpin vortices are in phase, very elongated low and high-speed streaks are observed. This corresponds also to the translative instability (Pierrehumbert & Widnall, 1982) in the context of a mixing layer. On the other hand, and if the initial spanwise perturbation is out of phase, the flow will give rise to hairpin vortices staggered in the streamwise direction. It corresponds to the so-called staggered mode in the boundary layer (Herbert, 1988), and is found also in the context of the mixing layer (see below). Now, the short low-speed streaks between the legs of the hairpins are out of phase, and should reconnect longitudinally, giving rise to long oscillating low speed streaks. This was observed both experimentally in a turbulent boundary layer by Kline *et al.* (1967), and numerically in a plane channel flow by Moin & Kim (1982).

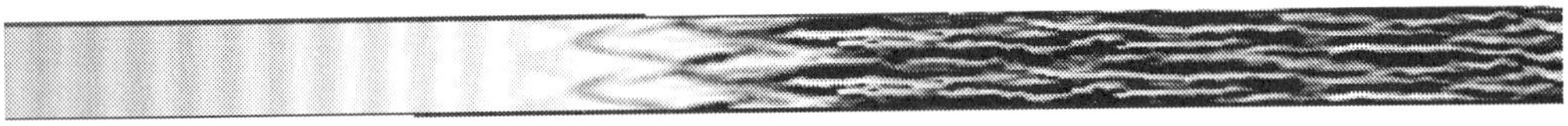

FIG. 2 - LES of a spatially-developing boundary layer at Mach 0.5 above a flat plate: horizontal slice of the streamwise-velocity-fluctuation field (from Ducros, 1995).

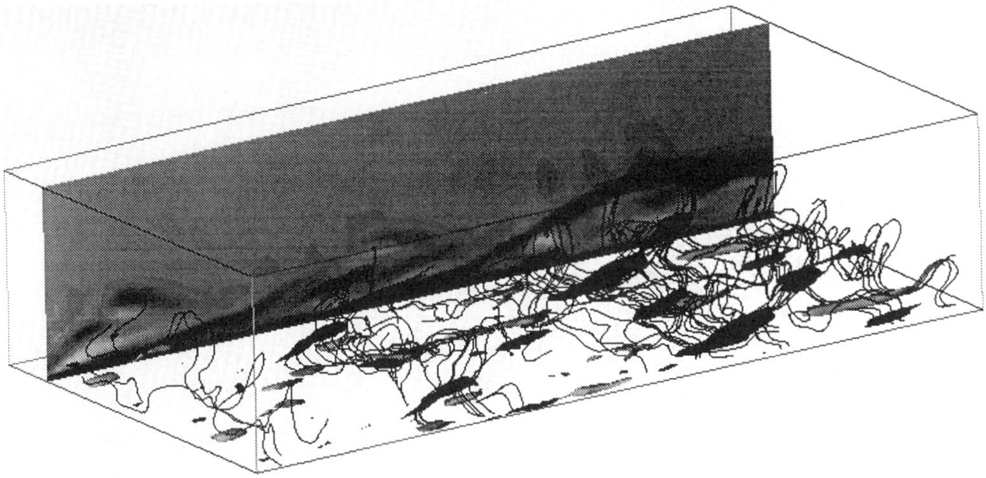

FIG. 3 - Same as Figure 2: Zoom on the turbulent region. Isosurfaces of streamwise vorticity (light grey and dark grey), plus vortex lines showing hairpin vortices. A vertical slice shows iso-ω_z contours (spanwise vorticity), to visualize the ejections (from Comte *et al.*, 1994).

Now, we will show that developed turbulence in shear flows preferentially organizes itself into the staggered mode since turbulence is more likely to provide initial perturbations, which are not totally in phase in the longitudinal direction. Figures 2 and 3 show a large-eddy simulation based upon the filtered structure-function model (see section 3.3.2.) of a spatially-developing boundary layer at Mach 0.5 performed by Ducros(1995) (see also Comte *et al.*, 1994). The Reynolds number, based on the upstream displacement thickness, is of 1000. The resolution is of $650 \times 32 \times 20$ grid points. A laminar profile is imposed at the inflow boundary, with two types of perturbations: a deterministic forcing at the fundamental mode (2D Tollmien-Schlichting waves of amplitude $8\ 10^{-3}\ U_\infty$) plus three-dimensional white noise of amplitude $5\ 10^{-3}\ U_\infty$.

Figure 2 displays the whole computational domain. The two-dimensional T.S waves can be seen in the upstream part of the domain. Further downstream, three-dimensionality first appears in the form of Λ-shaped vortices. Their staggered arrangement is clear. It results from the growth of oblique waves of streamwise period double that of the T.S. modes, that is, a sub-harmonic mode. Small-scale turbulence appears first at the tip of the Λ-vortices, where the streamwise velocity profile is locally inflexional. This causes the development of a Kelvin-Helmholtz-type instability of growth-rate much larger than that of the T.S waves, yielding very fast breakdown into turbulence. Elongated streaks are present downstream. Figure 3 is a zoom on the turbulent region, one clearly sees clearly the spatial correlation between the ejections and the head of the Λ-shape vortices, which correspond to vortex lines concentration.

2.3 Sensitivity to initial conditions

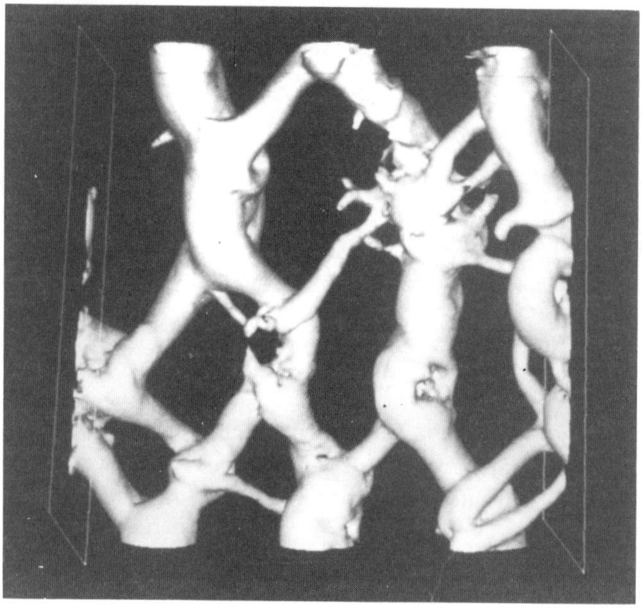

FIG. 4 - Top-view of the low-pressure in the LES of a mixing layer forced by a 3D random perturbation, showing a helical pairing (courtesy, J. Silvestrini, Grenoble).

Three-dimensional direct numerical simulations performed by Comte *et al.* (1992) at a Reynolds number of 100 have demonstrated the importance of the initial conditions nature on the further development of the mixing-layer coherent vortices. If a (small) quasi-two-dimensional perturbation is superposed upon the basic hyperbolic-tangent velocity profile, then quasi-two-dimensional large coherent Kelvin-Helmoltz vortices form and merge. They exhibit slight spanwise oscillations as in the translative instability proposed by Pierrehumbert & Widnall (1982). Between the big rollers, weak longitudinal hairpin vortices are stretched. On the opposite, for a three-dimensional perturbation (still of low amplitude), the growth of unstable oblique modes yield local roll-ups and pairings occuring in only certain regions along the span. This gives rise to a large-scale vortex lattice similar to that experimentally observed by Nygaard & Glezer (1990). This is shown on figure 4, presenting the results of a large eddy simulation of a temporal mixing-layer with four most-amplified Kelvin-Helmoltz wave-lengths initially, and using one of the subgrid-model presented in section 3.2 (the spectral-cusp model). Here, pseudo-spectral numerical methods (collocation methods; see Orszag, 1971), are used to solve the governing equations of motion. A top-view of the low pressure field is displayed here, which, as stressed above, is a very good indicator of vortices. This constitutes a numerical evidence of Pierrehumbert & Widnall's helical pairing instability. This three-dimensional structure is accompanied by an important production of longitudinal vorticity due to intense stretching mechanisms.

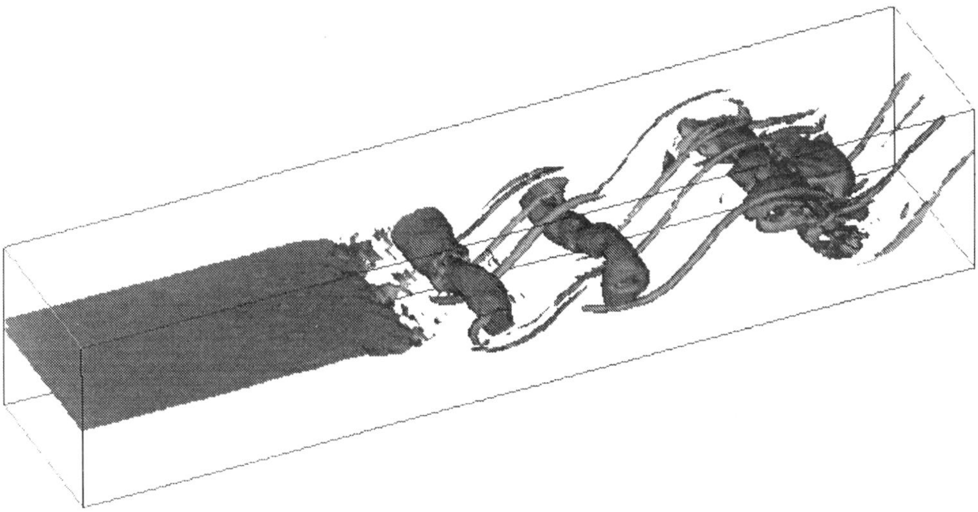

FIG. 5 - Large-eddy simulation of a spatially developing mixing-layer: structure-function model, quasi-two-dimensional initial conditions. Vorticity modulus iso-surface $\omega = 0.6\omega_i$ (courtesy J. Silvestrini, Grenoble).

We next present large-eddy simulations of spatially growing mixing-layer using the structure-function model (see section 3.3) and investigate the influence of initial perturbations. Its consists of two flows of parallel but different velocities (of same sign) U_1 and U_2, separated by a splitter plate upstream, in a configuration similar to the experiment by Brown & Roshko (1974). These computations have been performed by

J. Silvestrini (1994, private communication, Grenoble). In order to simulate this flow we have developed a three-dimensional numerical code combining, for the space derivatives computation, high-order difference schemes in the longitudinal direction and pseudo-spectral methods in the spanwise and shear directions (see Gonze, 1993). Compact difference schemes at the sixth order are used. Their precision is close to spectral methods. The resolution is $n_x, n_y, n_z = 240, 48, 48$. The trailing edge of the splitter plate upstream is approximated by a sharp hyperbolic-tangent mean velocity profile,

$$\bar{u}(y) = \frac{U_1 - U_2}{2} \tanh \frac{2y}{\delta_i} + \frac{U_1 + U_2}{2} , \tag{10}$$

Two random perturbations are superposed on it: the first one is two-dimensional (independent of the spanwise direction) of kinetic energy $\epsilon_{2D} \left(\frac{U_1 - U_2}{2}\right)^2$, and the second one is three-dimensional of kinetic energy $\epsilon_{3D} \left(\frac{U_1 - U_2}{2}\right)^2$, with $\epsilon_{2D} = 4 \times 10^{-4}$ and $\epsilon_{3D} = 4 \times 10^{-5}$. The upstream perturbation is therefore quasi-bidimensional. Figure 5 shows an iso-surface of the vorticity modulus, whose threshold is equal 60% of the upstream maximum spanwise vorticity ω_i. The present simulation exhibits features similar to the quasi-two-dimensional low-Reynolds number simulation by Comte *et al.* (1992): formation of quasi-two-dimensional Kelvin-Helmoltz vortices and quasi-two-dimensional pairing. However, as opposed to the direct numerical simulation of Comte *et al.* (1992), in which the maximum longitudinal vorticity was approximately equal to $0.1\omega_i$, here viscous effects do not prevent longitudinal vorticity stretching; for the present computational length, the longitudinal vorticity maximum is of the order of $2\omega_i$ and very neat longitudinal vortices are visible. They look strikingly like the sketch of the same vortices drawn by Bernal & Roshko (1986) from their experimental measurements.

We next consider a three-dimensional perturbation corresponding to $\epsilon_{3D} = 4 \times 10^{-4}$ (and $\epsilon_{2D} = 0$). Figure 6 visualizes the low-pressure regions. These show both the Kelvin-Helmoltz rolls and Λ-shaped vortices stretched in-between. The strength of these hairpin vortices is slightly higher than in the quasi-twodimensional case. The Kelvin-Helmoltz rolls are highly distorted in the spanwise direction. The spanwise width of the computational domain is, however, to narrow to allow for an helical pairing to take place; the width is equal to approximately twice the spacing between consecutive big rollers. Higher resolution computations with extended spanwise length should then be performed.

3 Large-eddy simulations: formalism

The reader is refered to the review article by Lesieur & Métais (1995) for more details about recent developments on LES of turbulent flows. As stressed above, in order to increase the Reynolds number in the numerical simulations, it is necessary

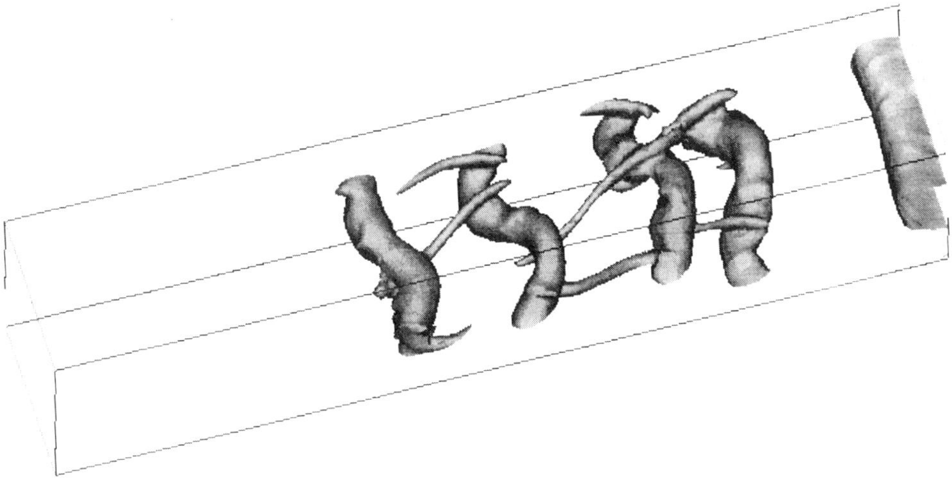

FIG. 6 - Large-eddy simulation of a spatially developing mixing-layer: structure-function model, three-dimensional initial conditions. Low-pressure iso-surface (courtesy J. Silvestrini, Grenoble).

to introduce a subgrid-scale model representing the action of scales smaller than the computational mesh Δx upon the explicitely resolved scales. The formalism of the large-eddy simulations for incompressible flows is the following: one considers a spatial filter G of width Δx, which filters out the *subgrid-scales* of wavelength $< \Delta x$. Thus the filtered field is defined as

$$\bar{u}_i(\vec{x}, t) = \int u_i(\vec{y}, t) G(\vec{x} - \vec{y}) d\vec{y} \quad , \tag{11}$$

and the subgridscale field is the departure of the actual flow with respect to the filtered field:

$$u_i = \bar{u}_i + u'_i \quad . \tag{12}$$

Because of the commutative properties of the filter and the derivatives, the continuity equation $\partial \bar{u}_j / \partial x_j = 0$ for the filtered field still holds. One considers now the incompressible Navier-Stokes equations in a frame rotating with constant angular velocity Ω about the x_3 axis and we apply the filter G to the equations. The equations are written here for a constant density flow. We then obtain:

$$\frac{\partial \bar{u}_i}{\partial t} + \frac{\partial}{\partial x_j}(\bar{u}_i \bar{u}_j) = -\frac{1}{\rho_0} \frac{\partial \bar{p}}{\partial x_i} + 2\epsilon_{ij3} \Omega \bar{u}_j + \frac{\partial}{\partial x_j} \left\{ \nu \left(\frac{\partial \bar{u}_i}{\partial x_j} + \frac{\partial \bar{u}_j}{\partial x_i} \right) + T_{ij} \right\} \tag{13}$$

T_{ij} is the subgrid-scale stress:

$$T_{ij} = -\left(\overline{u'_i u'_j} + \overline{\bar{u}_i u'_j} + \overline{u'_i \bar{u}_j} + \overline{\bar{u}_i \bar{u}_j} - \bar{u}_i \bar{u}_j\right) \quad , \tag{14}$$

Similarly to the Reynolds equations, an eddy-viscosity assumption is often performed for modelling the subgrid fluxes, namely:

$$T_{ij} = \nu_t \left(\frac{\partial \bar{u}_i}{\partial x_j} + \frac{\partial \bar{u}_j}{\partial x_i}\right) + \frac{1}{3} T_{ll}\, \delta_{ij} \quad . \tag{15}$$

We here present several commonly used subgrid-scale models. Most of these are based upon an analogous mixing length assumption consisting in writing $\nu_t \sim \Delta x\ v_{\Delta x}$, where $v_{\Delta x}$ is a characteristic velocity scales of the motions of size $\approx \Delta x$. Furthermore, the models generally depend upon one constant which can be analytically derive if it is assumed that the cut-off spectral wavenumber $k_C = \pi/\Delta x$ is located within a Kolmogorov kinetic energy cascade such that:

$$E(k) = C_K \epsilon^{2/3} k^{-5/3} \tag{16}$$

where $E(k)$ is the kinetic energy spectrum and C_K is the Kolmogorov constant.

3.1 Smagorinsky's model

The first of these subgrid models was proposed by Smagorinsky (1963) for numerical studies related to a two-layer quasi-geostrophic atmosphere. It consists in writing

$$\nu_t \sim \Delta x\ v_{\Delta x} \sim (\Delta x)^2 \frac{\partial v}{\partial x} = (C_S \Delta x)^2 \bar{S}^{1/2} \quad , \tag{17}$$

where $v_{\Delta x}$ is a characteristic velocity difference on the computational grid mesh, determined through the local strain

$$\bar{S} = \frac{1}{2}\left(\frac{\partial \bar{u}_i}{\partial x_j} + \frac{\partial \bar{u}_j}{\partial x_i}\right)\left(\frac{\partial \bar{u}_i}{\partial x_j} + \frac{\partial \bar{u}_j}{\partial x_i}\right) \quad . \tag{18}$$

For $\pi/\Delta x$ situated within an isotropic Kolmogorov cascade, it is obtained:

$$C_S = \frac{1}{\pi}\left(\frac{3C_K}{2}\right)^{-3/4} \sqrt{\frac{< \bar{S} >^{3/2}}{\left\langle \bar{S}^{3/2} \right\rangle}} \quad , \tag{19}$$

which gives $C \approx 0.18$ for a Kolmogorov constant of 1.4, assuming a strain ratio of unity.

This model was used in particular by Deardorff (1970) and Moin & Kim (1982) for the channel flow, and since then in many of the large-eddy simulations of engineering flows. Even for three-dimensional isotropic turbulence Smagorinsky's model behaves acceptably, although the kinetic energy spectrum decays faster than with the structure-function model (see section 3.3 and Métais & Lesieur, 1992). However, Smagorinsky's model turns out to be too dissipative in the presence of a boundary, and does not work for transition in a boundary-layer on a flat plate, starting from a laminar profile to which a small perturbation is superposed.

3.2 Spectral eddy-viscosity models

We assume that we work in the Fourier space, in the context of three-dimensional isotropic turbulence. The filter consists in a sharp cut-off filter clipping all the modes larger than k_C, where $k_C = \pi/\Delta x$ is the cut-off wavenumber. The concept of k-dependent eddy-viscosity was introduced by Kraichnan (1976) for three-dimensional isotropic turbulence, in the following way: if $T_{>k_C}(k,t)$ is the kinetic energy transfer across the cutoff k_C, corresponding to triadic interactions such that $k < k_C, p$ and (or) $q > k_C$ (see Lesieur, 1990, for details), one poses

$$\nu_t(k|k_c) = -\frac{T_{>k_C}(k,t)}{2k^2 E(k,t)} \quad , \tag{20}$$

in such a way that the kinetic energy spectrum $E(k,t)$ in the resolved scales ($k \leq k_C$) satisfies:

$$[\frac{\partial}{\partial t} + 2(\nu + \nu_t(k|k_c))k^2]\ E(k,t) = T_{<k_C}(k,t) \quad , \tag{21}$$

where $T_{<k_C}(k,t)$ is the kinetic energy transfer corresponding to resolved triads such that $k, p, q \leq k_C$. The kinetic energy transfers can be derived from the E.D.Q.N.M. theory, one of the statistical closures of isotropic turbulence (see Lesieur, 1990). We assume first that k_C lies within a Kolmogorov cascade, and $k << k_C$. Then the transfer $T_{>k_C}(k,t)$ across k_C can be analytically determined with the aid of expansions in powers of the small parameter k/k_C. This yields:

$$\nu_t^{\infty} = 0.441\ C_K^{-3/2} \left[\frac{E(k_C)}{k_C}\right]^{1/2} \quad , \tag{22}$$

where $E(k_C)$ is the kinetic-energy spectrum at the cutoff k_C. For k close to k_C, $T_{>k_C}(k,t)$ has to be evaluated using the E.D.Q.N.M. approximation. It still scales on $[E(k_C)/k_C]^{1/2}$, but with a coefficient rising rapidly in the neighbourhood of k_C (cusp behaviour). Notice that the scaling by $[E(k_C)/k_C]^{1/2}$ may be obtained by a mixing-length argument, where the eddy-viscosity is proportional to the product of k_C^{-1} by the characteristic turbulent velocity $v(k_C)$. The latter is itself proportional to $\sqrt{k_C E(k_C)}$. The analysis allows also to introduce a spectral eddy-diffusivity, with the same behaviour as the eddy-viscosity (plateau and cusp), and a turbulent Prandtl number which, from the E.D.Q.N.M. theory, is approximately constant and equal to 0.6.

This spectral cusp eddy-viscosity may be used in large-eddy simulations performed in spectral space, provided the kinetic-energy spectrum at the cutoff may be determined. Good results are obtained for decaying isotropic turbulence (see Lesieur & Rogallo, 1989, and Métais & Lesieur, 1992). The model works well even if the large scales are not isotropic, as was shown by Métais & Lesieur (1992) and Batchelor *et al.* (1992) for stably-stratified turbulence.

However, such an eddy-viscosity is difficult to employ if one wants to work directly in physical space, when the geometry of the problem prevents spectral methods to be used. An alternative method is to define an *average* spectral eddy-viscosity, using energy conservation arguments:

$$\int_0^{k_C} 2\nu_t^{av}(t) k^2 E(k,t) dk = \epsilon(t). \tag{23}$$

It yields (see Métais & Lesieur, 1992)

$$\nu_t = \frac{2}{3}\, C_K^{-3/2} \left[\frac{E(k_C)}{k_C}\right]^{1/2} , \tag{24}$$

This is in fact very close to Smagorinsky's model.

3.3 Structure-function model

This model, due to Métais & Lesieur (1992), is an attempt to use the spectral eddy-viscosity in physical space, while taking into account the intermittency of turbulence. The eddy-viscosity given by eq. (24) can be determined locally in physical space, if a local kinetic-energy spectrum $E_{\vec{x}}(k_C)$, with $k_C = \pi/\Delta x$ can be defined. This can be done using the local second-order velocity structure function

$$\bar{F}_2(\vec{x}, \Delta x, t) = \left\langle \|\vec{u}(\vec{x},t) - \vec{u}(\vec{x}+\vec{r},t)\|^2 \right\rangle_{\|\vec{r}\|=\Delta x} \tag{25}$$

For a uniform grid, the average is done on the six closest points surrounding $\vec{x}$ on the computational grid. If the grid spacing is not uniform, extrapolations can be performed using the fact that, in a three-dimensional Kolmogorov cascade, the second-order velocity structure function between two points a distant r apart scales like $r^{2/3}$. Using isotropic turbulence relations, such as

$$\bar{F}_2(\vec{x}, \Delta x, t) = 4 \int_0^{k_C} E(k) \left(1 - \frac{\sin(k\Delta x)}{k\Delta x}\right) dk , \tag{26}$$

it is obtained within a Kolmogorov spectrum

$$\nu_t(\vec{x}, \Delta, t) = C_F\, \Delta x\, [\bar{F}_2(\vec{x}, \Delta x, t)]^{1/2} , \tag{27}$$

with

$$C_F = 0.105\ C_K^{-3/2} \quad . \tag{28}$$

Notice that in the limit $\Delta x \to 0$ and for a regular mesh, we can write:

$$\bar{F}_2(\vec{x}, \Delta x, t) \approx \frac{1}{3}\Delta x^2 \left(\frac{\partial \bar{u}_i}{\partial x_j}\right)\left(\frac{\partial \bar{u}_i}{\partial x_j}\right) = \frac{1}{6}\Delta x^2\ [\bar{S} + \vec{\bar{\omega}}.\vec{\bar{\omega}}] \quad , \tag{29}$$

where $\bar{S}$ is defined by eq. (18) and $\vec{\bar{\omega}}$ is the large-scale vorticity vector. Equations (27) and (29) yield:

$$\nu_t(\vec{x}, \Delta x, t) \approx \frac{C_F}{\sqrt{6}}\ \Delta x^2\ [\bar{S} + \vec{\bar{\omega}}.\vec{\bar{\omega}}]^{1/2} \quad . \tag{30}$$

From equations (19) and (28), one gets $C_F \approx 2C_S^2$. Thus, we finally obtain

$$\nu_t(\vec{x}, \Delta x, t) \approx \sqrt{\frac{2}{3}}\ (C_S\Delta x)^2\ [\bar{S} + \vec{\bar{\omega}}\vec{\bar{\omega}}]^{1/2} \quad , \tag{31}$$

Therefore, as compared to Smagorinsky's model, the turbulent viscosity given by the structure-function model acts more efficiently in regions of high vorticity and low strain. Conversely, its action is less pronounced in regions of high strain and very low vorticity. It is important to remark that the structure-function model constant C_F has to vary like the square of Smagorinsky's constant. Therefore, if one uses Smagorinsky's model with a "non-universal" constant equal to αC_S (where C_S is given by eq. 19), comparisons with the structure-function model have then to be performed with a modified constant in eq. (27) equal to $\alpha^2 C_F$.

This model works very well for isotropic turbulence, where it gives a good Kolmogorov spectrum (see below). The structure-function model gives also good results for free-shear flows (see the wake of figure 10), and backstep flow (see figure 12). It works well also for rotating turbulence (see Bartello, Métais & Lesieur, 1994). It permits to simulate the transition on a periodic flat plate at Mach 4.5 (see Ducros *et al.*, 1994). However, it still does not work for transition on a wall at low Mach, because the eddy-viscosity reacts to the presence of TS waves enough to prevent them from developing into turbulence.

3.3.1 Selective structure-function model

In this model developed by David (1993), the eddy-viscosity is turned on only when the flow presents a sufficient degree of three-dimensionality. The latter is measured in terms of the mean angle between the vorticity at a given grid point and the vorticity of the closest neighbouring points. If this angle exceeds the most probable value obtained in simulations of isotropic turbulence at an equivalent resolution, the subgrid model is then turned on. This angle is of the order of 20^0 for resolutions ranging from 32^3 to 64^3. Otherwise, it is only molecular dissipation which acts. This model has been applied successfully to the simulation of a low-Mach number boundary layer over a heated wedge and a compression ramp at Mach 2.5 (see David, 1993). It has also been used for other flows of industrial interest such as the incompressible flow behind a swept backward-facing step (see Moinat, 1994; Moinat & Lesieur, 1994).

3.3.2 Filtered structure-function model

A model very close to the selective structure function, the filtered structure-function model, was developed by Ducros (1995) to simulate the transition in a weakly-compressible ($M_\infty = 0.5$) spatially-developing boundary layer on an adiabatic flat plate (see figures 2 and 3). In this model, a high-pass filter is applied to the velocity field, in the form of an iterated Laplacian. The kinetic-energy spectrum at the cutoff is calculated with the aid of this filtered field, allowing to evaluate the eddy-viscosity.

3.4 Dynamic model

This model is due to Germano (1992). A new filter $< . >$ (called the test filter, generally of width $2\Delta x$) is applied to $\bar{u}$, giving a "test-field" $< \bar{u} >$. The evaluation of the energetic transfers between the test field and $\bar{u}$ permits in principle to recalculate a time-varying Smagorinsky's constant which is local in space. One of the problem is that this procedure yields negative values for $C_S(\vec{x}, t)$ and requires an extra averaging (see Lesieur & Métais, 1995).

An analogous approach has been proposed in spectral space (Lesieur & Rogallo, 1989): a LES of decaying isotropic turbulence using the spectral eddy-viscosity (with cusp) was carried out, with a cutoff k_C. Then the explicit transfers across $k_C/2$, added to the modeled transfers across k_C, permitted to recalculate the constants of the model (level of the plateau and shape of the cusp). Unfortunately, the method, once iterated, did not converge.

4 Large-eddy simulations: validation

Ought to the Reynolds number limitation of the direct numerical simulations, only limited validations can be performed by comparing DNS and LES results. The confrontation with laboratory experiment measurements and with existing theories of turbulence is a good way of testing the performance of a given subgrid-scale model.

4.1 Three-dimensional isotropic turbulence

For instance, we have tested the ability for the structure function model to reproduce a well define Kolmogorov inertial range in istropic turbulence. Figure 7 shows the compensated kinetic-energy spectrum $\epsilon^{-2/3}k^{5/3}E(k,t)$ in a decaying isotropic LES using the structure funtion model, at a resolution of 96^3 (pseudo-spectral methods): the compensated spectrum is fairly constant (with a Kolmogorov's constant C_K of 1.4) from $k = 10$ to $k = 40$. The kinetic-energy spectrum obtained with Smagorinsky's model ($C_S = 0.2$) is steeper than the SF spectrum and close to a k^{-2} slope. On the

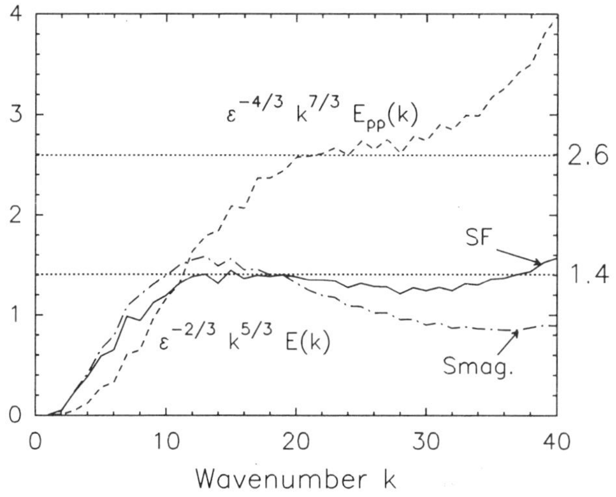

FIG. 7 - LES of decaying turbulence showing the compensated kinetic-energy spectra obtained with the structure-function model (SF) and Smagorinsky's model (Smag.) The compensated SF pressure spectrum is also plotted.

same figure, we have reported the pressure spectrum E_{pp}, compensated according to Batchelor's law

$$E_{pp}(k,t) = C_P \epsilon^{4/3} k^{-7/3} \quad . \tag{32}$$

This law can be obtained using the quasi-normal approximation for the determination of the second-order moment for pressure $< \hat{p}(\vec{k})\hat{p}(\vec{k}') >$. The constant C_P was calculated by Monin & Yaglom (1975). They found $C_P = \alpha\, C_K^2$ with

$$\alpha = \frac{7}{3}\left(\frac{27}{55}\right)^2 \frac{\Gamma\left(\frac{1}{3}\right)^2}{\Gamma\left(-\frac{4}{3}\right)} \approx 1.32 \quad . \tag{33}$$

On the figure, there is a tiny plateau of the compensated pressure spectrum at the right value C_P corresponding to $C_K = 1.4$. Higher resolution computations should be performed to investigate the nature of the pressure spectrum within the inertial range of three-dimensional isotropic incompressible turbulence.

As previously mentionned, low pressure regions are good indicator of the coherent vortices. This is confirmed through the examination of the pressure probability density function (pdf). Figure 8 shows the pressure pdf's as given by a direct numerical simulation of decaying three-dimensional isotropic turbulence at a resolution of 128^3 and the structure-function LES with 96^3 collocation points corresponding to figure 7. Both pdf's exhibits a very asymmetric shape, with an exponential fit in the lows. The amplitude is slightly lower for the LES. Conversely, the high pressure one is very close to gaussian. The skewed pdf is a clear signature of highly concentrated coherent vortices: the flow is composed of very localized regions of strong depression. Furthermore, as shown by Métais & Lesieur (1992), in the case of LES, these low pressure regions are better tracers of the coherent vortices than the high vorticity regions which are very scattered. Similarly skewed pdf's have been observed in the mixing layer direct numerical simulations by Comte *et al.* (1992). These have also been measured in experimental swirling turbulent flows by Abry *et al.* (1994) and in turbulents jets by Y. Gagne, E. Villermaux and M. Marchand (1994, private communication).

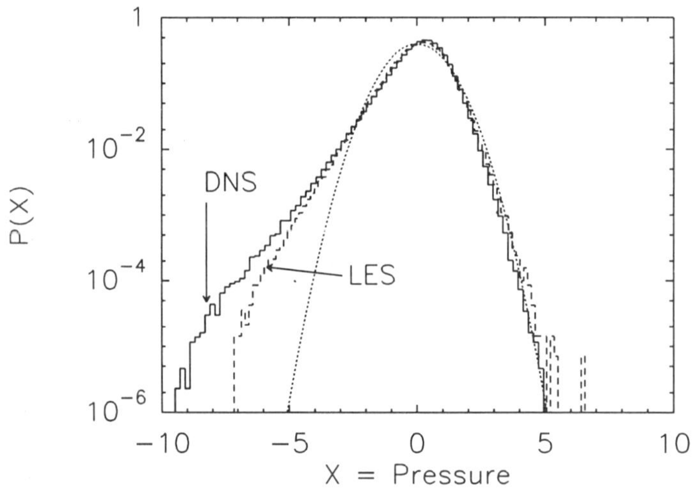

FIG. 8 - Probability density function $P(X)$ with X = pressure field, direct simulation (solid line); large-eddy simulation (dashed line). The pdf's are compared to a gaussian distribution of same variance (dotted line).

4.2 Free-shear flows

Large-eddy simulations results of the temporally and spatially growing mixing-layer were previously discussed in section (2.3). We here focus on the three-dimensional coherent vortices exhibited by the plane-wake. We first focus on the temporally-growing plane wake. The streamwise direction as well as the spanwise direction are periodic. The initial conditions consist of a low-amplitude random perturbation superposed upon a Gaussian mean-velocity profile given by:

$$\overline{u}(y) = U_m \exp\left[-(\log 2)\,\frac{y^2}{r_m^2}\right] \tag{34}$$

where r_m is the half-deficit velocity width. We here consider a quasi-two-dimensional perturbation resulting from the superposition of two random perturbations, a purely two-dimensional one (independent of the spanwise direction) of kinetic energy $\epsilon_{2D}U_m^2$, and a three-dimensional one of kinetic energy $\epsilon_{3D}U_m^2$, with $\epsilon_{2D} = 10^{-4}$ and $\epsilon_{3D} = 10^{-5}$. Pseudo-spectral methods are used with a resolution of $48 \times 48 \times 48$ collocations points We compare the coherent vortices respectively obtained in a direct numerical simulation (figure 9a) at Reynolds $R_e = U_m r_m/\nu = 200$, and in large-eddy simulations using various subgrid-scale models (zero molecular viscosity). Except for the viscosity, the initial conditions are identical in all the cases. We successively consider the structure function model (figure 9b), and Smagorinsky's model (figure 9c). In each case, we visualize an iso-surface of the vorticity modulus: $\omega = 0.5\omega_i$ (ω_i vorticity maximum associated with the initial mean-velocity profile) at time $t = 76r_m/U_m$. The length of the computational domain is twice the most amplified wavelength predicted by the linear stability analysis: $7.96r_m$ (Sato & Kuriki, 1961). Therefore, the Karman-street consists of two pairs of alternate sign vortices (see figure 9a). At low Reynolds number (figure 9a), the Karman rolls exhibit some oscillations in the spanwise direction but the longitudinal vorticity component remains small with respect to the spanwise one ($\approx$ 10%). The three-dimensionality is greatly enhanced in the case of large-eddy simulations. Figure 9b shows that the primary vortices of the Karman street are still present but vorticity stretching leads to the formation of intense counter-rotating vortices superposed onto the spanwise ones. These longitudinal structures are located within the braids connecting consecutive Karman vortices of anti-parallel vorticity. At that time, $\omega_{max} = 2.22\omega_i$ where ω_{max} is the maximum intensity of the longitudinal vorticity component. We have checked that both the primary Karman vortices and the secondary longitudinal hairpin vortices correspond to pressure lows (see e.g. Métais *et al.*, 1995). Smagorinsky's model turns out to be more dissipative than the structure-function model, yielding slightly lower values for the longitudinal vorticity component. It is important to note that the LES preserve both the coherence of the Karman rollers and of the thin longitudinal hairpin vortices whose diameter is close to Δx.

We next consider a flow configuration closer to experimental ones: the spatially-growing wake. We have used the same code as in the case of the previously-described

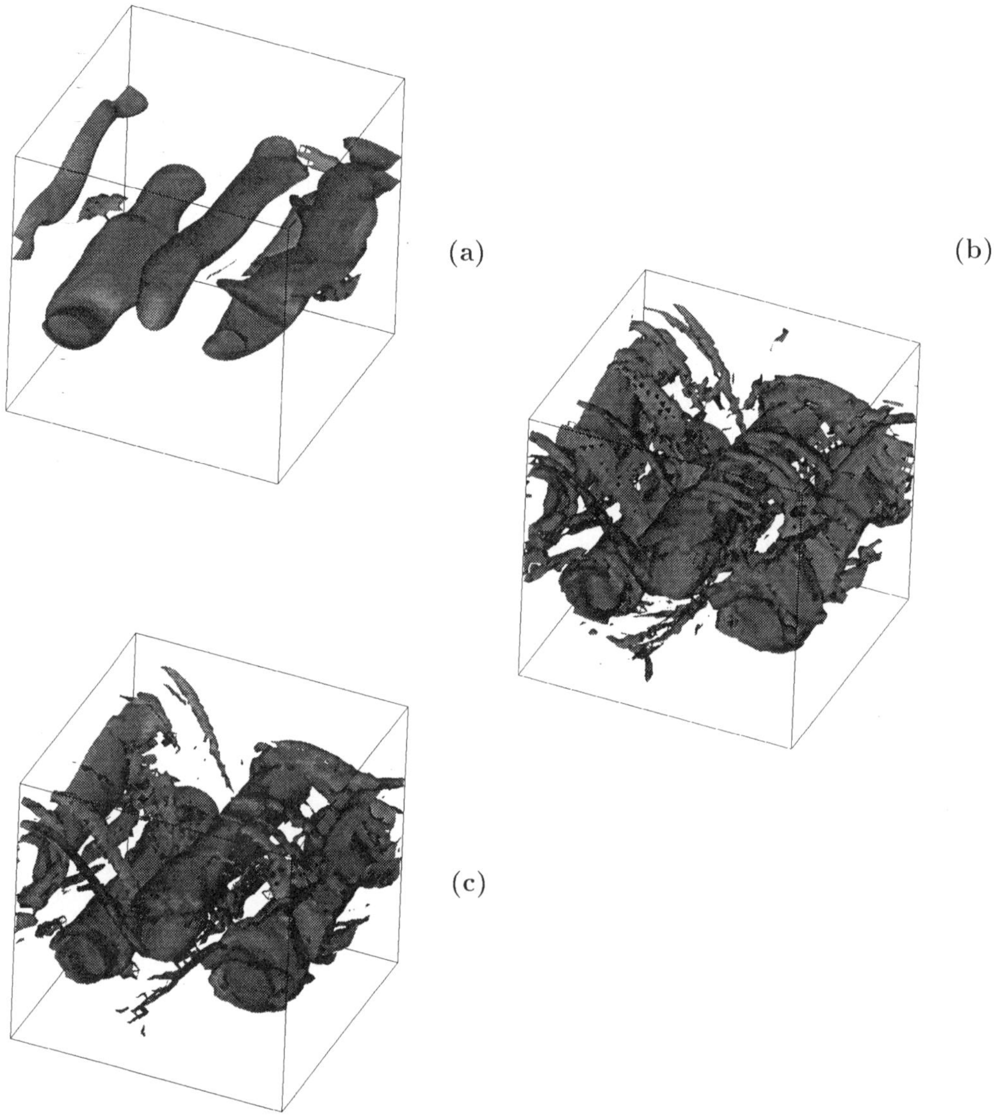

FIG. 9 - Vorticity modulus isosurface $\omega = 0.5\,\omega_i$ at $t = 76 r_m/U_m$: (a) Direct numerical simulation at $R_e = U_m r_m/\nu = 200$; Large-eddy simulations (zero-molecular viscosity), (b) structure-function model, (c) Smagorinsky's model.

spatially-growing mixing-layer (see figure 5).

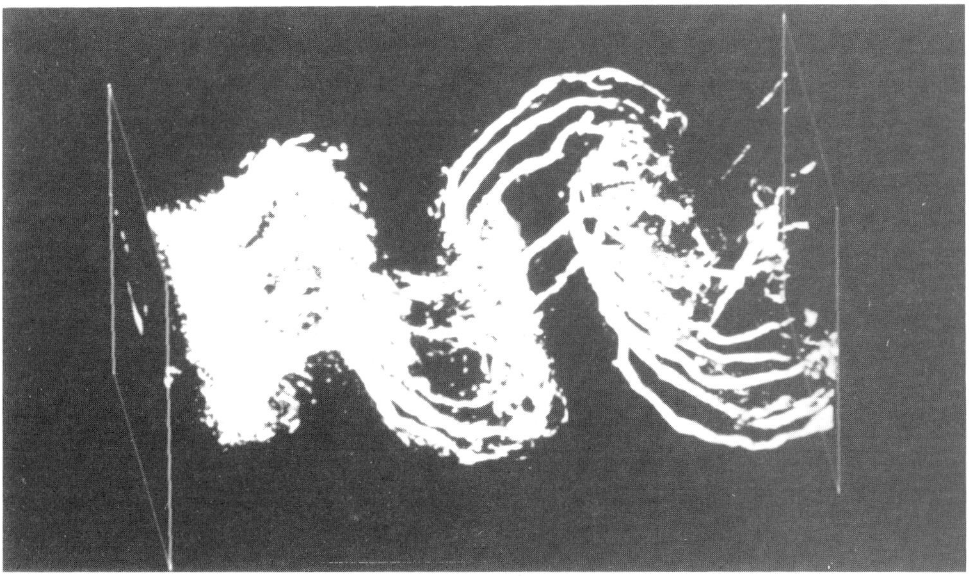

FIG. 10 - Large-eddy simulation of a spatially developing plane wake: structure-function model. Vorticity modulus iso-surface $\omega = \omega_i$ (courtesy M.-A. Gonze, Grenoble).

Figure 10 shows the spatial development of a plane wake obtained in a large eddy simulation based upon the structure function model (see Gonze, 1993). It displays an iso-surface of the vorticity modulus, whose threshold is equal to the upstream maximum spanwise vorticity ω_i. In order to simulate the near-wake, the obstacle is replaced at the inflow by an upstream deficit velocity profile close to a top-hat shape. To the mean profile is superposed a three-dimensional (white-noise) random perturbation of weak amplitude, which is regenerated at each time step. The resolution is $n_x, n_y, n_z = 160, 80, 80$. The calculation shows again a Karman street stretching intense longitudinal hairpins (see figure 10). The stretching of these hairpin vortices is even more intense than in the temporal case and the longitudinal vorticity maximum is $\omega_{max} = 4\omega_i$.

These longitudinal hairpin vortices are not easy to observe in experimental high-Reynolds number flows. They are, however, clearly displayed in cavitating wakes. Indeed, as previously mentioned, these intense vortical structures are pressure troughs, and therefore correspond to cavitating regions. These regions, clearly visualized in the cavitating turbulent-wake experiments such as the one performed by Franc *et al.* (1982) at a Reynolds number of 1.8×10^4, exhibit striking similarities with the numerically-simulated structures.

4.3 Separated flows

We here focus on the large-eddy simulations of the flow behind a backward facing step. These are based upon the TRIO code (finite-volume code) developed at the Atomic Energy Agency (CEA) in its center of Grenoble, and originally designed for industrial modelling purposes. The code has been used at a high Reynolds number (Silveira-Neto *et al.*, 1991, 1993) in large-eddy simulations using the structure-function model. The flow configuration is shown on figure 11. H and W respectively designate the step and channel heights, and we define the aspect ratio as W/H. The Reynolds number is UH/ν, where U is the velocity at the inlet. Although the flow geometry is simple, several features of complex turbulent flows are present such as separation, curvature, re-circulation and re-attachment.

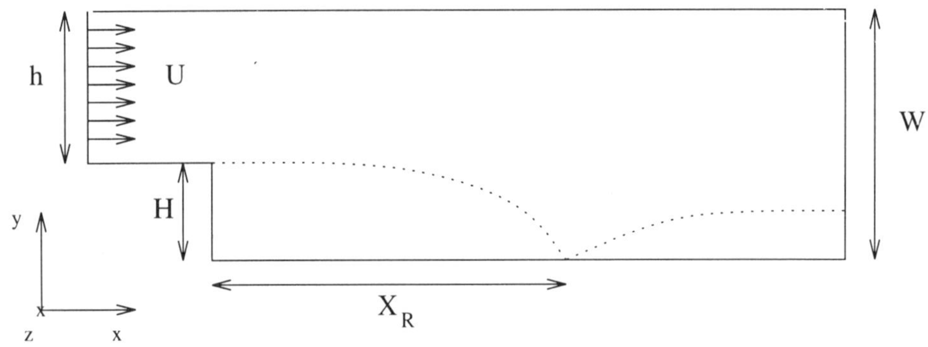

FIG. 11 - Schematic side view of the backward-facing step.

We here briefly recall some of the main results obtained by Silveira-Neto *et al.* (1993). At first, the case of a "high" step of aspect ratio 1.25, and at a Reynolds number of 6000 was investigated. Secondly, the "low" step case (aspect ratio 2.5, Reynolds 38 000), was chosen to fit the same step geometry as in the laboratory experiments by Eaton & Johnston (1980).

In the LES of the flow over the high-step, the resolution is $130 \times 25 \times 40$. Figure 12 displays the three-dimensional coherent vortices obtained in that case (see Silveira-Neto *et al.*, 1993, for other views). The flow topology resembles the one show on figure 5. The incoming flow, taken uniform, separates downstream of the step (left of the figure), and the resulting vortex sheet rolls up into quasi-two-dimensional Kelvin-Helmholtz billows which travel downstream and pair. Again, longitudinal hairpin vortices are stretched between the main spanwise rolls. Here the streamwise vorticity does not exceed approximately 0.4 ω_{max} where ω_{max} is the maximum spanwise initial vorticity modulus. A spanwise section of the longitudinal vortices yields a typical mushroom-like structure, analogous to what had been observed experimentally and numerically in mixing layers. In fact, calculations at longer times and higher resolution show that the

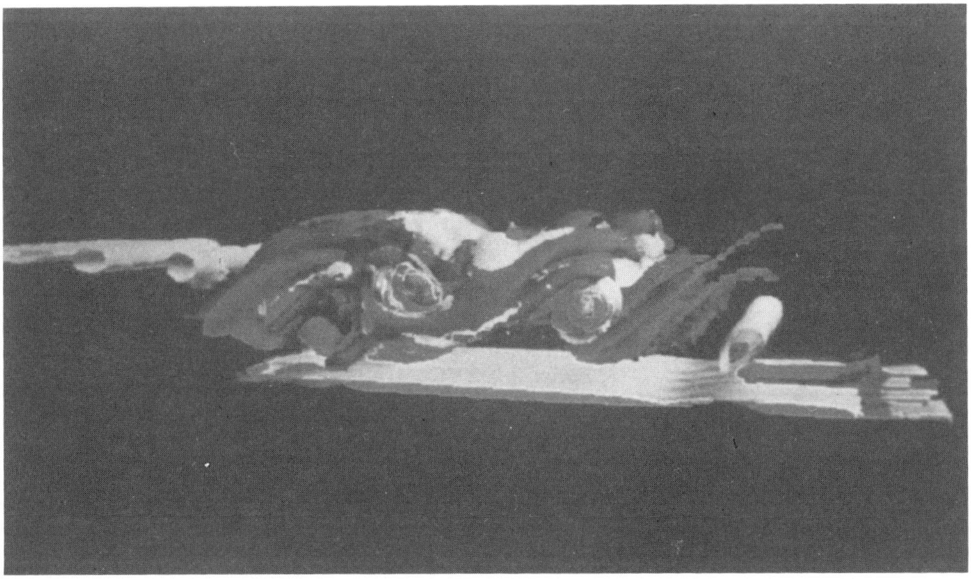

FIG. 12 - LES of the flow over a backward-facing step (high-step): spanwise and longitudinal vorticity isosurfaces.

flow evolves eventually towards highly-distorted Kelvin-Helmholtz billows more of the helical pairing type (see Comte *et al.*, 1992), with Λ-shaped vortices shed downstream.

In the "low" step LES (still using the structure-function model), visualizations of the vorticity field indicate the presence of intense longitudinal vortices. The low-pressure field shows large depressions which might undergo some sort of helical pairing (see figure 10 from Silveira-Neto *et al.*, 1993). In these simulations at high Reynolds number, the reattachment length is equal to 8 step heights, in very good agreement with the experiments. On the contrary, classical industrial models such as the $K-\epsilon$ model underestimate seriously this quantity, with a prediction of 6.

Comparisons of statistical data with Eaton & Johnston's (1980) experiment show that the structure function behaves better than Smagorinsky's model (see Silveira-Neto *et al.*, 1993, for details). Simulations without any subgrid-scale model were also carried out. These are not direct numerical simulations, since the numerical viscosity tends to damp the energy cascade towards the smallest scales. These simulations gives the poorest agreement with the experimental data. This indicates that some of the results given by numerical codes which rely on numerical diffusion to provide the necessary damping have to be considered with caution.

The LES version of the TRIO code is presently used to investigate the detached flows in more complex geometries In particular, a swept backstep study is undergone relating to external aerodynamics of cars in the region of the windshield ridge (see Moinat, 1994; Moinat & Lesieur, 1994). It has also been utilized to investigate a stably-stratified flow above a backward-facing step (see Fallon, 1994)

4.4 Compressible flows

Large-eddy simulations are also a powerful tool to understand the role played by coherent vortices in compressible flows. The structure function model and its derivatives (selective and filtered version) have been extensively used to study the transition to turbulence in weakly and highly compressible boundary layers (see Comte *et al.* 1994, and figures 2, and 3 for the low-Mach number spatially-growing boundary layer over a flat plate; Normand & Lesieur, 1992, and Ducros, Comte & Lesieur, 1994, for the high-Mach number temporal boundary-layer; David, 1993, for the heated wedge at Mach 0.5, and the compression ramp at Mach 2.5).

5 Influence of rotation on coherent vortices

The description of turbulent or transitional shear flows in a rotating frame is of upmost importance for turbomachinery and environment, and poses a severe challenge to classical industrial modelling methods. We here present in details the effects of rotation on shear-flows and on three-dimensional coherent vortices. In particular, our aim is to demonstrate how linear stability analysis, theoretical models and numerical simulations may complement each other.

Within geophysical and engineering flows, the Rossby number which characterizes the relative importance of inertial forces over Coriolis force varies importantly. Typical values of the Rossby number are ≈ 0.05 in the mesoscale oceanic eddies, ≈ 0.3 for the large synoptic atmospheric perturbations, and ≈ 2.5 for the atmospheric wake of a small island. Rotating turbulence finds numerous industrial applications in turbomachinery: the turbulent characteristics of the flow in blade passages of radial pumps and compressor impellers determine the efficiency of these devices. Turbulence is also of great importance for the cooling by the fluid inside the blades. Depending upon the magnitude of the radial velocity, the Rossby number within rotating machines can range from values close to unity to very small ones (≈ 0.05).

Lets consider a shear flow (free-shear or wall-bounded) of mean velocity, $\vec{u} = (\bar{u}, 0, 0)$ (x is the longitudinal direction, y the shear direction, and z the spanwise direction). We take a rotation vector $\vec{\Omega} = (0, 0, \Omega)$ oriented along the span (positive or negative). The vorticity vector associated with the mean velocity profile $\vec{\omega} = \left(0, 0, -\frac{d\bar{u}}{dy}\right)$ can be parallel or anti-parallel to $\vec{\Omega}$. We will refer to the first case as the cyclonic case, when the second will be called anticyclonic. In rotating channel flows, the cyclonic and anticyclonic walls are respectively nominated suction and pressure sides or trailing and leading sides. The effects of rotation on shear flows are drastically different depending on its cyclonic or anticyclonic nature. The laboratory experiments have shown that cyclonic rotation is always stabilizing. On the cyclonic side of channel flow, as compared to the non-rotating case, the turbulence energy production decreases with increasing rotation rate and fast rotation can lead to the total suppression of turbulent transition. Conversely, the anticyclonic side is destabilized (three-dimensionalized) for moderate rotation rates (high enough Rossby numbers). Destabilization and stabilization effects are also observed in rotating free-shear flows like mixing layer, wakes and jets. We here concentrate on the rotating mixing layer case and we will show,

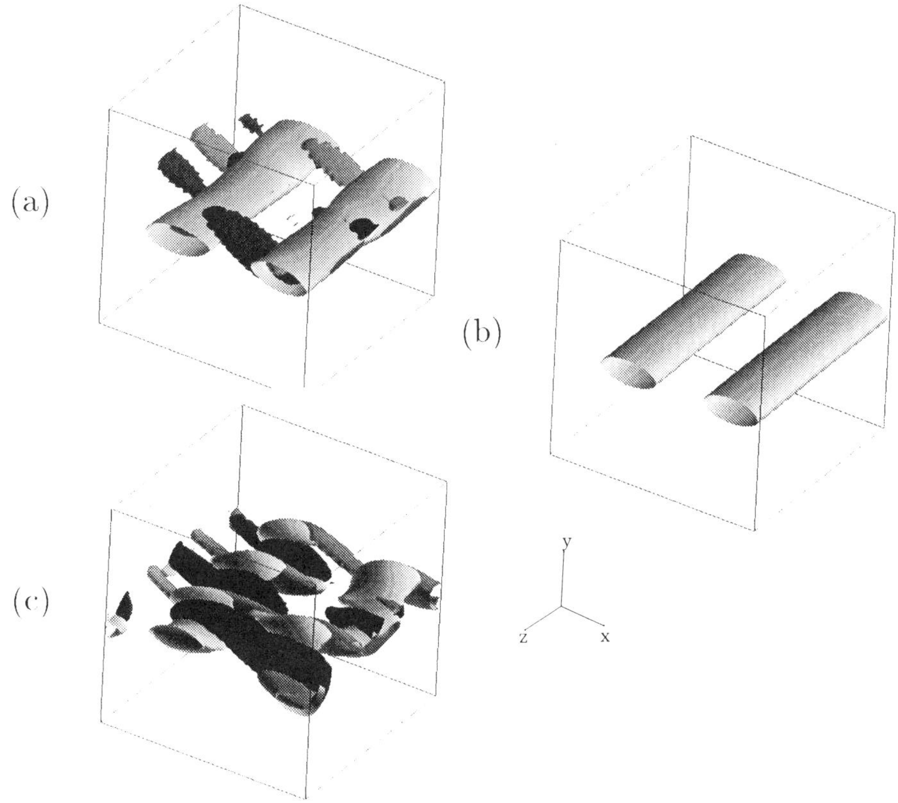

FIG. 13 - Mixing layer. Relative vorticity isosurfaces at $t = 17.8\delta_i/|U_0|$. a) non-rotating case: ω_z, light gray; ω_l (longitudinal vorticity) coloured by the sign of ω_x, black $\omega_x < 0$, dark gray $\omega_x > 0$. b) $R_o^{(i)} = -1$. c) $R_o^{(i)} = -5$.

through numerical simulations, how the rotation modifies the three-dimensional flow topology. The reader can refer to Lesieur *et al.* (1991), Métais *et al.* (1992), and Métais *et al.* (1995) for more details. A solid-body rotation does not have any influence on two-dimensional flow in a plane perpendicular to the rotation axis, since the Coriolis force is then proportional to the gradient of the stream function, and may be included into the pressure. Therefore, the phenomena observed in the laboratory experiments can only be explained by considering the influence of rotation on the growth of three-dimensional perturbations. In order to describe the early stage of the flow development, a three-dimensional linear-stability analysis of planar free-shear flows has been carried out by Yanase *et al.* (1993). For cyclonic rotation and for strong anticyclonic rotation, it was found that the flow was two-dimensionalized and the instability concentrated around the Kelvin-Helmoltz mode (which is not affected by the rotation). For moderate anticyclonic rotation $R_o < -1$, in addition of the Kelvin-Helmoltz instability, a new instability was discovered consisting of the strong amplification of a purely longitudinal mode. However, the flow quickly becomes non-linear and, in this régime, the previous analysis is no longer valid. Lesieur *et al.* (1991) have developed a phenomenological theory allowing to understand the non-linear stretching mechanims in the presence of solid-body rotation. They have emphasized the importance of considering the absolute vorticity, since, in the presence of rotation, Kelvin's circulation theorem directly applies to it. In particular it was predicted that cyclonic and rapid anticyclonic rotation will inhibit three-dimensional stretching. It was also predicted that the stretching of absolute vortex lines would be the most efficient in regions were the vorticity associated with the solid-body rotation annihilates the one due to the mean shear (regions of weak-spanwise absolute vorticity corresponding to a local Rossby ≈ -1).

We consider a temporal mixing layer associated with an hyperbolic-tangent mean-velocity profile: $\bar{u}(y) = U_0 \tanh y/\delta$ where $2U_0$ is the velocity difference accross the layer, and $\delta = \delta_i/2$, with δ_i initial vorticity thickness. We define the Rossby number as, $R_o^{(i)} = -U_0/2\Omega\delta$. $R_o^{(i)}$ is positive for cyclonic rotation (U_0 and Ω of opposite sign) and negative for anticyclonic rotation. The Reynolds number $Re^{(i)} = |U_0|\delta/\nu$ is here taken equal to 50. Initially, a low-amplitude random noise is superposed upon the ambient velocity profiles. Two different types of perturbations are considered. Firstly, a quasi-two-dimensional one consisting of the superposition of a purely two-dimensional perturbation (z independent) of kinetic energy $\epsilon_{2D}U_0^2$ and a three-dimensional perturbation of energy $\epsilon_{3D}U_0^2$, with $\epsilon_{2D} = 10\epsilon_{3D} = 10^{-4}$. The perturbation peaks at the fondamental Kelvin-Helmoltz mode. This case will be referred to as the "forced transition" case. The second type of perturbations which has been considered is purely three-dimensional, with $\epsilon_{3D} = 10^{-4}$ and $\epsilon_{2D} = 0$. The noise is now a white-noise which does not favour any mode, and the most amplified one can freely emerge: this case will be called the "natural transition" case. Now we look at the three-dimensional flow structure, in the forced transition case. We focus on the relative vorticity iso-surfaces at $t = 17.8\delta_i/|U_0|$ obtained in the non-rotating case ($R_o^{(i)} = \infty$), and for anticyclonic rotation at $R_o^{(i)} = -5$ and $R_o^{(i)} = -1$.

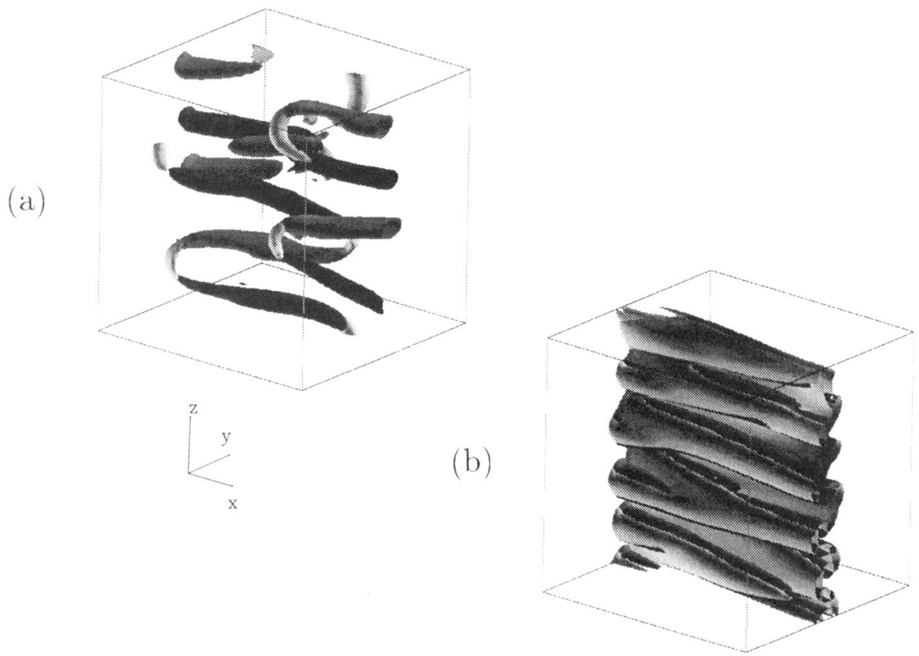

FIG. 14 - Mixing layer. Relative vorticity isosurfaces at $t = 26.8\delta_i/|U_0|$. (a) forced-transition; (b) natural transition. ω_z, light gray; ω_l (longitudinal vorticity) coloured by the sign of ω_x, black $\omega_x < 0$, dark gray $\omega_x > 0$.

(1) $R_o^{(i)} = \infty$ (figure 13a). Here, one observes quasi-two-dimensional Kelvin-Helmoltz billows, slightly distorted in the spanwise direction. Weak longitudinal vortices are stretched between the primary rolls: those are identified through isosurfaces of weak longitudinal vorticity.

(2) $R_o^{(i)} = -1$ (figure 13b) displays the spanwise vorticity field with the same iso-contour value than in the non-rotating case. Anticyclonic and cyclonic flows are similar at this Rossby number, and a strong two-dimensionalization is observed in both cases. The longitudinal vortices have disappeared. Furthermore, the two-dimensionalization tendency can be observed in the cyclonic case even for large positive $R_o^{(i)}$. This agrees well with both the predictions of the linear stability analysis by Yanase *et al.* (1993) and the phenomelogical theory proposed by Lesieur *et al.* (1991).

(3) $R_o^{(i)} = -5$ (figure 13c). The Kelvin-Helmholtz vortices are now highly distorted and exhibit strong oscillations along the spanwise direction. The longitudinal vorticity is much higher than in the non-rotating case: we observe the simultaneous formation

of Kelvin-Helmholtz vortices and longitudinal hairpin vortices which are stretched in-between. As time goes on, this produces an important increase of the longitudinal vorticity component: by the end of the run the longitudinal vorticity is approximately twice the one associated with the initial mean velocity profile. By the end of the run (figure 14a), the Kelvin-Helmholtz vortices have been totally dislocated and the flow is entirely composed of hairpin-shaped longitudinal vortices. A similar sequence had been proposed in this case by Lesieur *et al.* (1991), using the weak absolute vorticity stretching mechanism: weak absolute vorticity in the stagnation region between the Kelvin-Helmholtz rollers would be stretched longitudinally between the latter, yielding longitudinal alternate vortices which should destroy the primary vortices. It is worth noting that these vorticity structures originate from the growth of the longitudinal mode predicted by the linear stability analysis (see Yanase *et al.* 1993).

The signature of this mode is even clearer when considering the natural transition case. In that case, the linear growth gives initially rise to absolute vortex lines oscillating along the span and in phase in the streamwise direction. Afterwards, we have checked that the dynamics is dominated by a strong quasi-horizontal stretching of weak *absolute* spanwise vorticity corresponding to a phenomenological theory proposed by Lesieur *et al.* (1991). The isovorticity structures are completely longitudinal in that case (see figure 14b), and exhibit some analogies with the Görtler vortices observed in the boundary layer over a concave wall. Solid body rotation thus yields a very efficient mechanism to create intense longitudinal vortices in rotating anticyclonic shear layers, thanks to a linear longitudinal instability followed by a vigorous stretching of absolute vorticity.

Acknowledgemnts:

I am indebted to my collaborators and graduate students who contributed greatly to the theoretical results and numerical simulations presented here: P. Comte, C. Flores, D. Grand, M.-A. Gonze, M. Lesieur, J.J. Riley, A. Silveira-Neto, J. Silvestrini and S. Yanase. Some of the computations were carried out at the IDRIS (Institut du Développement et des Ressources en Informatique Scientifique, Paris).

References

ABRY, P., FAUVE, S., FLANDRIN P. & LAROCHE, C. 1994 Analysis of pressure fluctuations in swirling turbulent flows. *Journal de Physique*, **4**, pp 725–733.

BATCHELOR, G.K. 1967 *An introduction to fluid dynamics*, Cambridge University Press.

BATCHELOR, G.K., CANUTO, C. & CHASNOV, J. 1992 Homogeneous buoyancy generated turbulence *J. Fluid Mech.*, **235**, pp 349–378.

BARTELLO, P., MÉTAIS, O. & LESIEUR, M. 1994 Coherent structures in rotating three-dimensional turbulence. *J. Fluid Mech.*, **273**, pp 1–29.

BERNAL, L.P. & ROSHKO, A. 1986 Streamwise vortex structure in plane mixing layer *J. Fluid Mech.*, **170**, pp 499–525.

BRACHET, M.E. 1991 Direct simulation of three-dimensional turbulence in the Taylor-Green vortex *Fluid Dyn. Research*, **8**, pp 1–8.

BRADSHAW P. & KOH , M. 1981 A note on Poisson's equation for pressure in a turbulent flow *Phys. Fluids*, **24**(4), p 777.

BREIDENTHAL, R. 1981 Structure in turbulent mixing layers and wakes using a chemical reaction, *J. Fluid Mech.*, **109**, pp 1–24.

BROWN, G.L. & ROSHKO, A. 1974 On density effects and large structure in turbulent mixing layers *J. Fluid Mech.*, **64** , pp 775–816.

COMTE, P. 1989 Etude par simulation numérique de la transition à la turbulence en écoulement cisaillé libre. *PhD dissertation*, National Polytechnic Institute, Grenoble.

COMTE, P., LESIEUR, M., & LAMBALLAIS, E. 1992 Large and small-scale stirring of vorticity and a passive scalar in a 3D temporal mixing layer. *Phys. Fluids A*, **4**(12), pp 2761–2778.

COMTE, P., MÉTAIS, O., DAVID, E., DUCROS, F., GONZE, M.-A. & LESIEUR, M. 1994 Subgrid-scale models based upon the second-order structure function of velocity. *in Direct and Large-Eddy Simulation I*, P.R. Voke et al. (eds), Kluwer Academic Publishers, pp 61–72.

COUDER, Y. 1984 Two-dimensional grid turbulence in a thin liquid film *J. Phys. Lett.*, **45**, pp 353–360.

COUDER, Y, & BASDEVANT, C. 1986 Experimental and numerical study of vortex couples in two-dimensional flows *J. Fluid Mech.*, **173**, pp 225–251.

DAVID, E. 1993 Modélisation des écoulements compressibles et hypersoniques: une approche instationnaire, *PhD dissertation*, National Polytechnic Institute, Grenoble.

DEARDORFF, J.W. 1970 A numerical study of three-dimensional turbulent channel flow at large Reynolds number *J. Fluid Mech.*, **41**, pp 453–480.

DUCROS, F. 1995 Simulations numériques directes et des grandes échelles de couches limites compressibles. *PhD dissertation*, National Polytechnic Institute, Grenoble.

DUCROS, F., COMTE, P. & LESIEUR, M. 1994 Ropes and lambda-vortices in direct and large-eddy simulations of a high-Mach number boundary layer over a flat plate. Turbulent Shear Flows 9, *in press*, Springer-Verlag.

DRAZIN, P.G. & REID, W.H. 1981 *Hydrodynamic stability*, Cambridge University Press.

EATON, J. K. & JOHNSTON, J. P. 1980 Turbulent flow reattachment: an experimental study of the flow and structure behind a backward-facing step. *Stanford University, Rep.* MD-39.

FALLON, B. 1994 Simulations des Grandes Echelles d'Ecoulements Turbulents Stratifiés en Densité. *PhD dissertation*. National Polytechnic Institute, Grenoble.

FOUILLET, Y. 1991 Contribution à l'étude par expérimentaion numérique des écoulements cisaillés libres: effets de compressibilité, *PhD dissertation*, National Polytechnic Institute, Grenoble.

FRANC, J.P., MICHEL, J.M. & LESIEUR, M. 1982 Structures rotationnelles bi- et tri-dimensionnelles dans un sillage cavitant. *C.R. Acad. Sc. Paris.*, **295**, pp. 773–777.

HERBERT, T. 1988 Secondary instability of boundary layers. *Ann. Rev. Fluid Mech.*, **20**, pp 487–526.

GERMANO, M. 1992 Turbulence: the filtering approach. *J. Fluid Mech.*, **238**, pp 325–336.

GONZE, M.-A. 1993 Simulation numérique des sillages en transition à la turbulence. *PhD dissertation*, National Polytechnic Institute, Grenoble.

KLEBANOFF, P.S., TIDSTROM, K.D. & SARGENT, L.M. 1962 The three-dimensional nature of boundary layer instability. *J. Fluid Mech.*, **12** , pp 1–34.

KLINE, S.J., REYNOLDS, W.C., SCHRAUB, F.A. & RUNSTADLER, P.W. 1967 The structure of turbulent-boundary layers. *J. Fluid Mech.*, **30**, pp 741–773.

KONRAD, J.H. 1976 An experimental investigation of mixing in two-dimensional turbulent shear flows with applications to diffusion-limited chemical reactions. *PhD dissertation*, California Institute of Technology.

KRAICHNAN, R.H. 1976 Eddy viscosity in two and three dimensions. *J. Atmos. Sci.*, **33**, pp 1521–1536.

LASHERAS, J.C. & MEIBURG, E. 1990 Three-dimensional vorticity modes in the wake of a flat plate. *Phys. Fluids A*, **2**(3), pp 371–380.

LESIEUR, M. 1990 *Turbulence in Fluid* (second revised edition), Kluwer Academic Publishers.

LESIEUR, M., COMTE, P. & MÉTAIS, O. 1994 Numerical simulations of coherent vortices in turbulence, *to appear in Applied Mechanics Review.*

LESIEUR, M & MÉTAIS, O. 1995 New trends in large-eddy simulations of turbulence. *Annual Rev. Fluid Mech.*, **28**, *in press.*

LESIEUR, M & MÉTAIS, O. 1994 Turbulence and coherent vortices, *in Computational Fluid Dynamics, Les Houches, Session LIX*, M. Lesieur, P. Comte & J. Zinn-Justin, eds., *in press*, North-Holland.

LESIEUR, M. & ROGALLO, R. 1989 Large-eddy simulation of passive scalar diffusion in isotropic turbulence. *Phys. Fluids A*, **1** (4), pp 718–722.

LESIEUR, M., YANASE, S. & MÉTAIS, O. 1991 Stabilizing and destabilizing effects of a solid-body rotation upon quasi-two-dimensional shear layers. *Phys. Fluids A* **3** (3), pp 403–407.

LIN, S.J. & CORCOS, G.M. 1984 The mixing layer: deterministic models of a turbulent flow. Part 3. The effect of plane strain on the dynamics of streamwise vortices. *J. Fluid Mech.*, **141**, pp 139–178.

MÉTAIS, O. & LESIEUR, M. 1992 Spectral large-eddy simulations of isotropic and stably-stratified turbulence, *J. Fluid Mech.*, **239**, pp 157–194.

MÉTAIS, O., FLORES, C., YANASE, S., RILEY, J.J. & LESIEUR, M. 1995 Rotating free-shear flows Part 2: numerical simulations *J. Fluid Mech. in press.*

MÉTAIS, O., YANASE, S., FLORES, C., BARTELLO, P. & LESIEUR, M. 1992, Reorganization of coherent vortices in shear layers under the action of solid-body rotation. In *Turbulent Shear Flows VIII*, Eds U. Schumann et al, Springer-Verlag, pp 415–430.

MICHALKE, A. 1964 On the inviscid instability of the hyperbolic tangent velocity profile. *J. Fluid Mech.*, **19**, pp 543–556.

MOIN, P. & KIM, J. 1982 Numerical investigation of turbulent channel flow. *J. Fluid Mech.*, **118**, pp 341–377.

MOINAT, P. 1994 Application des techniques de simulation des grandes échelles à un écoulement en géométrie complexe: la marche en dérapage. *PhD dissertation*, National Polytechnic Institute, Grenoble.

MOINAT, P. & LESIEUR, M. 1994 Large-eddy simulation of the turbulent flow over a swept backward-facing step. *Proceeding of the International Symposium on Turbulence, Heat, and Mass Transfer*, pp 18.1.1–18.1.6

MONIN, A.S. & YAGLOM, A.M. 1975 *Statistical Fluid Mechanics*, Vol. 2, M.I.T. Press, 874 pp.

NORMAND, X., COMTE, P. & LESIEUR, M. 1992 Numerical simulation of a spatially-growing mixing layer *La Recherche Aérospatiale*, **6**, pp 45–52.

NORMAND X. & LESIEUR, M. 1988 Direct and large-eddy simulation of transition in the compressible boundary layer. *Theor. and Comp. Fluid Dynamics*, **3**, pp 231–252.

NYGAARD, K.J. & GLEZER A. 1990 Core instability of the spanwise vortices in a plane mixing layer. *Phys. Fluids A*, **2**(3), pp 461–464.

ORSZAG, S.A. 1971 Numerical simulations of incompressible flows within simple boundaries: accuracy. *J. Fluid Mech.*, **49**, pp 75–112.

PIERREHUMBERT, R.T. & WIDNALL, S.E. 1982 The two- and three-dimensional instabilities of a spatially periodic shear layer. *J. Fluid Mech.*, **114**, pp 59–82.

SATO, H. & KURIKI, K. 1961 The mechanism of transition in the wake of a thin plate placed parallel to a uniform flow. *J. Fluid Mech.*, **11**, pp 321–352.

SILVEIRA-NETO, A., GRAND, D., MÉTAIS, O. & LESIEUR, M. 1991 Large-eddy simulation of the turbulent flow in the downstream region of a backward-facing step. *Phys. Rev. Letters*, **66**(18), pp 2320–2323.

SILVEIRA-NETO, A., GRAND, D., MÉTAIS O. & LESIEUR M. 1993. A numerical investigation of the coherent vortices in turbulence behind a backward-facing step. *J. Fluid Mech.*, **256**, pp 1–25.

SMAGORINSKY, J. 1963 General circulation experiments with the primitive equations. *Mon. Weath. Rev.*, **91**(3), pp 99–164.

WINANT, C.D. & BROWAND, F.K. 1974 Vortex pairing, the mechanism of turbulent mixing layer growth at moderate Reynolds number. *J. Fluid Mech.*, **63**, pp 237–255.

YANASE, S., FLORES, C., MÉTAIS, O. & RILEY, J. J. 1993 Rotating free-shear flows. I. Linear stability analysis. *Phys Fluids A*, **5**(11), pp 2725–2737.